A MATHEMATICAL PANORAMA

A MATHEMATICAL PANORAMA

TOPICS FOR THE LIBERAL ARTS

William P. Berlinghoff, Bodh R. Gulati, and Kerry E. Grant

Southern Connecticut State College

D. C. Heath and Company   Lexington, Massachusetts   Toronto

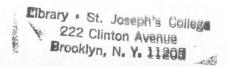

Cover photograph courtesy of Ken Buck

To Our Families

PREFACE

Any person who seeks to be liberally educated will benefit from the processes of logic and abstraction typified by mathematical thinking. For students not specializing in mathematics or science, the specific mathematical topics investigated are less important than the thought patterns they evoke. The student using this book will encounter a broad variety of mathematical concepts explored in sufficient depth to give a reasonable glimpse of the panorama of mathematics.

The material in this book grew out of a course for freshman nonscience majors taught at Southern Connecticut State College over a period of four years. The units have been thoroughly classroom-tested by a wide variety of students.

The three major facets of our approach to this course are as follows:

1. We believe that coverage of many topics maximizes the chance of catching a student's interest and enthusiasm. Furthermore, the chapters in this book are completely independent of each other; any subset of them can be taught in any order. This not only lends flexibility to the text, but also offers the student who falls behind an opportunity to make a fresh start several times each semester.

2. The wide variety of student backgrounds in a freshman course makes it difficult to require significant mathematical skill or sophistication as a prerequisite. We assume only that the student has some acquaintance with a few basic algebraic techniques and elementary geometric and set-theoretical ideas.

3. The lack of certain mathematical skills must not be confused with a general lack of ability to do college work. We believe that every college text writer has a responsibility to deal with significant ideas and to use a mature vocabulary.

Each chapter (with the exception of Chapter 1, which is described below) conveys an appreciation of a major area of mathematics by focusing on one or two important ideas in that area. The chapter starts with elementary concepts and gradually develops them through exposition, worked-out examples, and many exercises to reach some nontrivial result(s) in that area. Both the results themselves and the methods used to reach them have been carefully chosen to typify their respective areas. In this way the student will gain an appreciation for mathematics that is far broader than the scope of the particular topics covered.

Chapter 1 outlines the history of mathematics, concentrating on ideas rather than names and dates. It is intended primarily to provide a historical context in which the people and facts mentioned in the other chapters can be placed.

Chapter 2 develops the terminology and concepts necessary for investigating the twin existence problems of an Euler path and a Hamilton circuit in a graph. The problems are easily understood and formally quite similar. The radical dissimilarity of results, by contrast, provides an instructive lesson in the nature of mathematics.

Chapter 3 discusses the basic arithmetic of matrices and compares it with the familiar algebra of the number system. The main purpose is to examine solutions of simultaneous linear equations utilizing the Gauss-Jordan elimination method.

Chapter 4 provides a significant example of the unifying effect of mathematical abstraction. Its main theme is "equivalence relation"; the different types of numbers are shown to be equivalence classes of various sets.

Chapter 5 begins with the elementary notions of set and one-to-one correspondence, and progresses through Cantor's Theorem and the Continuum Hypothesis. A philosophical perspective on modern mathematics in general is provided.

Chapter 6 explains the empirical basis of probability, introduces some basic concepts in probability theory, and transmits a feeling for the questions that arise in everyday life. An elementary introduction to discrete random variables and their mathematical expectation is provided in the last two sections.

Chapter 7 introduces the descriptive nature of statistics in terms of frequency distributions, histograms, measures of central tendency, and measures of variation. The normal distribution, which plays an important role in both probability and statistics, is preceded by an intuitive introduction to continuous random variables.

Chapter 8 investigates the topic of perfect numbers in detail. The student learns that analytical observation, educated guesswork, and trial-and-error investigation are the sources of mathematical conjecture, but that patterns can be deceiving and proof is the unavoidable finale in mathematical discovery.

Chapter 9 uses observations about operation table patterns to suggest general logical arguments. The main topical theme of the unit is the understanding, proof, and elementary application of Lagrange's Theorem. More importantly, it also serves to illustrate the interrelationship between induction and deduction in mathematics.

Chapter 10 provides the student with the basic terminology of the components of an axiom system. In addition, an introduction to some of the basic metamathematical concepts and methods is presented.

Appendices A and B provide concise reference units for students who need to review (or learn) the elementary language of logic and sets. Finally, Appendix C is a deliberately opinionated essay on the nature of mathematics. It can be used to provoke class discussion or short papers at the beginning and/or end of the course, thus providing a vehicle for unifying the student's view of mathematics as a whole.

There are many people to whom we owe a debt of gratitude for their encouragement and assistance. Notable among these are the editors and staff of D. C. Heath and Company, whose patient prodding and amiable advice helped this project grow. For their conscientious reading of the manuscript and their helpful suggestions we especially thank the reviewers: Donald Catlin, University of Massachusetts at Amherst; George Kosan, Hillsborough Community College; and Richard Crouse, University of Delaware. We wish to thank Harper and Row for their permission to utilize some of the material published in *Finite Mathematics, an Introduction* and *College Mathematics with Applications to Business and Social Sciences* by one of us.

We are grateful to our colleagues at Southern Connecticut State College, particularly Professors Leo Kuczynski, J. Phillip Smith, and Robert Washburn for their helpful suggestions; to our families for their seemingly inexhaustible patience with the multiple moods of three often exasperated and harried authors; and to the many students whose questions and comments made this a better book.

William P. Berlinghoff
Bodh R. Gulati
Kerry E. Grant

TO THE STUDENT

A few words of advice and perspective are in order. The mathematics in this book is not highly technical, but the concepts are frequently challenging and require patience and perseverence to be mastered. We hope you will find our style and exposition lucid. But do not expect to read this, or any mathematics text, like a novel.

Expect to read and reread thoughtfully. Build a habit of having paper and pencil at hand to answer questions raised in the text, to work through exercises, and to create examples of your own. Examine your own understanding frequently. Do not go on to new material until you understand the old, or at least know what you do not understand. And do not hesitate to ask questions of your instructor.

Mathematics has played a central role in every known culture. It is intertwined with philosophy, science, business, art, music, and recreation. It is both content and form, practical and aesthetic, profound and profane. While we make no claim to encompass all of mathematics in this book, we do believe you can gain from our book a significant appreciation of many aspects of mathematics.

CONTENTS

6   ELEMENTARY PROBABILITY   193

7   INTRODUCTION TO STATISTICS   237

8   NUMBER THEORY   285

9   FINITE GROUPS   317

1

A BRIEF HISTORY
OF MATHEMATICS

1.1   INTRODUCTION

The investigation of any field of human achievement may be placed in perspective by relating individual accomplishments to the overall development of that field. For this reason it is appropriate in an introductory mathematics course to present a brief outline of the history of mathematics. No attempt is made here to detail mathematical theories; we merely offer a historical skeleton of names, places, discoveries, and inventions to provide a context for the material found elsewhere in the book. The required process of selection is unavoidably subjective, at least in part; however, choices and opinions advocated by recognized mathematical historians are followed wherever possible. We confine our attention predominantly to the mathematics of Western civilization, because mathematical development in the Far East was largely independent of the West until modern times, and the genealogy of mathematics as we know it today can be traced almost entirely through Europe and the Near East. This emphasis is not intended to minimize the mathematical achievement of the Orient, but merely to indicate the relatively small influence it has had on present-day mathematics.

The chronological sequence will be divided into eight major parts, determined mainly by the characteristics of the developments during each period. In more recent times this corresponds roughly to a division by centuries, but the lengths of the intervals prior to the Renaissance vary greatly. Of course, the partitioning is by no means absolute. Ideas and patterns of thought tend to overlap each other, intermingling with the passage of time; so this separation of history into pieces is only an approximation to aid in the analysis of individual accomplishments.

1.2   FROM THE BEGINNING TO 600 B.C.

Somewhere in prehistoric times, perhaps during the Middle Stone Age, two general concepts began to emerge from the countless diverse phenomena of the physical world. Thus were born *quantity* and *form*, the twin progenitors of two great families of thought whose true kinship would remain obscure until the seventeenth century A.D. Although accounts of this period are based largely on conjecture, it is generally believed that ideas about quantity started from attempts to compare collections of objects by counting, and slowly evolved into a variety of primitive number systems. The earliest of these were quite simple, often based on the idea of two or three things, with collections containing more than five or six things simply classified by "much" or "heap" or some other equally precise expression, until the necessity for exchange and barter gave rise to more extensive number systems. The emergence of form began as primitive art with plaited rushes, woven patterns in cloth, and similar ornamentation on pottery and buildings. The mathematical aspects of form did not become apparent for a while; what we would now call geometric figures began simply as decorative designs.

At the beginning of the historic period, about 5000 B.C., mathematics was well into its second stage of development. The quantitative needs of early societies had become so widespread and constant that it was desirable to develop general methods for calculation and to record these rules and results for future use. Since the Babylonians wrote on almost indestructible clay tablets and the Egyptian papyrus stayed well preserved in the dry climate of northern Africa, there are sufficient relics extant to provide a fairly detailed picture of this work in the early civilizations of the Near East. The earliest evidence of organized mathematical knowledge seems to indicate the existence of an Egyptian calendar in 4241 B.C. and possibly a Babylonian one before that. By 3000 B.C. the Sumerians had a workable mercantile arithmetic, and texts from the Third Dynasty of Ur (2100 B.C.) indicate a well-developed positional number system based on 60. Texts from the First Babylonian Dynasty, during the reign of King Hammurabi, show that by 1950 B.C. the Babylonians had an established algebra capable of handling linear and quadratic equations in two variables, and even some equations of higher degrees. Their geometry consisted of formulas for simple areas and volumes, and included a recognition of the rule for triangles which we now call the Pythagorean Theorem. Thus, the Babylonians may be credited with the first great period of mathematics.

Egyptian progress was not far behind that of its neighbors. The earliest source still in existence is the Ahmes Papyrus,* written in 1650 B.C. It is a practical handbook containing methods of solving types of linear equations, material on fractions with numerator 1 (a unique feature of Egyptian mathematics), measuring techniques, and problems in elementary series. The scribe Ahmes stated that he was copying an earlier work which had been written about 1800 B.C., so it may be considered a compilation of the mathematical knowledge of that time.

---

* This is also known as the Rhind Papyrus, named after A. Henry Rhind, a nineteenth-century English archaeologist who brought the manuscript to England.

FIGURE 1.1   Pythagorean Theorem

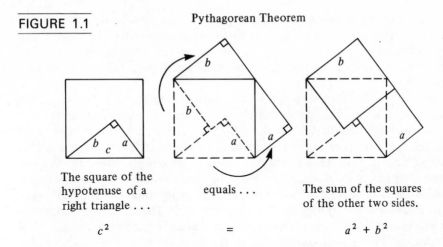

The square of the
hypotenuse of a   equals . . .   The sum of the squares
right triangle . . .   of the other two sides.

$$c^2 \qquad\qquad = \qquad\qquad a^2 + b^2$$

Our knowledge of early Chinese and Indian mathematics is comparatively poor. These peoples wrote on bark or bamboo, and hence all manuscripts were highly susceptible to decay. These natural inconveniences were sometimes compounded by man's perversity. For example, in 213 B.C. the Chinese Emperor Shi Huang-ti of the Ch'in Dynasty ordered all existing books burned and had protesting scholars buried alive so that he might be considered the creator of a new era of learning. Nevertheless, transcriptions of several ancient treatises have been preserved, among them the "Chou-pei," a dialogue concerning astronomy and mathematics. It was written shortly before 1100 B.C., and some of the material in it is attributed to a much earlier period. The Chou-pei contains material on mensurational geometry, the computational principle of the Pythagorean Theorem, some elementary trigonometry, and a discussion of instruments for astronomical measurements. We know even less about Indian mathematics of this period. All that can be said is that there exists evidence of a workable number system used for astronomy and other calculations, and of a practical interest in elementary geometry.

The outstanding feature of all pre-Hellenic mathematics is the complete absence of deductive reasoning. No attempt was ever made to justify a statement; the rules were given because they worked. Trial-and-error methods were the sources of knowledge, and successful results were noted and passed on to succeeding generations as formulas. Not until the blossoming of Greek civilization did mathematics come of age.

1.3 600 B.C. TO A.D. 400

With the advent of the first millennium before Christ, sweeping changes occurred in the lands around the Mediterranean Sea. The Age of Iron brought with it increased travel and trade, new towns sprang up along the coasts of Asia Minor and Greece, and the economic supremacy of the landlord gave way before the rising stature of wealthy merchants. Exchange of goods was accompanied by exchange of ideas and wealth begot leisure, so that by the sixth century B.C. there were men whose affluence allowed the luxury of intellectual speculation. With the first "Why?" mathematics entered its third phase of development, a science studied for its own sake.

Credit for that first impulse to speculate is attributed to Thales of Miletus (c. 640 to c. 546 B.C.), a merchant whose travels to Babylon and Egypt acquainted him with Oriental mathematics. Until this time geometry had been confined to its literal meaning, measure of the earth, and its propositions were simply rules for accomplishing this task. Thales, however, chose six statements, including "A circle is bisected by any of its diameters," "When two lines intersect, the vertical angles are equal," and "The sides of similar triangles are proportional,"* and he demonstrated that they followed logically from previous ones. The statements themselves were well known, but his approach to them was a radical departure from traditional mathematics. Thales also contributed to the fields of astronomy and number theory. As founder of the Ionian School, he influenced some of the finest minds of ancient Greece.

The most interesting member of the Ionian School was Pythagoras, whose life is as mysterious as his work was brilliant. Both the date and the place of his birth are uncertain, but it is generally held that he lived from about 570 to 500 B.C. Much of his life is obscured by myths, for he quickly became a legendary figure in Greece. His followers were banded together in a secret society that worshipped the idea of Number and hoarded knowledge as if it were gold. Pythagoras founded his own school at Crotona, a town of Magna Graecia on the southeast coast of the Italian peninsula. There he taught a philosophy based on the immutable elements of nature, embodied in and represented by whole numbers. Despite their indulgence in number mysticism, the Pythagoreans contributed greatly to number theory, the theory of music, astronomy, and geometry. Pythagoras was the first to insist on assumptions (axioms or postulates) as the basis for proof, and he offered the first proof of the theorem about right triangles that still bears his name. Strangely enough, one of the most significant ideas to emanate from the Pythagorean School was the verification of a concept that would completely destroy its own philosophy. The Pythagoreans discovered the existence of incommensurable line segments, which in our terminology means that they proved the existence of irrational quantities, or quantities that cannot be expressed in terms of whole numbers. They attempted to suppress this repugnant notion, but word leaked out and soon people were shopping for a new philosophy of nature.

Another disturbing figure in the turbulent world of Greek thought was Zeno of Elea, a philosopher of the early fifth century B.C. who speculated on the nature of motion or change of any kind. His contribution to mathematics consists of four paradoxes which he posed for the thinkers of his day. They involved two diametri-

---

* For a detailed discussion of all six propositions, see David Eugene Smith's *History of Mathematics* (Boston: Ginn and Company, 1923).

FIGURE 1.2

cally opposite viewpoints of infinity and motion, with two paradoxes for each viewpoint. We give here one example of each type:

> "*The Achilles*—Achilles running to overtake a crawling tortoise ahead of him can never overtake it, because he must first reach the place from which the tortoise started; when Achilles reaches that place, the tortoise has departed and so is still ahead. Repeating the argument we easily see that the tortoise will always be ahead.

> "*The Arrow*—A moving arrow at any instant is either at rest or not at rest, that is, moving. If the instant is indivisible, the arrow cannot move, for if it did the instant would immediately be divided. But time is made up of instants. As the arrow cannot move in any one instant, it cannot move in any time. Hence it always remains at rest."*

Any attempt to resolve these involves a consideration of limits, and although Zeno's contemporaries were unsuccessful in their attempts to unravel his verbal tangles, a seed had been planted that would eventually blossom into calculus. It would, however, require a growing season of 2000 years.

Plato and Aristotle exemplify the type of thought which was the Greeks' greatest contribution to mathematics. Their development of logical principles and axiomatic methods of demonstration put mathematics on a foundation that was considered unshakable until our own century. Under their guidance mathematics partook of the glory of the Golden Age as a philosophical science second to none. In this period arose the "problems of antiquity," perhaps the most famous problems of all time. They were geometric construction problems, allowing only an unmarked straightedge and a collapsing compass as tools for their solution. With these one was asked to:

1. Divide a given angle into three equal parts (Trisection of the Angle).
2. Find the side of a cube whose volume is twice that of a given cube (Duplication of the Cube).
3. Find the square whose area equals that of a given circle (Quadrature of the Circle).

---

* Quoted from E. T. Bell's *Men of Mathematics* (New York: Simon and Schuster, Inc., 1937), p. 24. The other two paradoxes may also be found there.

All three remained open questions until modern times, when the constructions were finally shown to be impossible. Nevertheless, the very fact of their existence bred a tantalizing kind of annoyance which led scholars to a host of new discoveries.

One of these scholars was Eudoxus (c. 408 to c. 355 B.C.), a pupil of Plato, physician, legislator, and mathematician. He is credited with developing a theory of proportions that overcame the difficulties of dealing with incommensurable quantities. He introduced the "method of exhaustion," a way of dealing with areas and volumes which is remarkably similar to some of the basic concepts of integral calculus.

As Alexander the Great set out to conquer the world in 334 B.C., Western civilization began to change. Greek culture and thought mingled with that of the Orient, and the mathematical center of the Western world shifted to Alexandria. That city's period of ascendancy began with Euclid (c. 300 B.C.), "the most successful textbook writer the world has ever known" and "the only man to whom there ever came or ever can come again the glory of having successfully incorporated in his own writings all the essential parts of the accumulated mathematical knowledge of his time."* His works were so comprehensive that they superseded all previous writings, and for this reason very few pre-Euclidean Greek manuscripts were preserved. Most of the information regarding work prior to the third century B.C. has had to be reconstructed from second-hand source material. Euclid's greatest work, the *Elements*, is a thirteen-book treatise arranged as follows:

> Books I–IV—plane geometry, including the Pythagorean Theorem;
> Books V and VI—Eudoxus' theory of proportions and its applications to similarity of figures;
> Books VII–IX—number theory, including the Euclidean Algorithm;
> Book X—geometric classification of quadratic irrationals and of their quadratic roots;
> Books XI–XIII—solid geometry, culminating in a proof of the existence of exactly five regular (Platonic) bodies.†

Some of this material is undoubtedly Euclid's own work, and he himself freely acknowledges his indebtedness for other parts, but the exact division is difficult to ascertain. However, even if he had contributed nothing original, the *Elements* would be equally significant, for it was a remarkably successful attempt to organize all of mathematics into a system logically deduced from a single axiomatic foundation.

Greek mathematics reached its peak with the flourishing of the School of Alexandria. Shortly after Euclid came Archimedes (287–212 B.C.), an astronomer and pioneer in physics and applied mathematics as well as a speculative mathema-

---

* Smith, *History of Mathematics*, Vol. I, pp. 103–104.

† For a more detailed outline of Euclid's *Elements*, see Dirk Struik's *A Concise History of Mathematics* (New York: Dover Publications, Inc., 1948), pp. 59–60.

tician. He studied at Alexandria for a short time and then returned to Syracuse, where he had been born. His scientific defense of that city against the hordes of Marcellus and his subsequent death at the hands of an impetuous Roman soldier are well known and will not be recounted here. Far more important are his works, which include major contributions to number theory and algebra, especially the treatment of infinite series, and an enormous amount of geometrical work, in which he developed some of the underlying principles of integral and differential calculus, anticipating Newton and Leibniz by 1900 years.

A contemporary of Archimedes was Apollonius of Perga (c. 260 to c. 210 B.C.), "the great geometer." Seven of his masterful eight books on conic sections have survived intact. In these he systematically developed many basic properties of the ellipse, the parabola, and the hyperbola, investigating such topics as congruence and similarity criteria for these curves, tangent figures, and inscribed and circumscribed polygons. The world was not to see another synthetic geometer of his stature until Jacob Steiner in the nineteenth century.

With the passing of Apollonius, the tide of Greek mathematics crested and ebbed slowly into oblivion with the rest of Greek civilization. Only two great waves appeared above the ripples of minor writers from this period. The astronomer Ptolemy (Claudius Ptolemaeus, c. A.D. 85 to c. 165) wrote a comprehensive treatise on astronomy known as the *Almagest*, in which the frontiers of computational mathematics were extended to plane trigonometry and the beginnings of spherical trigonometry, including tables of angle chord values and stereographic projection. About a century later Diophantus of Alexandria (c. 275 A.D.) wrote the *Arithmetica*, an unusual blend of Greek and Oriental mathematics, of which six books have survived. It is a milestone in the development of number theory, containing a treatment of indeterminate equations and problems requiring fractional solutions. It is the first treatise in which a type of algebraic symbolism is used systematically. In this regard Diophantus was several centuries ahead of his contemporaries.

The Oriental mathematics of this period in both India and China remained mostly computational and devoid of proof. With the exception of the book-burning incident in 213 B.C., the Chinese made slow but steady progress in the field of manipulative arithmetic, and the same may be said of India.

1.4 A.D. 400 TO 1400

Despite its productivity in many other areas, the Roman Empire was mathematically barren, and its conquest of the Mediterranean world did little for speculative science. Whatever activity there was resulted from surviving Greek and Oriental influences that remained substantially intact despite individual tragedies such as the death of Archimedes. With the fall of Rome and the dissolution of Latin political domination in 476, Western civilization entered a period of intellectual stagnation. The only person worthy of note here is Anicius Manlius Severinus Boethius (c. 475–524), a Roman citizen, statesman, philosopher, and mathematician, whose works include an arithmetic, a geometry, and a treatise on music (considered to be part of mathematics at the time). Although his texts lacked originality and were not especially rich in content, they were considered authoritative by the monastic schools for many centuries. This high regard may well have resulted from the author's martyrdom for Christianity rather than from their intrinsic merit. Indeed, the progress of ideas in western Europe reached its nadir in the sixth century.

Meanwhile, Hindu mathematics was beginning to bear fruit. The influences of Babylonian and Greek science blended with native Indian ideas and there emerged significant results in algebra and arithmetic. Hindu mathematicians of the fifth century worked with plane and solid numbers, obtaining both theoretical and computational results, including an approximation of π as $62,832/20,000$, or 3.1416. In the next century this work was extended to a treatment of indeterminate algebraic equations, following the methods of Diophantus. The Hindus, however, restricted their problems to allow only integral solutions, both positive and negative, and these are the conditions under which we now discuss Diophantine equations.

By far the most important achievement of Hindu mathematics was the development of the numeration system we use today, a place system based on 10 and including a symbol for zero. Decimal systems and place systems had been developed previously by other peoples, but this was the first combination of the two concepts. Evidence of its use dates from A.D. 595, but the zero symbol in Indian arithmetic cannot be traced back beyond the ninth century with any certainty, although the Babylonians had a somewhat similar notational device long before the Christian era.

Starting with Mohammed's Hegira in 622, the Moslems became the dominating influence in Western mathematics. The soldiers of Islam swept across North Africa, into western Asia, and through parts of Europe. Greek and Hindu mathematics were absorbed and merged by Arab scholars, who translated most of the important manuscripts into Arabic, contributing little that was original, but refining existing material and preserving the continuity of thought. The Mohammedan caliphs of the eighth and ninth centuries were great patrons of the exact sciences, especially astronomy and mathematics, and so scientific learning spread throughout the Arab world. Among the Moslem scholars of this period, one deserves special mention. He is Mohammed ibn Musa al-Khowarizmi (c. 825), who wrote two significant books, one on arithmetic and the other on algebra. Only the second still exists in the original Arabic, but both were translated into Latin by twelfth-century European scholars. The title of the first in translation was *Algorithmi de numero Indorum* (literally, "Al-Khowarizmi on Indian Numbers"). This was one of the means by which Europe was introduced to the Hindu number system, and is also the source of the word "algorithm." The second book was entitled *al-jabr w'al muqabalah* (literally, "Restoration and Opposition") and was devoted entirely to the study of linear and quadratic equations. By latinization, the key word of the title became "algebra," and, because of the widespread popularity of this text in Europe, the term soon became synonymous with the science of equations.

One more person merits brief mention here. Omar Khayyám, who lived in northern Persia at the end of the eleventh century, is known principally as the author of the *Rubáiyát*, but he was also an excellent mathematician. He wrote on Euclid, instituted a revised Persian calendar, and penned a treatise on algebra in which he discussed the determination of roots of cubic equations as intersection points of two conic sections.

It was not until the latter part of the eleventh century that the Greek mathematical classics started to filter into Europe, and with its slow emergence from feudalism the West began to stir from intellectual somnolence. During the twelfth century the great mathematical works that had been translated into Arabic a few centuries before were translated from Arabic into Latin, especially in Spain, where many Jewish scholars were employed for this purpose after the defeat of the Moors in 1085. As commercial intercourse between East and West expanded, powerful trading centers were established on the coasts of Italy. European merchants began to visit the Orient to seek and put to practical use whatever scientific information they could find. The first of these to do significant mathematical work was Leonardo of Pisa (c. 1170 to c. 1250), more commonly known as Fibonacci. His main work, the *Liber Abaci*, was instrumental in acquainting western Europe with the Hindu-Arabic numeration system. He also wrote on algebra and geometry, and investigated the infinite sequence which still bears his name.*

Not all mathematics of this period was done solely for its practical value. Scholastic philosophers speculated on the infinite, and their ideas influenced mathematicians of both the seventeenth and nineteenth centuries. Clerical scholars also worked in the fields of geometry and algebra. Outstanding among these men was Nicole Oresme (c. 1323–1382), Bishop of Lisieux, who was also an excellent economist. His mathematical work includes the first known use of fractional exponents and the location of points by numerical coordinates.

As cities sprang up throughout Europe, the church schools began to assume a new role. They became universities, empowered to grant degrees recognized by both Church and State. The first universities were at Paris, Oxford, Cambridge, Padua, and Naples, all chartered in the thirteenth century. This innovation held out great promise for the progress of learning during the fourteenth century, a promise only partially fulfilled, for although other universities were established, the Hundred Years' War (1337–1453) and the Black Death (1347–1351) staggered the emerging European culture and postponed the revival of mathematical creativity.

---

* 0, 1, 1, 2, 3, 5, 8, 13, ... Each term is the sum of the two preceding terms.

1.5   THE FIFTEENTH AND SIXTEENTH CENTURIES

The Renaissance of European art and learning began in earnest during the fifteenth century. When Constantinople fell in 1453, many Greek scholars migrated to the cities and new universities of the West. This coincided with the invention of movable-type printing, an invaluable aid in the dissemination of information. The ever-increasing demands of trade, navigation, astronomy, and surveying spurred mathematical study and at the same time confined it somewhat. The dominating theme of fifteenth- and sixteenth-century mathematics was computation, and great strides were taken in the achievement of accuracy and efficiency in technique.

The leading mathematician of the fifteenth century was Johannes Müller (1436–1476), also known as Regiomontanus. Besides translating many classical Greek works and studying the stars, he wrote *De triangulis omnimodis*, the first treatise to be devoted solely to trigonometry. This book differs little from the trigonometry of today except in notation, and it marked the beginning of trigonometry as a study independent of astronomy.

The first printed books on mathematics were a commercial arithmetic that appeared in 1478 and a Latin edition of Euclid's *Elements* that appeared in 1482. However, mathematics derived its first real benefits from printing with the appearance in 1494 of *Summa de Arithmetica* by Luca Pacioli, a Franciscan monk. The book was written in Italian (rather than Latin) and was a complete summary of all arithmetic, algebra, and trigonometry known at that time, ending with the remark that the general cubic equation was insoluble with existing mathematical machinery. As if in direct response to this challenge, mathematicians at the University of Bologna attacked the problem and disposed of it within the first quarter of the next century.

The foremost scientist of this age was Leonardo da Vinci (1452–1519), a man of universal brilliance whose fields of activity included painting, sculpture, biology, architecture, mechanics, and optics. His mathematical work was centered about geometry and its applications to art and the physical sciences. The application of mathematics to art was by no means confined to da Vinci's work. Several famous artists of the sixteenth century were excellent geometers, notably Albrecht Dürer. He wrote the first printed work dealing with higher plane curves, and his investigation of perspective and proportion is reflected in his paintings and the artistic works of his contemporaries.

The father of English mathematics was Robert Recorde (c. 1510–1558), a medical doctor, educator, and public servant. He published four books on mathematics, written in English dialogue with clarity, precision, and originality: *The Ground of Artes*, an arithmetic that was to go through at least 29 editions; *The Castle of Knowledge*, the first English exposition of the Copernican theory in astronomy; *The Pathwaie to Knowledge*, an abridged version of Euclid's *Elements*; and *The Whetstone of Witte*, an algebra, in which the equality sign "$=$" appears for the first time.

In the second half of the sixteenth century, a French lawyer by the name of François Viète (1540–1603) began to devote his leisure time to mathematics, and he advanced algebraic technique considerably. He was the first to use alphabetical coefficients in solving equations, and his consistent symbolism, including the signs + and −, helped him to pioneer in the development of general methods for solving equations. By applying these algebraic methods, Viète extended and generalized the study of trigonometry. Not until the eighteenth century would there be another algebraist of comparable ability.

The last name of importance connected with the sixteenth century is that of Simon Stevin of the Netherlands, developer of the theory of decimal fractions in 1585. Thus ended 200 years of preparation for a scientific explosion that signaled the dawn of the third great mathematical era.

1.6 THE SEVENTEENTH CENTURY

Two previously destructive trends began to bear fruit in the seventeenth century. The first and more obvious of these was the frequent occurrence of political and religious wars both large and small. Ever since the time of the Crusades, Europe had been in turmoil, and hardly a year had passed without open conflict somewhere. As has been the case throughout history, war and its ever-insistent need for newer and better weapons pressed the best minds of the age to compete in devising machines for battle. Once the weapons were made, however, the truly great minds that had designed them turned to the investigation of machines in general; Leonardo da Vinci's work is a prime example of this. Slowly, Europe embarked upon the first stages of mechanization, and in an effort to develop the necessary engineering skill, its scientists and scholars turned to the study of motion and change. The second trend was just as powerful but began more subtly. The religious reformation championed by Martin Luther on the Continent and by Henry VIII and his daughter Elizabeth in England unleashed an intellectual revolt against authority and tradition that had been brewing for many years. The general skepticism rejuvenated philosophy, and the explorers of ideas pushed ahead in all directions. This composite age of science and rationalism produced some of the greatest men of history—astronomers Galileo and Kepler; philosophers Hobbes, Locke, and Spinoza; and authors Dryden, Milton, and Shakespeare. Mathematics experienced a period of growth unparalleled until modern times, and the number of people who contributed significantly to the advance of science becomes so large from this time on that we shall henceforth be compelled to confine ourselves to creative mathematicians of the first rank.

The Tudors had just relinquished the British crown to the Stuarts and William Shakespeare was in his prime when John Napier reached the peak of his scientific career. Born in 1550, he was a Scottish contemporary of Galileo and Kepler, and a true product of his time. He spent much of his life alternating between attacks on the Catholic Church* and the contemplation and design of military devices, envisioning the use of such futuristic fantasies as submarines and self-propelled cannon. But his claim to immortality was established only a few years before his death, with the publication of *Mirifici Logarithmorum Canonis Descriptio*, in which he set forth the theory of logarithms. This work was followed by another on logarithms, one on computing rods, and one on algebra. Napier's logarithm system was a bit unwieldy, but several years after his death in 1617 it was perfected by a friend and colleague, Henry Briggs, who had originally suggested the use of 10 as the base. Oddly enough, the development of logarithms preceded that of exponential functions, the seemingly easier inverse theory, by about 50 years.

Not to be outdone by its neighbors across the Channel, France produced four brilliant mathematicians within half a century. The first of these was the philosopher-scientist René Descartes (1596–1650). His many years of study and contemplation convinced him that all scientific investigations are related and the key to that relation is mathematics. In his famous *Discourse on Method*, published in 1637, Descartes stated, "The long chains of simple and easy reasonings by means of which geometers are accustomed to reach the conclusions of their most difficult demonstrations, had led me to imagine that all things, to the knowledge of which man is competent, are mutually connected in the same way, and that there is nothing so far removed from us as to be beyond our reach, or so hidden that we cannot discover it, provided only we abstain from accepting the false for the true, and always preserve in our thoughts the order necessary for the deduction of one truth from another."† He explained that his method was a fusion of logic, the "Analysis (geometry) of the Ancients," and the "Algebra of the Moderns," and set forth four basic rules of procedure, saying, "In this way I believed that I could borrow all that was best in both Geometrical Analysis and in Algebra, and correct all the defects of one by the help of the other." To say that Descartes succeeded in unifying all of science would be a bit of an exaggeration, but he did achieve a remarriage between the two great families of quantity and form in one of the three appendixes to the *Discourse on Method*, simply titled "La Geometrie." This was the first publication of analytic geometry. In it Descartes applied the methods of algebra to geometry, expressing and classifying various curves and other geometric figures by means of algebraic equations relative to a coordinate system. He used this algebraic approach to investigate and settle a number of geometric questions, including some classical problems which hitherto had been insoluble.

A countryman and acquaintance of Descartes was Pierre de Fermat (1601–1665), called by some the greatest pure mathematician of the seventeenth century. He was certainly one of the foremost scientific amateurs in history. Fermat was a quiet, unobtrusive lawyer and civil servant who indulged in mathematics sheerly for the fun of it, publishing little but exhibiting his creativity in exchanges of letters

---

* The most popular of these was *A Plaine Discovery of the Whole Revelation of Saint John*, first published in Edinburgh in 1593.

† This and subsequent quotes from *Discourse on Method*, trans. John Veitch (London and Washington: L. Walter Dunne, 1901).

FIGURE 1.3

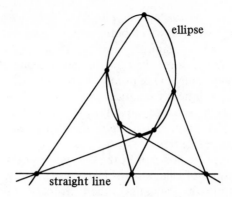

Pascal's Theorem, for an ellipse

with Descartes, Mersenne, Pascal, and others. He invented analytic geometry independently of Descartes, conceived the tangential approach to differential calculus before either Newton or Leibniz was born, and was one of the creators of the mathematical theory of probability. Despite these monumental achievements, Fermat is best known for his work in number theory on the properties of primes. Ironically, his name has been permanently associated with a statement that he neither originated nor proved. In his copy of a translation of Diophantus, one of Fermat's many marginal jottings was next to a problem asking for values x, y, and a to satisfy the equation $x^2 + y^2 = a^2$. In this note he states that for any power greater than 2 there are no integer solutions, claiming, "I have discovered a truly marvellous demonstration [of this] which this margin is too narrow to contain."* Unfortunately, he apparently never wrote it down anywhere else, and "Fermat's Last Theorem" still ranks among the most puzzling unsolved problems of mathematics.

The third man in this group was Gérard Desargues (1593–1662), soldier, engineer, and geometer. During his lifetime much of his work was eclipsed by the general interest in Descartes' writings, but two centuries later his treatise on conics was republished and was immediately hailed as a classic in pure geometry. Desargues introduced the geometric treatment of points at infinity and worked extensively with perspective, thus becoming the founder of modern projective geometry.

Completing this quartet of famous Frenchmen is Blaise Pascal (1623–1662). From the age of twelve he regarded geometry as recreation, and by sixteen he had proved one of the most beautiful and far-reaching results in all of geometry,† applying it to consolidate and extend much of the previous work in this field. He invented a computing machine when he was nineteen, and in his twenties was

---

* *Œuvres de Fermat*, ed. P. Tannery and C. Henry (Paris, 1891–1912), Vol. III, p. 241.

† "If a hexagon is inscribed in a conic section, then the points of intersection of the three pairs of opposite sides are collinear." (See Figure 1.3 for an example of this theorem.)

recognized as a competent physicist. He was cocreator with Fermat of probability theory, a study to which both were led in a joint attempt to answer questions from members of the gambling nobility. He also investigated the properties of a special planar curve called the cycloid, and was one of the first to use mathematical induction as a method of proof. The tragedy of Pascal is that at the age of twenty-five he became a nearly fanatical Jansenist, and considered mathematics a trifle to be toyed with only occasionally. Most of his later life was devoted to a study of philosophy and religion, from which emerged two literary classics, "Pensées" and "Provincial Letters."

During the first part of the seventeenth century the final building block for the foundation of calculus was fashioned in Italy. Bonaventura Cavalieri, a Jesuat* professor of mathematics at the University of Bologna, proposed the "principle of indivisibles" in 1629. This principle sets forth a criterion for comparing areas and volumes of certain geometric figures, based on the assertion that a planar region can be considered as composed of an infinite set of parallel line segments and a solid figure as composed of an infinite set of parallel planar regions. His work circulated throughout European scientific circles, and in the latter part of the century two men combined it with Cartesian geometry to erect a towering mathematical structure with a very shaky ground floor.

One of these was Isaac Newton (1642–1727), whose achievements as a physicist are well known. When he was a boy, Newton showed little formal mathematical ability, and his early scholastic record was somewhat less than outstanding, but after a brief absence from school he returned to his studies with renewed enthusiasm. He entered Trinity College at Cambridge as a student at the age of nineteen, and in eight years had become Lucasian professor of mathematics. In the decade between 1666 and 1676, Newton developed his "theory of fluxions" in three treatises. Although they were not published until many years later, these papers provided a basis for a subsequent work, *Philosophiae Naturalis Principia Mathematica*, an axiomatic development of physics in which Newton first set forth a complete mathematical formulation of his famous laws of motion. The *Principia* quickly assumed the role of an indispensable prerequisite for future scientific and technical progress, and it ranks among those few mathematical works that have radically affected the history of civilization.

Among Newton's contemporaries and successors, however, acceptance and acclaim were far from unanimous. The man of genius had known what he wanted, and he allowed intuition to suppress a few logical scruples, accepting the theory primarily because it worked. But some of his peers were justifiably skeptical. None of these critics was as caustically witty and incisive as George Berkeley, Anglican Bishop of Cloyne. In *The Analyst*, published in 1734, he contended that scientists who criticized faith in mysteries of religion had precisely the same difficulties in their own field: "And what are these fluxions? The velocities of evanescent increments. And what are these same evanescent increments? They are neither finite quantities, or quantities infinitely small, nor yet nothing. May we not call them the

---

* Not to be confused with "Jesuit." The Jesuats were a congregation within the Catholic Church from 1367 to 1668.

ghosts of departed quantities? Certainly . . . he who can digest a second or third fluxion . . . need not, methinks, be squeamish about any point in Divinity."

Newton's arch rival was Gottfried Wilhelm von Leibniz (1646–1716), Germany's universal genius. His extraordinary talents benefited law, diplomacy, religion, philosophy, physical science, and mathematics, in which he independently developed differential and integral calculus shortly after his British counterpart had done the same. This fact, however, was hotly disputed for many years, with charges of plagiarism from both camps of admirers flying back and forth across the English Channel, and engendering a bitter patriotic partisanship among the scientific community that was to last for many decades. Far less publicized but at least as important is Leibniz's work in combinatorial analysis. In his search for a "universal characteristic" unifying all of mathematical thought he became one of the founders of symbolic logic, a study that would not be investigated intensively until two centuries later.

1.7 THE EIGHTEENTH CENTURY

Calculus dominated mathematical development in the eighteenth century. Exploration of this powerful new theory proceeded in two directions—extension and application to other parts of mathematics and to physics, and examination of its logical foundation. Research during this era was carried on mainly by Royal Academies, subsidized by the "enlightened despots" of the age, while the universities played only a minor part in the production of ideas. The most prominent academies were at Berlin, London, Paris, and St. Petersburg. France held the lion's share of mathematical talent, but Switzerland also contributed significantly to the field.

A family named Bernoulli fled from Belgium in 1583 to avoid religious persecution, and ultimately settled in Switzerland. Their descendants provide a strong argument for the inheritance of intellectual ability; within three generations they had produced eight eminent mathematicians, of whom four achieved international renown! The first and best known of these were two brothers, Jacob (1654–1705) and Johann (1667–1748). Lured away from careers in theology and medicine by the fascination of Leibniz's pioneering work in calculus, Jacob and Johann Bernoulli entered into a fierce mathematical rivalry with each other and with Leibniz himself that produced much of the material now contained in elementary calculus texts, as well as results in the theory of ordinary differential equations. Much of their work centered about the properties of various special curves, including the catenary and the logarithmic spiral, and Johann Bernoulli is often considered the father of the calculus of variations because of his study of the brachistochrone.* Besides his contributions to calculus, Jacob also did outstanding work in geometry and wrote the first book devoted to the theory of probability.

---

* This is the curve along which a particle will slide from one point to another under the influence of gravity in the least possible time, friction being neglected.

Two of Johann's sons also achieved fame—Nicholaus (1695–1726) for his work in geometry, and Daniel (1700–1782) for his extensive publications in the fields of astronomy, mathematical physics, and hydrodynamics.

The most productive mathematician of the century was Leonhard Euler, born in Basel, Switzerland, in 1707. He was instructed in mathematics by Johann Bernoulli, and also studied theology, medicine, Oriental languages, astronomy, and physics. In 1727, he accepted a position at the St. Petersburg Academy, became head of the Berlin Academy in 1747, then returned to St. Petersburg 20 years later, where he remained until his death in 1783. He was an indefatigable worker whose labors were not significantly diminished by partial blindness at the age of twenty-eight, nor by complete blindness at fifty-nine. He wrote almost 900 important books and papers, including works on analysis, algebra, arithmetic, mechanics, music, and astronomy. Euler is responsible for much of our modern notation in algebra and calculus, and he put trigonometry into its present form; but perhaps his most consequential achievements resulted from his efforts to establish calculus as a purely analytic theory having no necessary dependence on geometry.

The latter part of the eighteenth century was a period of political turmoil. Great Britain was having a bit of difficulty disciplining her rowdy American colonies, and France's nobility had lost their knack of controlling the peasantry. The upheaval in France was so drastic it is incredible that scientific investigation not only endured but even thrived throughout the entire period. In fact, French mathematics retained its position of superiority. The Continental mathematicians had a strong advantage over the English in that the differential and integral calculus of Leibniz was much easier to understand and apply than Newton's unwieldy theory of fluxions, and their achievements indicate that they put it to good use.

An important breakthrough in calculus was accomplished by Jean d'Alembert (1717–1783) when he exorcised Newton's "ghosts of departed quantities" by introducing the notion of limit. His contemporaries, however, failed to appreciate the significance of this idea and it lay dormant for years.

"The lofty pyramid of the mathematical sciences," according to Napoleon Bonaparte, was Joseph Louis Lagrange (1736–1813), a brilliant but modest mathematician who was honored extravagantly by Napoleon and two foreign monarchs, and whose career was advanced immeasurably by the selfless support of Euler. Lagrange improved and organized much of Euler's calculus and worked extensively in the theory of equations, number theory, and mechanics. He was also responsible for inaugurating intensive mathematics curricula in France's two newly established schools, the École Normale and the École Polytechnique.

Pierre Simon Laplace (1749–1827) was the century's applied mathematician par excellence. The son of a peasant couple, he used his exceptional methematical talent to rise through the social strata until he was made a count by Napoleon. His most famous works are *Théorie analytique des probabilités* and the enormous *Mécanique céleste* (five volumes), which reviewed, unified, and greatly extended all previous work in the fields of probability and celestial mechanics. Despite an annoying tendency to borrow ideas without acknowledgment, Laplace is generally recognized as an outstanding creative scientist. The size of his works belies their succinctness; in the words of one of his translators,* "I never come across one of Laplace's 'Thus it plainly appears' without feeling sure that I have hours of hard work before me to fill up the chasm and find out and show how it plainly appears."

Brief mention must be accorded to Adrien-Marie Legendre and Gaspard Monge. Both of these men reached the peaks of their careers at the close of the eighteenth century, but their works differ radically from each other. Although he was responsible for a text which provided a thorough reorganization of Euclidean geometry, a major portion of Legendre's work was in analysis and applied mathematics, the predominant fields of that time. Monge, on the other hand, was strictly a geometer, one of the first modern mathematical specialists. He is best known for his development of descriptive geometry, and geometric ideas pervade all his works. This marks him as a herald of the next age.

1.8   THE NINETEENTH CENTURY

The modern era began with Carl Friedrich Gauss (1777–1855), "prince of mathematicians." Just as Archimedes' influence pervaded the science of the Hellenistic age and Newton overshadowed the post-Elizabethan period, so did Gauss dominate nineteenth century mathematics. He was a child prodigy who could do arithmetic at the age of three, and was familiar with infinite series by the time he was ten. The modern theory of numbers traces its ancestry back to his monumental *Disquisitiones Arithmeticae*, published in 1801. As a result of his writings on celestial mechanics Gauss became generally recognized as the leading mathematician of Europe. His work was clear and concise, and was characterized by rigorous proof. Although landmarks in almost every branch of mathematics bear his name, Gauss expressed a strong personal inclination toward one, saying, "Mathematics is the queen of the sciences, and the theory of numbers is the queen of mathematics."

With the dawn of the new century, mathematical creativity began to increase exponentially, and by 1900 had produced about five times as much original mathematics as had been accomplished in all previous ages. With this phenomenal wealth of material, mathematics was becoming a subject so vast that its totality defied the comprehension of a single mind. Except for those few men at the highest level of brilliance, such as Gauss, Riemann, Klein, and Poincaré, mathematicians were forced to confine their efforts to one major branch of the subject, such as algebra, geometry, or analysis. Many other changes were occurring in the mathematical landscape. The dominance of royally supported scientific academies declined rapidly, and research became an important function of the universities. Within the subject itself, mathematicians became increasingly self-critical. Their demands for a new rigor in all proofs and distrust of intuition brought forth the study of symbolic logic and axiomatics.

From this point on, a strictly chronological succession of biographies is no longer sufficient to outline mathematical progress; it must give way to a series of

---

* Nathaniel Bowditch (1773–1838), American astronomer.

narratives, each following some topical subdivision. There is an unavoidable over-
lapping of names, times, and ideas, but every effort has been made to preserve
perspective in the overall picture.

While physical science and engineering continued to reap the rewards of
calculus, mathematics itself began to benefit from the spirit of revolution that was
sweeping the Western world. In France the overthrow of the monarchy and the
subsequent Napoleonic period provided an ideal climate for the cultivation of new
ideas. Into this atmosphere was born Évariste Galois, a brilliant, temperamental
boy, who spent most of his life being thrown out of schools and into jails. Despite
frequent educational and political embroilments, he devoted much of his time to
algebra, which at that time was little more than generalized arithmetic, but his
writings went unnoticed. In 1832, shortly before his twenty-first birthday, Galois
became involved in a duel that cost him his life. The night before the "affair of
honor," he dashed off a hurried letter to a friend, sketching some of his recent
mathematical ideas. He wrote "I have made some new discoveries in analysis. . . .
Later there will be, I hope, some people who will find it to their advantage to
decipher all this mess." "This mess" was the theory of groups, the foundation of
modern algebra and geometry. Work along this line was done independently by
Niels Henrik Abel, a Norwegian mathematician of the same period who also died
before he was thirty.

The liberation of algebra from its dependence on arithmetic took a giant stride
forward with the discovery of quaternions by William Rowan Hamilton (1805–
1865), an Irish astronomer and mathematician. A child prodigy who at the age of
twelve had a working knowledge of twelve languages,* Hamilton spent much of his
early scientific career applying mathematics to physical theories, especially
mechanics and optics. In 1835 he turned his attention to algebra, and eight years
later discovered quaternions. Roughly speaking, the system of quaternions is a
generalization of the complex number system, and its multiplication was the first
worthwhile example of a noncommutative operation. General classes of algebras
were soon forthcoming from the predominantly geometric work of Hermann
Grassmann, and the subject was well on its way to abstraction. England was the
center of this nineteenth-century algebra with its geometric applications, which
flourished under the active guidance of Arthur Cayley (1821–1895), the originator
of matrix theory, and James Joseph Sylvester (1814–1897), who was also a prime
mover in the early development of American mathematics. Two other outstanding
names in group theory are Felix Klein (1849–1925) and Marius Sophus Lie (1842–
1899). Klein was a geometer who studied discrete groups, whereas Lie worked
with continuous groups.

With the appearance of Augustin Louis Cauchy (1789–1857) and his contemporaries, analysts in general became more cognizant of the need for strictly
logical demonstration. Cauchy gave a workable definition of limit and proceeded
to establish a firm foundation for calculus. He also developed the theory of
functions of a complex variable at about the same time that Gauss published his

---

* English, Latin, Greek, Hebrew, French, Italian, Arabic, Sanskrit, Syriac, Persian, Hindustani, and
Malay.

complex arithmetic. Bernhard Riemann* (1826–1866) of Germany also pioneered in complex number theory; much of his work had decidedly geometric overtones. The most important single contribution of Gauss, Abel, Cauchy, Riemann, and the other mathematicians of the early nineteenth century was their meticulous insistence on rigorous proof. Their work paved the way for Karl Weierstrass (1815–1897), a mathematician famous for his deliberate and painstaking reasoning. He clarified the notions of function and derivative, and eliminated all remaining obscurity from calculus. "With Weierstrass began that reduction of the principles of analysis to the simplest arithmetical concepts which we call the arithmetization of mathematics."†

This approach is typified by Leopold Kronecker (1823–1891), who asserted, "All results of the profoundest mathematical investigation must ultimately be expressible in the simple form of properties of the integers." He was a number theorist, but is best known for his prolonged ideological feud with Weierstrass, whose theories were based on the concept of infinite progressions. Kronecker, on the other hand, would not admit the mathematical existence of anything that was not actually constructible in a finite number of steps. Diametrically opposed to this view were Richard Dedekind (1831–1916) and Georg Cantor (1845–1918). Dedekind developed rigorously the concept of irrational number, thus enabling the real number system to become the basis for all of analysis. Cantor, in his *Mengenlehre* (Set Theory), based the concept of number on that of set, and proceeded to develop different types of infinity, or transfinite numbers, that behave somewhat like the whole numbers of elementary arithmetic. This, in Kronecker's opinion, was a dangerous travesty of mathematics, and he attacked both theory and person with such vehemence that Cantor suffered a series of breakdowns and ultimately died in a mental hospital. Set theory, however, has remained as a prominent but controversial part of mathematical thought.

The revolution in geometry was foreshadowed as early as 1733 by the work of Girolamo Saccheri. Ever since Euclid set forth his *Elements*, the fifth postulate (often called the "Parallel Postulate") had been questioned by those who thought it might be provable from the other four. Knowing that all previous attempts to establish this had failed, Saccheri proposed a radically different approach to the problem. He replaced the questionable postulate with its negation, hoping to arrive at two contradictory statements within the new system. Had he done this, it would have meant that the original fifth postulate is a necessary consequence of the other four, but the new system yielded no contradictions. Disappointed, Saccheri turned back just when one more step would have given him the discovery of the century, and his work was promptly forgotten.

Early in the nineteenth century, three men in three different countries used Saccheri's approach, but they had the insight to realize the meaning of their "failures" and the courage to publish their findings. Nicolai Lobachevsky in 1829, János Bolyai in 1832, and Bernhard Riemann in 1854 all published consistent non-Euclidean systems of geometry, each independently of the other two. Gauss

---

* Actually, George Friedrich Bernhard Riemann.
† Struik, *A Concise History of Mathematics*, p. 237.

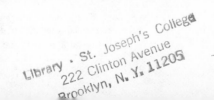

had had some of these ideas several decades before, but had refrained from publishing them for fear of criticism. These ideas conflicted with the prevailing Kantian
philosophy, which contended that space conception is Euclidean *a priori*, so they
remained obscure for several decades. But the logical floodgates had been opened.
No longer were postulates statements that were intuitively evident; they were
simply the initial assumptions whose choice was completely arbitrary, subject to
no preconditions. This was the start of formal axiomatics.

Now that geometry was no longer confined to visual images it expanded at a
fantastic rate. Grassmann's *Ausdehnungslehre* (Theory of Extension) gave the
world a fully developed geometry of *n* dimensions for metric spaces.* This work
established him as one of the founders of vector analysis (along with Hamilton).
Jacob Steiner (1796–1863), a purely synthetic geometer who detested algebra and
analysis, developed much of projective geometry. Felix Klein, on the other hand,
unified all geometries by means of modern algebra, declaring in his "Erlanger
programm"† that every geometry is the study of invariants of a set with respect to a
particular group of functions. This theory was greatly extended by Cayley and Lie.

The trend toward unification of mathematics was personified by Henri Poincaré
(1854–1912), last of the great universal mathematicians. His almost superhuman
memory and powers of logical apprehension enabled him to produce valuable
contributions to arithmetic, algebra, geometry, analysis, astronomy, and mathematical physics. In addition, he wrote popularizations of mathematics and was
actively interested in the psychology of creativity. He profoundly influenced the
development of topology, a relatively new branch of mathematics formerly called
analysis situs and sometimes referred to as "rubber sheet geometry."

The formalistic treatment of algebra in England and the abstract axiomatic
approach to geometry on the Continent triggered a sudden interest in logic and the
foundations of mathematics, an interest that was redoubled after the appearance of
Cantor's controversial set theory. The first significant mathematical studies of
logic were *The Mathematical Analysis of Logic* (1847) and *The Laws of Thought*
(1854), both by George Boole (1815–1864). In these works he exhibited a completely symbolic approach to logic and laid the foundation for future extensions of
the field. In 1884, Gottlob Frege (1848–1925) published *Die Grundlagen der
Arithmetik*, which offered a derivation of arithmetical concepts from formal logic
and thus greatly stimulated efforts to unify logic and mathematics.

1.9   THE TWENTIETH CENTURY

In 1900, the members of the International Congress of Mathematicians, assembled
in Paris, listened as one of their foremost colleagues lectured on mathematics in the
newborn century. David Hilbert (1862–1943) had just completed his now-famous
Grundlagen der Geometrie (Foundations of Geometry), a complete renovation of

---

* "Metric space" is a generalization of the usual intuitive concept of physical space. It is a set of points
on which some idea of distance is defined.

† Inaugural address upon elevation to professorship at the University of Erlangen in 1872.

Euclid's *Elements* using modern axiomatic methods. Hilbert outlined twenty-three unsolved problems, a challenge for the new century. His insight was so accurate that every one of these problems has led to important new results, and the solution of even a part of one of Hilbert's problems carries with it international recognition for the solver. Most of the problems have now been solved, some within the past decade; a few remain important open questions in contemporary mathematics.

Even Hilbert, however, could not foresee how mathematics would mushroom in the twentieth century. The phenomenal growth rate that began in the 1800s has continued, with mathematical knowledge doubling every fifteen or twenty years. Thus, far more original mathematics has been produced since the end of World War II than in all previous history! There are approximately 300 periodicals published in various parts of the world that devote a major part of their attention to the publication of articles on mathematics. The abstracting journal *Mathematical Reviews* alone publishes each year about 8000 synopses of recent articles containing new results. The present century is justifiably called the "golden age of mathematics." Quantity alone, however, is not the key to the unique position that the twentieth century occupies in mathematical history. It is essential to understand that beneath this astounding proliferation of knowledge there is a fundamental trend toward unity, a unity even more profound and productive than that envisioned by Descartes and Leibniz.

The basis for this unity is abstraction. Although the non-Euclidean geometries of the nineteenth century paved the way for an abstract axiomatic treatment of mathematics as a whole, many of the fundamental interrelationships among the major branches of the subject did not begin to appear until the 1940s. The recent recognition of these unifying theories and of the vast unexplored fields they unlocked has led some prominent mathematicians to regard the end of World War II as the beginning of a new era in mathematics.*

As we approach the present, accurate evaluation of the relative significance of new mathematical results becomes almost impossible, so we have made little effort to compare individual contributions with one another, leaving that task to a later generation. Hence, although all mathematicians mentioned here have achieved prominence through acclaim from the modern scientific world, no claim is made for the completeness of the listings, nor is the topical coverage intended to represent a comprehensive survey of twentieth-century mathematics. The primary purpose of this section is to indicate briefly the scope and power of contemporary mathematical activity.

As the result of Boole's work and the recognition of formal axiomatics which followed the birth of the non-Euclidean geometries, interest in the logical foundations of mathematics began to spread rapidly. The most notable successor to Boole's initial efforts in mathematical logic is the *Principia Mathematica*, a monumental two-volume work that appeared during the years 1910–1913, in which the philosopher-mathematicians Bertrand Russell (1872–1970) and Alfred North Whitehead (1861–1947) attempted to fulfill Leibniz' dream by expressing all of

* See, for example, Jean Dieudonné, "Recent Developments in Mathematics," *American Mathematical Monthly*, Vol. 71, No. 3 (March 1964).

mathematics in a universal logical symbolism. Hilbert, too, dreamed of unifying mathematics, and labored for many years to find a single provably consistent set of axioms upon which all mathematics could be based. Thus, mathematics entered a fourth phase. Starting only with an animal skin, it had first fashioned some serviceable garments and then acquired an elaborate wardrobe. Now, elegantly attired, mathematics began looking into its own pockets for unseen holes. The holes started to appear as Russell found inconsistencies in Cantor's theory of sets, a theory upon which all of mathematics could be based. This made Hilbert's goal all the more desirable, and hence the mathematical community was profoundly shocked when Kurt Gödel (1906–   ) proved in 1931 that this goal was unattainable. Gödel also laid the groundwork for one of the century's most spectacular mathematical discoveries. In 1964, Paul J. Cohen, basing his work on that of Gödel, proved that both the continuum hypothesis and the axiom of choice* are independent of the currently accepted axioms for set theory. Thus, Cohen has become the Saccheri of set theory by showing that these two statements are unprovable from the other axioms, and that the inclusion of their negations in set theory can lead to entirely new theories of mathematics as a whole.

Once they recovered from their initial surprise, mathematicians accepted the fact that mathematics was not provably consistent from within and continued to explore their own branches of the subject. In fact, the development of most parts of the subject was virtually unaffected by the sudden tremor that rocked its foundation. Algebra became far more general than it had ever been before, and similar tendencies toward abstraction in geometry led to extensive advances in the field of algebraic geometry, founded by Cayley in the previous century. Another extremely productive hybrid field of investigation is differential geometry, a synthesis of geometry and parts of analysis, in which there was a resurgence of interest engendered by attempts to reconcile the theory of gravity with certain electromagnetic phenomena. Analysis itself is undergoing an extraordinary metamorphosis during this century. When Henri Lebesgue (1875–1941) revolutionized the theory of integration in 1902, he opened the way for a much more abstract and unified treatment of analysis, as exemplified by the general theories of abstract spaces developed by E. H. Moore in 1906 and Maurice Fréchet in 1928. This new generality linked analysis with both algebra and topology.

According to at least one expert commentator on contemporary mathematics, "the main fact about our time which will be emphasized by future historians of mathematics is the extraordinary upheaval which has taken place in and around what was earlier called algebraic topology."† Algebraic topology began to be a major field of investigation with the work of Henri Poincaré at the end of the last century, and during the first half of this century it became the spawning ground for some of the most powerful tools in all of mathematics. Perhaps the outstanding

---

* Roughly speaking, the axiom of choice states that, given any collection of pairwise disjoint sets, we can choose exactly one element from each set in the collection. Although not recognized explicitly as an axiom until 1904, this assumption is basic to all of topology and much of modern analysis, and simplifies many arguments in algebra. For a discussion of the continuum hypothesis, see Chapter 5.

† Dieudonné, "Recent Developments," p. 243.

achievement in topology in the twentieth century to date took place in 1962, when the American mathematician John Milnor (1931–) disproved a famous conjecture that related algebraic topology to its older brother, combinatorial topology. Milnor's proof that spaces which are combinatorially equivalent need not be topologically equivalent won for him a Fields medal, an international mathematics award similar to the Nobel prize. Besides extending the frontiers of mathematical knowledge, the methods of algebraic topology became the basis for an even newer field, homological algebra, which has forged powerful bonds of unity among previously separate theories in algebra, analysis, and geometry. Although homological algebra and a close relative called "category theory" did not begin to appear in print until the 1940s and 1950s,* their techniques are already in the process of invading much of mathematics.

Along with any unifying trend in research there is normally a concomitant desire to consolidate and simplify previous results. One outstanding manifestation of this desire in mathematical history was the appearance of Euclid's *Elements*. In this century another attempt has been made to integrate all contemporary mathematics within a single framework. It was begun in the mid-1930s by a small group of young French mathematicians, who formed a secret society and wrote under the pseudonym Nicolas Bourbaki. The first volume of their encyclopedic endeavor *Éléments de Mathématique* appeared in 1939, and there are currently more than 30 volumes in the series. As the fame of the Bourbaki series spread, the Association of the Friends of Bourbaki exhibited its ubiquitous sense of humor by inventing a complete biographical myth about the "man" whose works they were writing. In this way they have managed to keep their exact membership secret, although some names have leaked out from time to time. The Bourbakians realized both the immenseness of their task and the fact that creative mathematics is primarily a young peoples' sport, so they agreed to retire from the society before the age of fifty and elect younger colleagues to replace them. Thus, Nicolas Bourbaki became a renowned but mysterious international scholar, always at the peak of his professional productivity, supplying the scientific world with a continuing series of modern, clear, accurate expositions in all fields of contemporary mathematics.†

The utilization of mathematical techniques in the physical and social sciences has been so widespread and penetrating in this century that it would be futile to attempt any summary of modern advances in applied mathematics. Nevertheless, there are several men whose works must be mentioned because of their profound influence on the world we live in. Albert Einstein (1879–1955) at the age of twenty-six revolutionized physical science with his theory of relativity, a radically different analysis of change based in part on non-Euclidean geometry. Einstein's work made him the best-known scientist of his day. John von Neumann (1903–1957) was a

---

* Henri Cartan and Samuel Eilenberg published the first book on a general theory of homological algebra in 1956. Category theory began with a 1945 paper by Eilenberg and Saunders MacLane, "General Theory of Natural Equivalences," *Transactions of the American Mathematical Society*, Vol. 58, pp. 231–294.

† For further information, see Paul R. Halmos, "Nicolas Bourbaki," *Scientific American*, Vol. 196 (May 1957), pp. 88–99.

Hungarian-born American who directed the development of some of the first electronic computers at Princeton's Institute for Advanced Study, helped to develop quantum theory in physics, and is considered to be the founder of game theory, a mathematical analysis of strategies. Von Neumann also made important contributions to the development of the atomic and hydrogen bombs and to long-range weather forecasting. Finally, we come the the father of automation, Norbert Wiener (1894–1964), whose work on information processes resulted in a field named by the title of his book, *Cybernetics, or Control and Communication in the Man and the Machine* (1948).

The unique feature of twentieth-century applied mathematics has been the invention and development of electronic computers. Their ability to perform routine calculations at speeds of up to 6,000,000 operations per second has radically altered problem-solving methods not only in the physical sciences, but in the social sciences as well. The social sciences are currently undergoing a fundamental shift of emphasis from qualitative to quantitative considerations, a transition made extremely difficult by the many variables that are usually present in social situations. The proper handling of these variables would be virtually impossible by traditional techniques, but computers can simulate and analyze complex situations involving hundreds of items, including people, keeping track of random effects and other pertinent data. Thus, the electronic computer is fast becoming an indispensable instructional and research tool in economics, sociology, psychology, education, and other related fields. Moreover, the "computer age" is affecting pure mathematics as well. It has created the need for a new branch of mathematical logic concerned with problems of machine design and control (such as coding), and has aided and encouraged the development of numerical analysis.

The growth of mathematics in the United States during this century has been phenomenal. Mathematical development in this country is usually considered to have begun with the extended visit of British mathematician James Sylvester at Johns Hopkins University late in the nineteenth century, but American mathematics did not become truly self-supporting until the emergence in 1913 of George David Birkhoff (1884–1944) as an internationally renowned mathematician. Birkhoff was the first American-educated mathematician to achieve such stature, and he in turn directed and guided many of the most productive American mathematicians of this century through his position as a professor, first at Princeton and then at Harvard from 1912 until his death. As the prestige of the United States has grown, more and more European and Asian mathematicians have migrated to the major American research centers, especially Princeton's Institute for Advanced Study (whose original faculty included Einstein and von Neumann). Today, the combined membership of the three major American mathematical societies* is approximately 30,000, as compared with a mere handful in 1913. Moreover, the employment focus of this mathematical talent has shifted and broadened over the

---

* The American Mathematical Society, the Mathematical Association of America, and the Society for Industrial and Applied Mathematics.

years. In the early part of this century, almost all mathematicians worked in academic settings. Now, however, a variety of industrial, commercial, and governmental needs, often occasioned by the ever-widening utilization of computers, have combined with the prospect of a continuing gradual decline in college enrollments to draw more and more mathematicians away from academe and into applied mathematical endeavors. In this way, mathematics is riding the historical pendulum, swinging back from the abstractions of the past hundred years toward applications of these theoretical results to "real" problems in other areas. Perhaps this remarriage of mathematics and reality will carry with it an appropriate social attitude change toward the subject, and mathematics will once again be viewed less as an area of irrelevant esoterica and more as a vital subject whose pursuit is essential to the progress of the human race.

FOR DISCUSSION OR ESSAY

In his book *The Decline of the West*, Oswald Spengler advances the "culture clue thesis," stated by C. J. Keyser in the following form: "The type of Mathematics found in any major Culture is a clue, or key, to the distinctive character of the Culture taken as a whole."* Discuss this idea, offering historical reasons for agreement or disagreement.

For Further Reading

Bell, E. T. *Men of Mathematics*. New York: Simon & Schuster, Inc., 1937.

Dantzig, Tobias. *Number, the Language of Science*, 4th ed. New York: Macmillan Publishing Co., Inc., 1954.

Keyser, C. J. *Mathematics as a Culture Clue*. New York: Scripta Mathematica, Yeshiva University, 1947.

Kline, Morris. *Mathematical Thought from Ancient to Modern Times*. New York: Oxford University Press, 1972.

———. *Mathematics in Western Culture*. New York: Oxford University Press, 1964.

Newman, James R., ed. *The World of Mathematics*, Vols. I–IV. New York: Simon and Schuster, Inc., 1956.

Smith, David Eugene. *History of Mathematics*, Vols. I and II. Boston: Ginn and Company, 1923.

Struik, Dirk J. *A Concise History of Mathematics*. New York: Dover Publications, Inc., 1948.

---

* C. J. Keyser, *Mathematics as a Culture Clue* (New York: Scripta Mathematica, Yeshiva University, 1947), p. 46.

2

GRAPH THEORY

2.1 INTRODUCTION

In this chapter we shall look at a relatively new branch of mathematics called graph theory. As will become clear, this topic has virtually no connection with the notion of graphs as encountered in high school algebra, where a graph is a geometric representation of relationships between numbers.

The origins of graph theory are traditionally traced to a paper presented by Leonhard Euler* in 1736. The paper was a mathematical treatment of a famous puzzle, known as the Königsburg Bridge Problem. Throughout the following years, few mathematicians took interest in graph theory, and the few problems that arose generally tended to fall into the category of mathematical games.

More recently, the techniques of graph theory have proved to be of considerable practical value in analyzing a number of applied problems. We shall discuss some of these later, but let us ease ourselves into the topic gently (and historically) by considering several puzzles. Many are quite old, so you may well recognize an old friend (or nemesis).

---

* See Section 1.7.

EXAMPLES 2.1

1. Can Figure 2.1 be traced in one continuous path without lifting one's pencil from the paper and without tracing any line segment more than once?

FIGURE 2.1

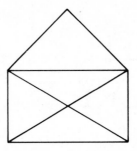

2. **(The Königsburg Bridge Problem)** Königsburg* was a major city of Prussia and was physically situated on both sides of a river and on two islands in the river, as indicated in Figure 2.2. Seven bridges interconnected the land masses, as indicated. The problem posed by the townspeople was to walk through the various parts of the city crossing each bridge exactly once. Can it be done?

FIGURE 2.2

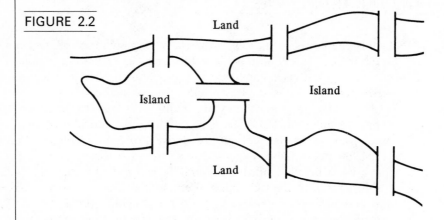

3. A city street crew wishes to install new street signs in a section of the city indicated in Figure 2.3. Signs should be posted at every intersection. What route should be taken by the crew so that each intersection is visited exactly once?

4. Another city street crew is sent to the same section of the city (Figure 2.3) to repair potholes. Since every street must be inspected, this crew would like to find a route that travels each street exactly once. Can it be done?

---

* Königsburg was renamed Kaliningrad after World War II. It is currently part of the Soviet Union.

FIGURE 2.3

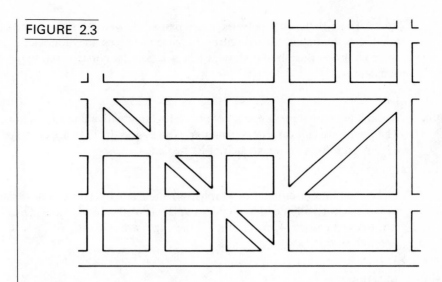

5. Can you draw a continuous curve in the plane of the page which crosses each segment in Figure 2.4 exactly once?

FIGURE 2.4

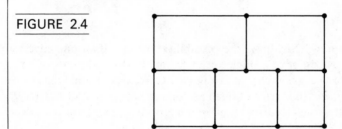

Before proceeding to the following section, you should try your hand at the puzzles described in Examples 2.1. For those you answer successfully, you are commended; for those you cannot answer, consider the possibility that there is *no* solution of the type called for. (Can you *prove* that there is no solution? This would also "solve" the problem, although in a different sense.)

2.2 TERMINOLOGY OF GRAPH THEORY

In this section we shall develop some of the basic terminology of graph theory. We shall see that all the problems posed in Section 2.1 can be rephrased in graph theory. It should be noted that graph theory, being a mere two centuries old, is still in a developmental stage, and the terminology is not yet universally agreed upon. Thus, there may be some discrepancies between terms we introduce here and those you find in the references. In some instances, the difference is deliberate, to avoid the confusion of dissimilar terms for similar concepts caused by historical accident.

DEFINITION 2.2.1

A **graph** is a nonempty, finite set of points together with a finite set of line segments. The points are called **vertices** (singular, **vertex**) and the lines are called **edges.** Each edge must have a vertex at each of its two ends.

To facilitate discussion of graphs, we shall label vertices with capital letters. Edges can normally be identified by the labels of their endpoints. Thus, in Figure 2.5, the vertices are A, B, C, and D; the edges are AB, AC, BC, and CD (or BA, CA, and so on).

FIGURE 2.5

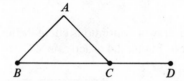

Definition 2.2.1 allows the possibility of more than one edge joining a given pair of vertices, or of a single vertex being "both" endpoints of an edge. A graph with these features is shown in Figure 2.6. We have numbered the multiple edges there, and can thus refer to them, for example, E_1F and E_2F (or F_1E and F_2E). Frequently, such a distinction is unnecessary, and multiple edges will not be numbered. The edge HH is an example of a loop.

FIGURE 2.6

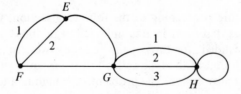

DEFINITION 2.2.2

An edge whose endpoints coincide as a single vertex is called a **loop.**

A loop is not an edge with no end, or only one end. Rather its two ends coincide. Remember that every edge must have two endpoints, both of which are vertices— these vertices need not be distinct.

Figure 2.7 shows three vertices, J, K, and L, an edge KL, and a line segment with one end at J. But the other end of that segment is not a vertex, and thus

FIGURE 2.7

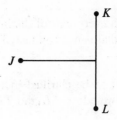

Figure 2.7 as labeled is *not* an example of a graph. Figure 2.8, on the other hand, is a graph, since every edge connects two vertices. Note that vertex *M* is not an

FIGURE 2.8

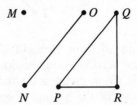

endpoint of any edge. Also, this graph is in several "pieces," although it is a single graph. To describe such a graph (and to help describe a number of problems) we consider the following:

DEFINITION
2.2.3

A **path** in a graph is a nonempty, finite sequence of successive edges that form a route from one vertex to another vertex (not necessarily distinct).

For example, in Figure 2.5, *AB*, *BC*, *CD* is a path from *A* to *D*. For brevity, we shall denote this path *ABCD*. (Note that this three-edge path is denoted by four letters.) Another path from *A* to *D* is given by *ACD*. Another is given by *ACBACD*. (A path may use some edges more than once.) Note that *ABD* does not designate a path from *A* to *D*, because there is no edge from *B* to *D*.

Where there are multiple edges between vertices, we can distinguish their use in paths by numbering, as indicated above. Thus, in Figure 2.6, a path from *E* to *H* is given by E_2FG_1H. If we are not concerned with which edge is used, we may omit the numerical subscripts.

You may have observed that our notation allows some apparent ambiguity: Does *XY* name the edge joining *X* and *Y* or a one-edge path from *X* to *Y*? The answer is "yes." In other words, every edge is a path. There are not really two *different* ideas represented by an ambiguous label; rather, one idea incorporates the other, and this is naturally represented in the notation.

We can classify the graph in Figure 2.8 as different from those in Figures 2.5 and 2.6 by the existence of paths among vertices.

DEFINITION
2.2.4

A graph in which every pair of distinct vertices is joined by a path is a **connected** graph. A graph that is not connected is **disconnected.**

Another point to be clarified in the terminology of graphs is exemplified in Figure 2.9. The two edges, *SU* and *TV*, have no vertex in common, although they

FIGURE 2.9

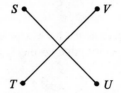

do intersect. This does not disqualify Figure 2.9 as a graph, because every edge does connect two vertices. You should note that this is another example of a disconnected graph. Connectedness is a function of the existence of paths—sequences of edges—and not of the geometric configuration. The distinction of the graph of Figure 2.9 from earlier graphs is given in the following:

DEFINITION
2.2.5

A **planar** graph is one in which two edges, or two parts of the same edge, have points in common only at a vertex.

To summarize: Figures 2.5 and 2.6 show connected planar graphs. Figure 2.8 is planar and disconnected. Figure 2.9 is nonplanar and disconnected. And Figure 2.10 is nonplanar but connected.

Let us now turn to the puzzles of Section 2.1 and rephrase them in terms of graph theory.

The figure (2.1) to be traced in Example 2.1.1 is essentially a graph already. We need only label the vertices. This is done in Figure 2.11. The problem of tracing translates into one of finding a path with the constraint that each edge must be used *exactly* once.

FIGURE 2.10

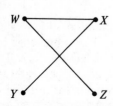

FIGURE 2.11

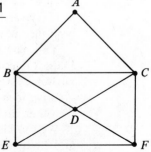

DEFINITION 2.2.6	An **edge path** in a graph is a path in which every edge is used exactly once (once and only once).

The tracing puzzle thus becomes the question: Is there an edge path for the graph of Figure 2.11?

Turning to the Königsburg Bridge Problem, we can translate Figure 2.2 into a graph containing all the significant features of the original. Observing that strolling about on one side or the other of the river or on one of the islands has no effect on the problem (the challenge is finding a way to cross the bridges appropriately), we may identify each of these land masses as a vertex and each bridge as an edge. This is done in Figure 2.12.

FIGURE 2.12

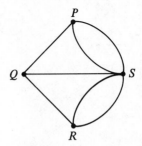

The problem of crossing each bridge exactly once translates into finding an edge path in this graph—the same problem as the tracing puzzle.

The street map of Examples 2.1.3 and 2.1.4 translates in an obvious way into a graph, with each intersection becoming a vertex and each street an edge. This is left as an exercise.

The problem for the pothole repair crew, to travel each street exactly once, is again the edge-path problem. The task for the street sign crew, on the other hand, is different but analogous. Visiting each intersection exactly once translates to finding a path that uses each vertex exactly once. Quite naturally, we define this as a vertex path.

DEFINITION 2.2.7	A **vertex path** in a graph is one in which every vertex is used exactly once.

As in the first puzzle, Figure 2.4 is already a graph, except for the labeling of the vertices. The problem, though, does not translate into finding a path of any sort. Here we are challenged to find some curve that *crosses* edges instead of following them, as in a path. We thus create a term to say precisely this.

DEFINITION 2.2.8	A **crossing curve** in a graph is a continuous curve that crosses each edge exactly once and passes through no vertex.

It should be noted that a crossing curve may have none, one, or both ends inside one of the regions (if any) enclosed by edges of the graph. It should also be noted that there is no constraint about a crossing curve crossing itself one or more times; it may not cross any edge more than once.

EXERCISES 2.2

1. Which of the drawings in Figure 2.13 represent graphs?

FIGURE 2.13

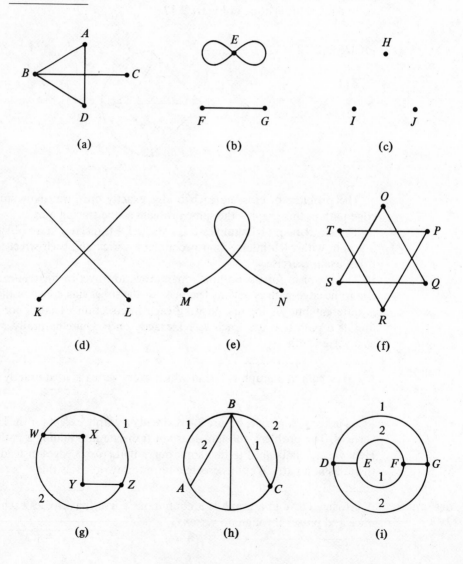

2. Which of the graphs in Figure 2.13 is connected?

3. Which of the graphs in Figure 2.13 is planar?

4. Which of the graphs in Figure 2.13 contains a loop?

5. In Figure 2.13(g):
   (a) Give a path from Z to X which uses no edge more than once.
   (b) Give a path from W to X which uses every edge at least once.
   (c) Give a path from W to Z which is an edge path.
   (d) How many different edge paths are there from W to Z?
   (e) How many different edge paths are there from X to Y?
   (f) How many different paths are there from X to Y?

6. In a certain graph, the following is a path: $AFEBBCEAGCB$. Determine whether the following are true, false, or cannot be determined without additional information.
   (a) This path contains 11 edges.
   (b) The graph contains at least 10 edges.
   (c) The graph contains at least 8 edges.
   (d) The graph contains at least 7 vertices.
   (e) There is no edge joining A and B.
   (f) There is a two-edge path joining A and B.
   (g) The graph contains multiple edges joining some pairs of vertices.
   (h) The graph is connected.
   (i) The graph is planar.

7. Answer Exercise 6 with the additional assumption that the path is an edge path.

8. Draw a graph that represents the street map given in Figure 2.3.

9. The Town of Mud Flats is built on the banks of a river, much like Königsburg (see Figure 2.14). Many years ago the river followed a different channel (indicated by dashed lines) and one of the old bridges was left standing for its scenic beauty. Draw a graph representing the current land masses as vertices and the bridges as edges. Can you find an edge path for this graph?

FIGURE 2.14

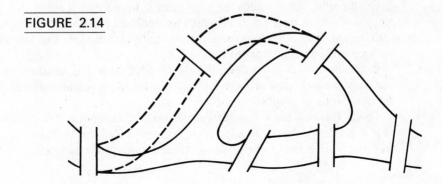

10. For each of the following sets of conditions, give an example of a graph with four vertices which satisfies all the conditions. If not possible, explain why not.
    (a) Disconnected, nonplanar.
    (b) Disconnected, with an edge path.
    (c) Disconnected, with a vertex path.
    (d) Connected, with no vertex path.
    (e) With an edge path, but no vertex path.
    (f) With a vertex path, but no edge path.
    (g) With an edge path and a vertex path.

11. Redo Exercise 10 with three vertices rather than four.

2.3   EDGE PATHS

In this section we turn to the first type of puzzle identified in the previous sections—that of tracing each edge (or each bridge, or each street) exactly once (that is, of finding an edge path).

By trial and error, you may have found an edge path in some of the puzzles in Section 2.1. Perhaps you were lucky and found an edge path with little effort; perhaps you found one only after many trials. In some instances, you certainly did not find an edge path, for none exists in some of the puzzles posed.

Unfortunately, failure to find an edge path does not solve the problem, for we cannot be sure, on the basis of trial and error, whether there is no edge path, or whether we have merely overlooked an edge path. We could record every path that uses no edge more than once between each pair of vertices in the graph, stopping when an edge path is found or, if none is found that uses every edge, establishing that no edge path exists. For any but the simplest graphs (where the existence or nonexistence of edge paths is already obvious), this is thoroughly impractical. For example, in the relatively simple "envelope" graph of Figure 2.11, there are 21 pairs of vertices (including such pairs as C, C) and over 50 different paths that use no edge more than once between each pair. Something better must be found.

On the other hand, when an edge path is found and it solves the particular puzzle at hand, it mathematically calls for analysis. Have we found the only edge path, or are there others? Is there an edge path between any two vertices or only between the pair we happened upon? Why?

Basically, why does one graph have an edge path and another not? Mathematically, we seek some property or combination of properties that will distinguish the two classes of graphs.

To do this, we use a common mathematical technique. We assume that we have a graph which has an edge path, and try to deduce whatever we can about that graph. Effectively, then, we are determining properties possessed by every graph that has an edge path. Hopefully, some of these properties will exist *only* in graphs with edge paths, thus allowing us to identify such graphs by those critical properties.

Suppose, then, that A is a vertex of our graph-with-edge-path; and suppose further that A is neither the beginning nor the end of the path, but some vertex along the way. Clearly, there must be at least two edges at A, one by which the path first comes to A and one by which it leaves. If the path does not return to A, then there are *exactly* two edges at A (Figure 2.15). On the other hand, if the path does return to A, it cannot do so by one of the two edges previously mentioned, since it is an edge path; so there must be a third edge to return by, *and a fourth edge* to leave by **(Figure 2.16)**.

FIGURE 2.15   **FIGURE 2.16**

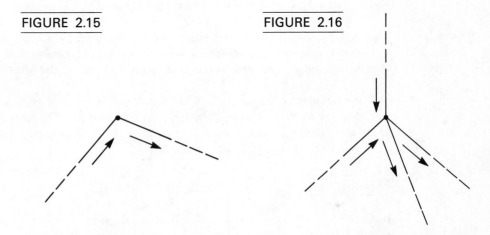

As before, if the path does not return to A, then there are exactly four edges at A. If the path does return, there must be a fifth *and a sixth* edge at A, to allow the path to come and go. In general, since the edge path neither starts nor ends at A, there must be an *even number* of edges at A.

There is one slight flaw in this analysis. We have assumed that when the path leaves A, it goes on to some *other* vertex (and comes to A from some other vertex). If there is a loop at A, the path leaves A and returns on the same edge, so that only one edge is involved in this leaving and returning.* From the localized viewpoint, however, we can incorporate this possibility in our earlier analysis by noting that there are still an even number of *ends of* edges at A.

This discussion motivates the next definitions.

DEFINITION 2.3.1

The **degree** of a vertex is the number of ends of edges incident with it.

---

* For this reason, among others, some authors exclude loops in defining what a graph is.

DEFINITION 2.3.2

An **even vertex** is one with even degree; an **odd vertex** is one with odd degree. An **isolated vertex** is one with degree zero.*

For example, in Figure 2.8, vertices P, Q, and R each have degree 2 and are thus even vertices; N and O each have degree 1 and are odd vertices; and M is an isolated vertex, having degree zero.

Our previous observations, then, yield the fact that in a graph with an edge path, a vertex that neither starts nor ends that path must be an even vertex. A similar analysis of a vertex that is one endpoint of the edge path (the start or finish, but not both) must be an odd vertex. If the edge path starts and ends at a single vertex, that vertex must be even. Thus, we have reasoned to the following:

THEOREM 2.3.1

If a graph has an edge path, then either all the vertices are even and the path starts and ends at one vertex; or there are exactly two odd vertices and all others are even and the edge path starts at one odd vertex and ends at the other.

For example, in Figure 2.17, $PRTSRQ_1S_2QP$ is an edge path starting and ending at vertex P. And, as our observations indicated and Theorem 2.3.1 states, each vertex in the graph is even. In Figure 2.18, $ACDAB$ is an edge path starting at one odd vertex, A, and ending at another.

FIGURE 2.17

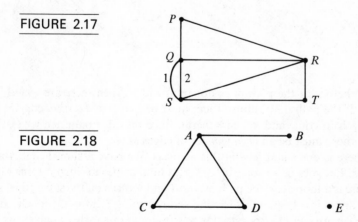

FIGURE 2.18

Since a graph with an edge path can have at most two odd vertices, we have a relatively easy way to rule out a large number of graphs as conclusively having no possible edge path. We need merely test each vertex to determine if it is odd; if we find three such vertices, there is no need to proceed any further. This is formalized in:

---

* Some authors exclude isolated vertices in defining a graph.

THEOREM
2.3.2

If a graph has more than two odd vertices, it has no edge path.

Does it follow that a graph with no odd vertices or two odd vertices *has* an edge path? Clearly not, as Figure 2.19 shows. The problem in this graph is connectedness.

FIGURE 2.19

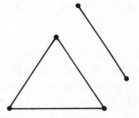

If we again examine our hypothetical graph-with-edge-path, the existence of that edge path guarantees our ability to get from one vertex to another—*with one exception*. That exception is the possible presence of isolated vertices in a graph with an edge path. Since no edges are incident with an isolated vertex, such a vertex has no effect on the existence of an edge path,* although it does cause the graph to be disconnected.

DEFINITION
2.3.3

A graph that is connected, or connected except for one or more isolated vertices, is a **nearly connected** graph.

For example, Figure 2.17 is connected, and thus nearly connected. Figure 2.18 is nearly connected, although Figure 2.19 is not. With this terminology, the following becomes obvious.

THEOREM
2.3.3

A graph that has an edge path is nearly connected.

If we combine the two properties we have identified, we do obtain a characteristic way to distinguish graphs with edge paths.

THEOREM
2.3.4

If a graph is nearly connected and has at most two odd vertices, then it has an edge path.

Since near-connectedness (or its lack) is obvious at a glance, this theorem effectively reduces the determination of the existence of an edge path to that of

---

* This is one reason some authors exclude isolated vertices.

counting the degree of vertices. (Actually, since only odd or even parity need be determined, it is literally as simple as performing the numerical analog of "She loves me, she loves me not.") The proof of Theorem 2.3.4 is somewhat lengthy. It is not difficult, although you should proceed slowly, making sure you understand each step before going on to the next.

PROOF   We shall prove the case of two odd vertices; the remainder is left as an exercise.

Step 1   Let us assume that the two odd vertices are labeled P and Q, respectively. We start at one, say P, and randomly construct a path, using no edge more than once, continuing as long as there are unused edges available.

Each time this path comes to a vertex other than P or Q, the path uses one edge of an available even number, so there must be another at that vertex not yet used, by which the path can continue. Coming to and leaving such a vertex uses two of an even number of edges (or ends of edges), leaving an even number yet to be used. Thus, the path cannot end at such a vertex.

One edge of an odd number of edges at P is used (the very first one of the path), leaving an even number; so if the path returns to P the situation is exactly like coming to an even vertex. Namely, the path cannot end at P.

Eventually, the path must end, since there are only a finite number of edges. Our analysis shows that the only place it can end is at Q. At this point, all the edges at Q will be used up—otherwise, the path could continue.

At this juncture, an odd number of edges at P have been used, all the edges at Q have been used, and an even number of edges (possibly zero) at every other vertex have been used. There may be edges not used in the path. If there are any, there are an even number remaining at any vertex where they exist.

Step 2   If the path constructed in step 1 has used all the edges, it is an edge path, and we are finished. If there are unused edges, then there must be some vertex which has *both* some edges that were used in the path and some that were not. (If this were not the case, the graph would not be nearly connected.) We choose one such vertex, and assume for reference that it is labeled X.

Step 3   Starting at X, we again construct a path, using no edge which has previously been used (in step 1 or this step), continuing as long as there are available edges. As was the case in step 1, this path must end, but cannot stop where an even number of edges are available. Thus, it must stop at X, where the starting edge left an odd number of edges.

Step 4   We now "cut and paste" the two paths, constructed in steps 1 and 3, where we know they share a common vertex, X. Specifically, we follow the original (step 1) path from P until we come to X; we then follow the path constructed in step 3 to its end at X, where we resume following the original path to its end at Q. The result is an expanded path from P to Q, using no edge more than once.

Step 5   If all edges have now been used, the expanded path is an edge path and we are finished. If not, the analysis of step 2 is applicable, so we choose an appropriate vertex, with both "used" and "unused" edges, and repeat steps 3 and 4.

Each time we do this, we use up more edges. Eventually, all edges in the graph must be included in the expanded path, and we have an edge path.

Theorems 2.3.1, 2.3.3, and 2.3.4 completely characterize which graphs have edge paths. We end this section with several definitions.

DEFINITION
2.3.4

A path that starts and ends at distinct vertices is an **open** path.

DEFINITION
2.3.5

A path that starts and ends at one vertex is a **closed** path.

DEFINITION
2.3.6

A connected graph that has a closed edge path is an **Euler graph**.

DEFINITION
2.3.7

A graph with no edges is called a **null graph.**

EXERCISES 2.3

1. In Figure 2.20(g), give the degree of each vertex.

2. For each of the seven graphs in Figure 2.20, determine which vertices are even and which are odd.

3. For each of the seven graphs in Figure 2.20, how many odd vertices are there?

4. (a) Which of the seven graphs in Figure 2.20 have no edge path?
   (b) Which have an open edge path?
   (c) Which have a closed edge path?
   (d) Which are Euler graphs?

5. In the text, only one case of Theorem 2.3.4 was proved—that of two odd vertices. Modify this proof to cover the case of no odd vertices.

6. Can you construct a graph with exactly one odd vertex? (*Hint:* study the reasoning of step 1 in the proof of Theorem 2.3.4.)

7. Generalize the reasoning of step 1 in the proof of Theorem 2.3.4 to show that every graph (connected or not) has an even number of odd vertices.

8. Apply the methods of this section to Examples 2.1.1, 2.1.2, and 2.1.4 to answer the questions posed there.

9. Show that if the people of Königsburg built one additional bridge from one land mass to another, they could find a path that crosses each of the bridges

---

FIGURE 2.20

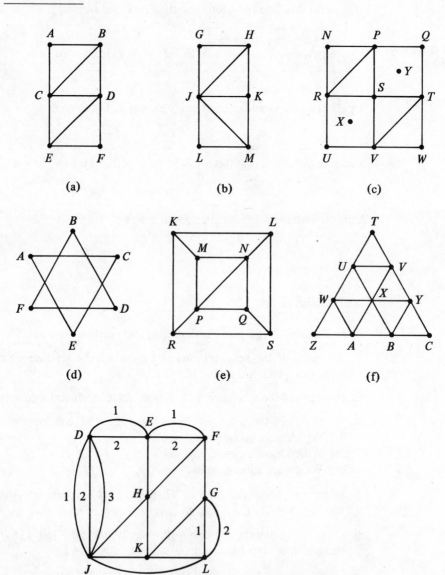

(a)

(b)

(c)

(d)

(e)

(f)

(g)

exactly once. (See Example 2.1.2 and Figures 2.2 and 2.12.) Show that the same would be true if they tore down one bridge.

10. How many new bridges should be built in Königsburg if the townspeople wish to be able to cross every bridge exactly once and return to their starting place? (See Exercise 9.) Where should the new bridge(s) be built?

11. If a connected graph has exactly six odd vertices, how many new edges must be added to change it into an Euler graph? (See Exercises 9 and 10.) Where should these edges be added? Apply this idea to the graph in Figure 2.20(e); what vertices can be joined by edges to convert the graph to an Euler graph?

12. If a connected graph has exactly n odd vertices (n is some natural number), how many new edges must be added to change it into an Euler graph? (See Exercise 11.)

13. Show that an Euler graph must have at least as many edges as it has vertices.

14. Show that by deleting certain edges from the graph of Figure 2.20(e) you can convert it into an Euler graph. (See Exercise 11.) What is the smallest number of edges that can be deleted to accomplish this? What is the largest number? (See Exercise 13.)

15. Give an example of a connected graph that is not an Euler graph and cannot be converted into an Euler graph by the deletion of any number of edges. (See Exercise 14.)

16. Consider carefully Definitions 2.2.3, 2.2.4, 2.2.6, 2.3.3, 2.3.6, and 2.3.7 to determine the correct answer to the following:
    (a) Is a null graph connected?
    (b) Is a null graph nearly connected?
    (c) Is there a path in a null graph?
    (d) Is a null graph an Euler graph?

17. A certain graph is not a null graph and is not an Euler graph. What is the smallest number of vertices this graph can have?

18. A certain graph is not a null graph and has no edge path. What is the smallest number of vertices this graph can have?

2.4   VERTEX PATHS

We turn now to the problem suggested by Example 2.1.3, that of finding a path in a graph which passes exactly once through each vertex. Noting the close parallel in form with the problem of the previous section, we shall approach it with the same method.

Specifically, let us suppose that we have a graph with a vertex path. What can we logically assert about such a graph? Perhaps the most obvious fact is that the graph must be connected; since the vertex path passes (exactly once) through *every* vertex of the graph, some part of the vertex path must be a path between any two vertices we choose. Thus, we have proved:

THEOREM
2.4.1

If a graph has a vertex path, it is connected.

It is not difficult to see that connectedness is not sufficient to guarantee that a graph has a vertex path, as Figure 2.21 shows. Although this graph is connected, a little reflection shows that it cannot have a vertex path. Vertex A has degree 1, and thus any vertex path must either start or end at A. But the same is true of vertices C and F; and a path cannot have three ends. Let us label such vertices, and state our observation as a theorem.

FIGURE 2.21

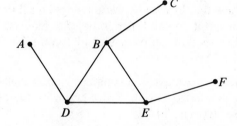

DEFINITION
2.4.1

A vertex of degree one is a **terminal vertex.** The edge at this vertex is a **terminal edge.**

THEOREM
2.4.2

If a graph has a vertex path, it has at most two terminal vertices.

We might note that a vertex path in a graph with one or two terminal vertices would have to be an open vertex path, or:

THEOREM
2.4.3

If a graph has a closed vertex path, it has no terminal vertices.

Rather than continuing to state pairs of theorems for open versus closed vertex paths, let us restrict ourselves to consideration of one type, leaving the very similar observations about the other for the exercises. For simplicity, and historical interest, we shall consider only closed vertex paths for the remainder of the section. Analogous to Euler graphs, we have:

DEFINITION
2.4.2

A **Hamilton* graph** is one with a closed vertex path.

Suppose, then, that we have a connected graph with no terminal vertices. Is it a Hamilton graph? Unfortunately, the answer is no, as we can observe in Figure 2.22. Here we see a graph that has no terminal edges and is connected, but is not a Hamilton graph. The problem is that it has two "halves" with only one vertex, T, connecting them. A closed vertex path must pass through vertices on either side of T and form a *closed* path. This requires passing through T *twice*, a contradiction to the definition of a vertex path, so no closed vertex path exists.

FIGURE 2.22

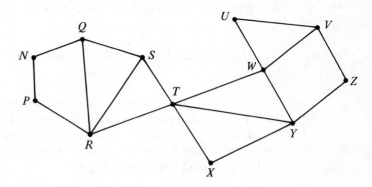

We can characterize T in the following definition, and identify the general connection with Hamilton graphs.

DEFINITION
2.4.3

A vertex in a connected graph is a **critical vertex** if its removal, together with all edges incident with it, leaves a disconnected graph.

THEOREM
2.4.4

A Hamilton graph contains no critical vertices.

Do we now have enough facts about Hamilton graphs to characterize them? If a graph is connected, has no terminal vertices or critical vertices, is it a Hamilton graph? Again, the answer is "no," as is shown in Figure 2.23.

FIGURE 2.23

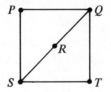

---

* See Section 1.8.

In this graph, a closed vertex path must go through edges PQ and SP, as the only way to reach (and leave) vertex P. But the same is true of edges SR and RQ, for vertex R, and of edges ST and TQ for vertex T. Thus, three edges at Q (and also at S) *must* be in any possible closed vertex path in order to reach every vertex; but this implies passing through Q (and S) more than once. Again, we define and record :

DEFINITION
2.4.4

An edge incident with a vertex of degree two is called a **Hamilton edge.** (We shall abbreviate as *H*-edge.)

THEOREM
2.4.5

In a graph, a closed vertex path must contain every *H*-edge.

THEOREM
2.4.6

In a Hamilton graph, no vertex is incident with more than two *H*-edges.

A fairly simple circumvention of this constraint is suggested in Figure 2.24. In part (a), multiple edges join the vertices; in part (b), loops have been added. The net effect in either case is to increase the degree of the vertices, thus technically eliminating *H*-edges, but the graphs still have the same basic drawback as Figure 2.23 in failing to be Hamilton graphs.

FIGURE 2.24

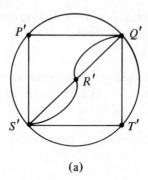 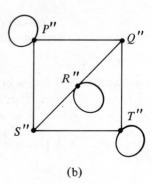

(a)   (b)

In general, multiple edges and loops do not provide useful additional edges to be used in a closed vertex path. They only serve to obfuscate our search for a characteristic of Hamilton graphs. So we shall ban them for the remainder of this section.

DEFINITION
2.4.5

A **simple** graph is one with no loops and no multiple edges joining any pair of vertices.

If we collect all our observations so far, we are ready to pose our characterization question again. If a simple, connected graph has no terminal vertices and no critical vertices and no vertex incident with more than two H-edges, is it a Hamilton graph? You guessed it. The answer is still "no."

Figure 2.25 shows a graph with all the conditions called for. Each H-edge is marked with a short cross-mark. Here, we can identify our new problem in terms of these edges, but it is not that there are too many at one vertex. Rather, the H-edges at A, B, C, and D form a closed path. By Theorem 2.4.5, each of these edges must be used in a closed vertex path; by definition, the other vertices (E, F, G, H) must also be used. But there is no way to do both of these things. We have another exclusion to add to our growing list.

FIGURE 2.25

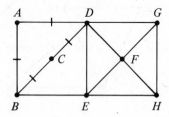

THEOREM 2.4.7

If the H-edges in a graph form a closed path among some, but not all, vertices, the graph is not a Hamilton graph.

Are we finished? If a simple graph passes all the tests implied in Theorems 2.4.1, 2.4.3, 2.4.4, 2.4.6, and 2.4.7, is it a Hamilton graph? Unfortunately, no. Figure 2.26 is a counterexample.

FIGURE 2.26

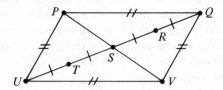

As before, H-edges are marked with a single cross-mark. Thus, any closed vertex path must contain $QRSTU$. Clearly, this cannot be accommodated in a vertex path which must reach both P and V and be closed.

One way to classify the problem is to note that since RS and ST are H-edges, PS and SV cannot be used. Effectively, then, PQ, QV, VU, and UP are "secondary" H-edges because they are the only edges actually available at P and V. (These are marked with double cross-marks.) Then the H-edges and "H_2-edges" together rule out a closed vertex path by either Theorem 2.4.6 or Theorem 2.4.7.

We could formally define "H_2-edges," "H_3-edges," and so on, and generalize Theorems 2.4.6 and 2.4.7 accordingly. But even then we would fail to cover every quality that characterizes Hamilton graphs.

Figure 2.27 shows a simple connected graph with no isolated or critical vertices and no H-edges. Yet there is no closed vertex path for this graph.

FIGURE 2.27

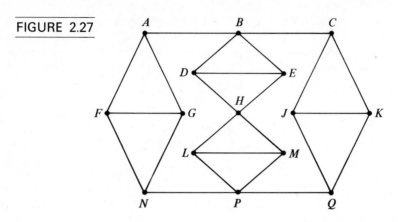

What we have experienced in this section is a taste of the frustration met by every individual who has tried to characterize Hamilton graphs. Although the problem is over a century old, no one has yet found any combination of properties whose presence in a graph will guarantee the existence of a closed vertex path.

This is not to say that our efforts have been wasted. We have identified a number of properties that are necessary for a Hamilton graph. Their absence, then, allows us to easily identify that a given graph is not Hamilton. And our observations about H-edges can be an aid to finding a closed vertex path if one exists, even though we have no theorems to guarantee existence of the desired path.

Similarly, although we cannot characterize whether or not a graph has an open edge path, we can modify the theorems on closed edge paths to provide useful information.

In all, determining whether a graph has a vertex path or not remains largely a matter of trial and error, although our theorems can greatly reduce the number of errors and can provide an organized guide as to what to try.

EXERCISES 2.4

1. Which graphs in Figure 2.28 have terminal vertices?

2. Identify the H-edges in Figure 2.28(b), (e), and (h).

3. Which graphs in Figure 2.28 are Hamilton graphs?

4. Which graphs in Figure 2.28 have open vertex paths?

FIGURE 2.28

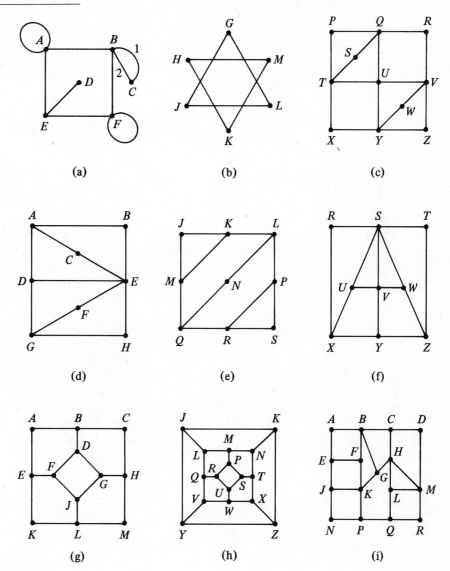

(a)   (b)   (c)

(d)   (e)   (f)

(g)   (h)   (i)

5. Can the street sign crew of Example 2.1.3 do their job by visiting each intersection exactly once? (See Figure 2.3.)

6. Show that every Hamilton graph has an open vertex path. Give an example to show that the converse is not true.

7. Prove that an open vertex path may fail to contain at most two *H*-edges in a graph. (*Hint*: See Theorem 2.4.5.)

8. Prove that if a graph has a vertex incident with five or more *H*-edges, it contains no vertex path. (*Hint*: See Theorem 2.4.6.)

9. Prove that a connected graph with more than two vertices which has a terminal vertex must have a critical vertex.

10. Show that if a closed path is both a vertex path and an edge path for a graph, the number of vertices must be equal to the number of edges.

2.5   CROSSING CURVES

We now turn to the type of problem posed in the puzzle of Example 2.1.5; that of finding a continuous curve crossing each line segment in a figure exactly once. In that puzzle, we asserted that the figure could be labeled in an obvious and unambiguous way to become a graph. This was a correct assertion for that, and many other, graphs, but this is not always the case, and the option may dramatically alter the answer to the crossing-curve question.

Figure 2.29(a) shows a simple figure which, with appropriate identification of vertices, is a graph. Two ways of making this identification are given in Figure 2.29(b) and (c). In part (b), the resulting graph is nonplanar and has a crossing curve, as indicated, but seems to inadequately convert Figure 2.29(a). In part (c), the resulting graph is planar (and has more edges) and appears a more honest graph representation of part (a), but there is no crossing curve possible.

We shall restrict ourselves to planar graphs. We shall also consider only connected graphs, on the basis that, for purposes of a crossing curve, a disconnected graph can be treated as several different connected graphs.

FIGURE 2.29

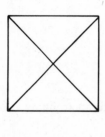

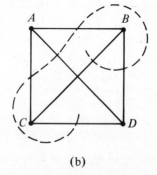

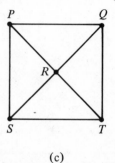

(a)   (b)   (c)

DEFINITION
2.5.1

A **polygonal graph** is a connected planar graph. Each region of the plane determined by the edges of a polygonal graph is a **face** of the graph.

It should be noted that the portion of the plane "outside" a polygonal graph is also considered a face. For example, in Figure 2.29(c), there are five faces.

A crossing curve, then, is one that crosses from one face to another, intersecting each edge exactly once. An exception occurs if a graph has a terminal edge. Here, the crossing curve remains within a given face when crossing such an edge. You should note that terminal edges pose no barrier to the existence of a crossing curve. A crossing curve will exist if we are able to reach every edge. A curve will be impossible if all the edges surrounding some face are used up (crossed once already) and the curve is "trapped" within that face while some edges, elsewhere, are not yet crossed.

Thus, the existence of a crossing curve is determined by the number of times a possible curve can enter or leave a face. We define a term to identify this number.

DEFINITION
2.5.2

The **order** of a face is the number of edges adjacent to it. A terminal edge is counted twice.

The peculiarity of this definition with regard to terminal edges can be explained in that a crossing curve, in crossing a terminal edge, leaves and enters the face simultaneously. Alternatively, we could think of each edge as having two sides; usually one side is in one face and the other is in a neighboring face, but both sides of a terminal edge lie in the same face.*

EXAMPLE 2.5.1

In Figure 2.30(a), there are three faces; one of order three, one of order four, and one (the "outside" face) of order five. In Figure 2.30(b), there are four faces: two of order three, one (contained by the loop) of order one, and one of order seven.

Let us consider Figure 2.30(a) and seek a crossing curve. Since the face enclosed by the path *BCEB* has order three, a crossing curve can enter this face twice and

---

* This peculiarity could be avoided by excluding terminal edges in the definition of polygonal graphs. This option also has its problems.

FIGURE 2.30

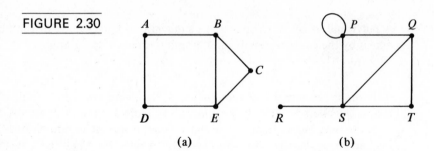

(a)   (b)

leave once, or leave twice and enter once. In other words, the start or finish of the curve must lie in this face. The same is true of the "outside" face of order five. Fortunately, the face of order four must be entered twice and exited twice, and the curve need have no end in this face. A crossing curve is thus theoretically possible—one such curve is indicated in Figure 2.31.

FIGURE 2.31

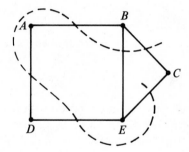

Turning to Figure 2.30(b), each of the four faces has an odd order, requiring one more entry than exit, or vice versa. Thus, a crossing curve must have one of its ends in each of four faces—an impossibility. There is no crossing curve for this graph.

If this "even versus odd" analysis is reminiscent of the discussion of edge paths, there is good reason. The problem of finding a crossing curve from face to face *across* edges is exactly analogous to finding an edge path from vertex to vertex *along* edges. We can formally make the transition via the following definition and theorem.

DEFINITION 2.5.3

The **dual** of a given polygonal graph is one in which each face of the given graph has a corresponding vertex in the dual, and each edge in the given graph has a corresponding edge in the dual such that an edge which separates two faces in the given graph corresponds to an edge in the dual connecting the vertices that correspond to those faces.

EXAMPLE 2.5.2

The graph of Figure 2.32(a), with faces labeled by lowercase letters, has as its dual the graph of Figure 2.32(b). The correspondence between edges and between

FIGURE 2.32

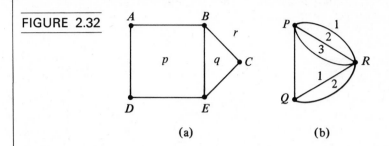

(a)   (b)

faces and vertices is given in Table 2.1. You should check that an edge separating two faces corresponds to an edge joining the corresponding vertices.

TABLE 2.1

Graph (a) Face	Dual (b) Vertex	Graph (a) Edge	Dual (b) Edge
p	P	AB	P_1R
q	Q	AD	P_2R
r	R	BC	Q_1R
		BE	PQ
		CE	Q_2R
		DE	P_3R

THEOREM 2.5.1 A polygonal graph has a crossing curve if and only if its dual graph has an edge path.

The proof of this theorem is contained in the previous discussion. We need only note that in the correspondence from a polygonal graph to its dual, the order of a face corresponds to the degree of the corresponding vertex, so that the discussion and proofs in Section 2.3 apply exactly to crossing curves via the correspondence of the dual graph.

For example, the crossing curve indicated in Figure 2.31 corresponds to the edge path $Q_2R_3P_2R_1PQ_1R$ in the dual graph (Figure 2.32). Conversely, any edge path in the dual can be translated to a crossing curve in the original graph.

We can, then, determine the existence of crossing curve by constructing the dual graph and applying the theorems of Section 2.3. More simply, we can translate the theorems of that section to theorems about crossing curves by replacing face for vertex, order of face for degree of vertex, and so on.

EXAMPLE 2.5.3

Consider the graph of Figure 2.33. Faces have been labeled and the order of each face identified in Table 2.2. Since there are only two odd faces, there is a crossing curve, with one end in face (a) and the other in face (f). The drawing of the crossing curve is left as an exercise.

FIGURE 2.33

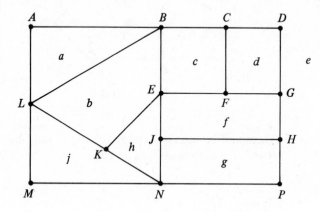

TABLE 2.2

Face	Order	Face	Order
a	3	f	5
b	4	g	4
c	4	h	4
d	4	j	4
e	10		

EXERCISES 2.5

1. Find a crossing curve for the graph of Figure 2.33.

2. Does the **graph** in Example 2.15 have a crossing curve? If so, draw one.

3. Which of the graphs in Figure 2.34 have crossing curves? For those which do, draw the crossing curve.

4. Construct a graph that is the dual of Figure 2.34(c).

5. A **self-dual** graph is one that, with the appropriate correspondence, is dual to itself. Give a correspondence of faces and edges in Figure 2.34(a) to show that it is self-dual. Show that Figure 2.34(b) is self-dual.

FIGURE 2.34

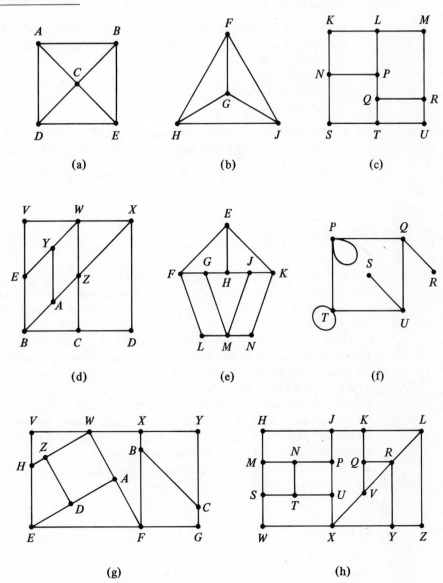

(a)   (b)   (c)

(d)   (e)   (f)

(g)   (h)

6. If possible, construct a self-dual graph with exactly three vertices. (See Exercise 5.) If impossible, explain why.

7. If possible, construct a self-dual graph with exactly four vertices which is different from Figure 2.34(b) (that is, not with four faces each of order three). (See Exercise 5.) If impossible, explain why.

8. Show that if a graph named graph 2 is the dual of graph 1, then graph 1 is the dual of graph 2.

9. In the process of forming dual graphs, face and vertex are duals of each other; crossing curve and edge path are duals, and so on. Identify the dual of each of the following:
   (a) Terminal edge.
   (b) Terminal vertex.
   (c) Isolated vertex.

10. Give definitions of "open crossing curve" and "closed crossing curve" which are dual to open edge path and closed edge path. (See Exercise 9.)

11. Construct a graph with exactly one odd face. If impossible, explain why.

2.6   EULER'S FORMULA

We have already noted some strong connections between graph theory and numbers in searching for edge paths and crossing curves. In both instances, the even or odd degree of vertices (or order of faces) largely determined the existence of the path (or curve).

There are many other numerical properties of graphs. For example, if we add up all the degrees of all the vertices in a graph, the result must be even, because we have effectively counted all the ends of all the edges, which must come in pairs. This fact allows us to prove that odd vertices, if any, must be even in number, and provides an alternative method of proof from that suggested in Exercise 2.3.7.

A more significant, and useful, relation exists among the number of vertices, edges, and faces in a polygonal graph. For a source of data, let us turn to the graphs of Figure 2.34 in the exercises at the end of the previous section. In Table 2.3, we have recorded the number of vertices, edges, and faces in each, denoted by V, E, and F, respectively.

TABLE 2.3

Graph	V	E	F
(a)	5	8	5
(b)	4	6	4
(c)	10	13	5
(d)	10	15	7
(e)	9	13	6
(f)	6	8	4
(g)	12	18	8
(h)	17	25	10

At first glance, there may not seem to be much of a pattern. For example, graphs with the same number of vertices do not have the same number of edges, and vice versa. Those with more vertices tend to have more edges and faces, but in apparently random fashion.

We can note that edges tend to be more abundant than either vertices or faces. If we look at the excess of edges over vertices or faces, a striking pattern emerges (see Table 2.4).

TABLE 2.4

Graph	V	E	F	$E - V$	$E - F$
(a)	5	8	5	3	3
(b)	4	6	4	2	2
(c)	10	13	5	3	8
(d)	10	15	7	5	8
(e)	9	13	6	4	7
(f)	6	8	4	2	4
(g)	12	18	8	6	10
(h)	17	25	10	8	15

You should now note that $E - V$ is in every instance exactly 2 less than F. Also, $E - F$ is exactly 2 less than V. If we state these two observations in the form of algebraic equations, we obtain

$$E - V = F - 2 \tag{2.1}$$

and

$$E - F = V - 2 \tag{2.2}$$

It is not hard to see that both contain the same information, and we have a single formula relating V, E, and F. This can be juggled algebraically in many ways. Regardless of the form, the relation is called Euler's Formula.* Of course, the fact that this formula is correct for eight examples is not proof that it is always correct. The formal statement and proof are not too difficult.

THEOREM 2.6.1 (Euler's Formula)

In any polygonal graph, the number of vertices, edges, and faces are related by: $V - E + F = 2$.

PROOF   Any polygonal graph can be constructed by starting with one vertex and expanding step by step with one of the following:

1. Add a new vertex and join it by an edge to some vertex already present; or
2. Add a new edge joining two vertices already present (including the possibility of adding a loop).

---

* This formula was reportedly known to earlier mathematicians, but was expostulated and proved by Euler in a brief paper.

For example, we can construct the polygonal graph in Figure 2.35(g) by start-ing with a single vertex in Figure 2.35(a) and adding vertices and edges according to these two steps. Each intermediate step is shown in Figure 2.35, and is identified with a "1" or "2," referring to the two types of construction identified above. Although this is not the only sequence of steps you might follow, you can see that this, or any, polygonal graph can be constructed with these two constructions alone.

Turning to the quantities involved in Euler's Formula, we start with $V = 1$, $E = 0$, and $F = 1$ (the plane is one face), and Euler's Formula is satisfied.

Each time construction 1 is used, we add a vertex and an edge, so V and E each increase by 1. Since the graph we are building is polygonal, it is planar. Thus, the new edge lies entirely within a face that already exists, and no new faces are created by this construction (that is, F remains unchanged). In all, $V - E + F$ does not change from its value before the step.

Each time construction 2 is used, E increases by 1 and V remains fixed. Since the added edge lies within some face (planar graph) and the graph is connected at each step (new vertices are always connected by a new edge, in step 1), this edge must "cut off" part of the "old" face, creating a new, additional face (that is, F increases by 1). In all, $V - E + F$ does not change.

Therefore, we start with $V - E + F = 2$, and build up the graph by a series of constructions that leave the quantity $V - E + F$ unchanged at each step. We must end with a graph satisfying Euler's Formula.

FIGURE 2.35

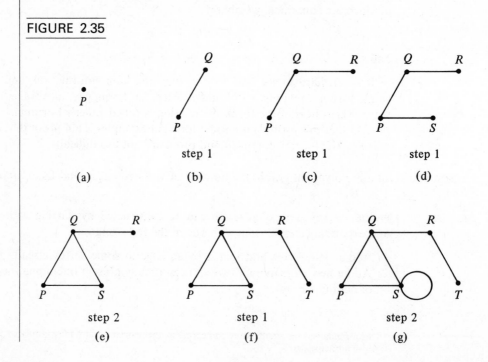

step 1   step 1   step 1

(a)   (b)   (c)   (d)

step 2   step 1   step 2

(e)   (f)   (g)

There are a number of applications and generalizations of Euler's Formula of importance in graph theory, geometry, and topology. We shall consider one such classical case.

Let us consider some solid geometric figures, such as cubes, prisms, pyramids, and so on. Specifically, we shall look at any solid whose edges are straight-line segments, whose faces are polygons (triangles, quadrilaterals, and so on), and which have no "holes" (for example, no square doughnuts allowed).

If we think of the solid as reduced to an open-latticework figure (such as a toothpick model), with elastic edges, we can flatten the solid into a plane, forming a polygonal graph.* For example, in Figure 2.36(b), we have a "graphic" representation of the cube indicated in Figure 2.36(a).

FIGURE 2.36

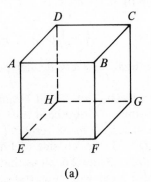
(a)

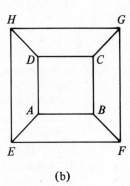
(b)

Note that the graph accurately represents the number of vertices, edges, and faces of the solid, plus the number of edges (and faces) meeting at each vertex (degree) and the number of edges surrounding each face (order). Note that one face of the cube (*EFGH*) corresponds to the "outside" face of the graph. Admittedly, the geometric shape of most of the faces is distorted.

Since the number of vertices, edges, and faces translates from solid to graph (and vice versa) unchanged, Euler's Formula must hold for such solids. This allows one to prove several important facts about geometric solids.

As one example, let us consider regular solids, that is, solids with each face an equilateral, equiangular polygon, and the same number of faces at each vertex. The cube is probably the most familiar example of a regular solid. Are there others? How many and what do they look like?

Euler's Formula and a little arithmetic can answer these questions. If we let D be the degree of each vertex (the degree must be the same for each if the solid is regular), then $D \cdot V$ counts the sum of all the degrees of all the vertices. This must equal twice the number of edges, as noted earlier. Similarly, letting O be the order

---

* It is precisely this connection that lends the name "polygonal" to polygonal graphs.

of each face, $O \cdot F$ must equal $2E$; this follows either by duality or by noting that each edge has two faces adjacent to it. Thus, we know that

$$D \cdot V = 2E \quad \text{or} \quad V = \frac{2E}{D} \tag{2.3}$$

and

$$O \cdot F = 2E \quad \text{or} \quad F = \frac{2E}{O} \tag{2.4}$$

Substituting these into Euler's Formula, we obtain

$$\frac{2E}{D} - E + \frac{2E}{O} = 2 \tag{2.5}$$

Dividing this expression by E, we have

$$\frac{2}{D} - 1 + \frac{2}{O} = \frac{2}{E} \tag{2.6}$$

as a formula that must hold for every regular polyhedron (solid with polygonal faces).

We can now try some values of D and O, and determine the corresponding values of E from this formula. O must be at least 3 (a triangle is the simplest plane figure) and D must be at least 3 (each vertex of a solid must join at least three faces). If we try $D = O = 3$, the left side of Eq. 2.6 becomes $\frac{2}{3} - 1 + \frac{2}{3}$, or $\frac{1}{3}$. E must equal 6. Equations 2.3 and 2.6 tell us that V and F are each 4 in this case, and we have described a solid with four triangular faces. This figure is called a *tetrahedron*.

The number of choices for D and O turns out to be quite limited. This follows from Eq. 2.6. Since E must be positive, and thus $2/E$ also, it follows that the left side of the equation must also be positive. If D or O are too large, $2/D$ and $2/O$ become so small that their sum is less than 1, and the left side is negative or zero. Thus, neither D nor O can be too large. By trial, there are only five possible pairs of values for D and O. These are given in Table 2.5, together with the resulting values of E, V, and F.

These five regular solids were known to the ancient Greek mathematicians. They could find no others. But they failed to prove that there were no others. It

TABLE 2.5

D	O	E	V	F	Name of Solid
3	3	6	4	4	tetrahedron
3	4	12	8	6	hexahedron (cube)
3	5	30	20	12	dodecahedron
4	3	12	6	8	octahedron
5	3	30	12	20	icosohedron

was not until the advent of graph theory that these simple numerical observations were made in a form that could be proved.

EXERCISES 2.6

1. The graph in Figure 2.37 is planar but not connected. If we stretch the definition of "face," we can identify three faces. Count the vertices and edges and compute $V - E + F$.

FIGURE 2.37

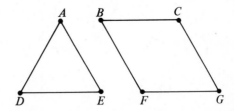

2. Construct several planar graphs that are disconnected into two parts. (See Exercise 1.) For each, compute $V - E + F$. Conjecture a formula for such graphs. Prove your conjecture if you can.

3. Repeat Exercise 2 for planar graphs disconnected into three parts.

4. Let P be the number of parts in a planar graph. (For a connected graph, $P = 1$; the graph of Figure 2.37 has $P = 2$; etc.) Determine a formula for planar graphs which relates V, E, F, and P. (See Exercises 2 and 3.)

5. In Eq. 2.6, if $O = 3$ and $D = 6$, the left side of the equation becomes zero. The right side, $2/E$, can be thought of as equaling zero if E is infinite. This corresponds to a figure with six ($D = 6$) equilateral triangles ($O = 3$) at each vertex, extending infinitely throughout the plane. This is called a regular **tesselation** of the plane with triangles (see Figure 2.38). Find any other values of O and D that have this property. Identify the corresponding tesselation.

FIGURE 2.38

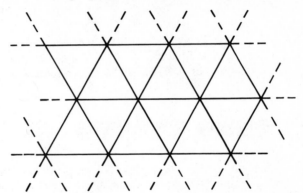

6. Verify Euler's Formula for the solids pictured in Figure 2.39.

FIGURE 2.39

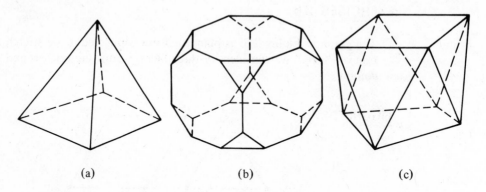

(a)   (b)   (c)

7. Calculate $V - E + F$ for the solids in Figure 2.40. These solids, with polygonal faces, have one "hole" through them. Note that no single face has a hole; rather the hole is surrounded by several polygons.

FIGURE 2.40

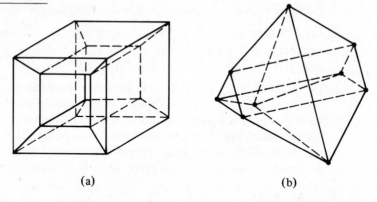

(a)   (b)

8. Let H be the number of holes in a solid. (For the solids in Figure 2.39, $H = 0$; for those in Figure 2.40, $H = 1$; and so on.) Determine a formula for solids which relates V, E, F, and H. (See Exercises 6 and 7.)

2.7   DIGRAPHS AND APPLICATIONS

We mentioned at the outset that graph theory has been applied in a number of areas. Some of these areas are internal to mathematics. As we saw in Exercises 2.5, solids have a formula to relate the number of vertices, edges, faces, and holes.

There are figures for which it is difficult to determine whether or not a hole should be counted as such. The known formula becomes a way of deciding this and classifying the solid. A similar technique aids topologists in classifying the dimension and shape of various figures.

Our discussion of edge paths and vertex paths can be applied in some fairly simple situations. In general, though, such problems are amenable to common-sense analysis, and can be figured out (or circumvented) without appealing to graph theory.

An area we have not pursued deals with the construction of planar graphs with specific conditions as to which vertices are to be joined by edges. This problem is solved theoretically, although practical problems may arise. In other words, theory may guarantee that a certain planar graph exists, but the actual drawing can be very tedious. Such graphs find application in the design of printed circuits.

Many of the applications of graph theory come from adding the additional ideas of direction to tracing paths in a graph.

DEFINITION 2.7.1

A **directed edge** in a graph is one that is assigned a direction. This is indicated by drawing the edge with an arrowhead.

DEFINITION 2.7.2

A graph in which every edge is directed is a **directed graph** or **digraph.**

DEFINITION 2.7.3

A **directed path** is a path in a directed graph. The direction of the path must coincide with the direction of each edge in the path.

EXAMPLE 2.7.1

Figure 2.41 shows two examples of digraphs. In 2.41(a), *ABED* is a directed path from *A* to *D*. There is no directed path from *D* to *A*. In Figure 2.41(b), there are two odd vertices (*P* and *S*), but there is no *directed* edge path, since each of *P* and *S* has two exiting edges and only one entering edge.

FIGURE 2.41

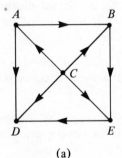

(a)

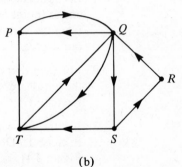

(b)

The comment about edge paths suggests that the theorems of Section 2.3 need some revision for directed paths. This is pursued in the exercises.

The great utility of the digraph comes from its applicability in so many areas. Vertices may represent individuals or groups and the directed edges their financial, social, or political influence on one another. The corresponding digraph can allow one to analyze the interrelations among the individuals or groups.

The vertices of a digraph may represent locations in a computer's memory, and the directed edges may represent the ways in which information can be transferred from one location to the other. Such a digraph can be extremely useful in data processing, where efficient manipulation of and access to stored information is essential.

Digraphs can represent the interlocking food chains in some ecosystems. They can model the results of athletic contests. They can represent the organizational structures of a business, either the official structure or the real one.

Graphs, directed or not, can be coupled with arithmetic to provide other applications. One obvious such utility is via assigning numbers to represent lengths of edges. The numbers might represent distance (this is exactly what a road map is) or they might measure distance in time, seniority, and so on.

Alternatively, numbers can be assigned to vertices. For example, each vertex of a digraph might represent a task to be accomplished in some project. Numbers would represent the time required for each task, and the directed edges would represent the required order of precedence among the tasks, where appropriate. The digraph is then a planning tool to determine the most efficient means to allocate resources and personnel. This is the heart of the management technique known as PERT (Program Evaluation and Review Technique).

We shall not pursue the theory of digraphs and arithmetic graphs. The references provide a good introduction to such additional topics and introduce a number of additional applications. Suffice it to say that relatively simple mathematical techniques in graph theory, arithmetic, and matrix theory provide a very powerful tool with many applications in science, technology, sociology, and business.

EXERCISES 2.7

1. Let us define the **in-degree** of a vertex in a digraph as the number of edges which end at (point to) that vertex, the **out-degree** as the number of edges which start from the vertex. For each digraph in Figure 2.41 determine the in-degree and out-degree of each vertex.

2. For each digraph in Figure 2.41, what is the sum of all the in-degrees? Of all the out-degrees? What is the connection between the two sums, and why? (See Exercise 1.)

3. Let the **net-degree** of a vertex be the in-degree *minus* the out-degree. (See Exercise 1.) Determine the net-degree of each vertex in each of the digraphs in Figure 2.41.

4. If a digraph has a closed directed edge path, what can you say about the net-degree of the vertices? (See Exercise 3.)

5. If a digraph has an open directed edge path, what can you say about the net-degree of the vertices? (See Exercise 3.)

6. State a theorem that characterizes which digraphs have directed edge paths. (See Exercises 4 and 5.)

2.8   CONCLUSIONS

In reflecting on what you have learned from this chapter, several things should be pointed out as significant. We have, purposely, seen some aspects of mathematics which are very typical.

The methods by which graphs with edge paths were identified in Section 2.3 is common in mathematics. Recall that we were looking for characteristic properties of graphs so that we could find an edge path. What we did was to assume that we had a solution (an edge path) and investigate the graph. In a sense, we assumed the "answer" and investigated the "question."

In Sections 2.3 and 2.4 we examined the problem of finding edge paths and vertex paths, respectively. The two *problems* seem very analogous, each dealing in exactly the same way with one of the two component entities of a graph. But the *results* were anything but analogous. Edge paths were completely characterized in terms of degree of vertex and connectedness, but vertex paths defied characterization in terms of any combination of properties we could find. We were left with only partial results about vertex paths, and hence a problem that remains unsolved, as it has been for some two centuries. Such disparity of solution or theory for apparently similar problems is not unusual in mathematics.

On the other hand, the problem of crossing curves, which at first seemed quite different from the path problems, turned out to be virtually identical to that of edge paths. This, too, is not uncommon in mathematics.

In fact, one may identify as a fundamental process in mathematics the attempt to abstract the essential structure from a given situation—mathematical or "real world." By stripping away misleading details, mathematics can identify a problem as what it really is, hopefully utilizing known facts from previously solved problems which are essentially the same. Even when solutions are not known, the identification of abstract similarities in two or more problems allows the crossover of techniques from one to another, leading to a possible solution of all the problems involved.

In many of these instances, as in the situation we have encountered, the similarities are so pervasive that one is the dual of the other. Typically, the connection between the duals is relatively simple, although the details that transfer may be many in number. In several areas of mathematics, the transfer of ideas via duality is a major source of new concepts, theorems, and so on.

Finally, the historical development of graph theory, from puzzles and mathematical diversion to a variety of applications, is again typical of much of mathematics.

At various times in the history of mathematics, topics such as negative numbers or entire branches of mathematics such as non-Euclidean geometry were considered mathematical curiosities, of value at best to clarify the understanding of "real" mathematics. But they later took on great practical significance as the best available mathematics to describe and analyze real situations.

For Further Reading

Biggs, N. L., E. K. Lloyd, and Robin J. Wilson. *Graph Theory 1736–1936*. Oxford: Clarendon Press, 1976.

Flores, Ivan. *Data Structures and Management*. Englewood Cliffs, N.J.: Prentice-Hall, Inc., 1970, Chap. 2.

Harary, Frank. *Graph Theory*. Reading, Mass.: Addison-Wesley Publishing Company, Inc., 1969.

Steen, Lynn Arthur, ed. *Mathematics Today, Twelve Informal Essays*. New York: Springer-Verlag, 1978.

3

MATRICES AND LINEAR SYSTEMS

3.1 INTRODUCTION

Historically, theory of matrices originated in the famous works of William R. Hamilton (1804–1865) and was developed by Arthur Cayley (1821–1895), James S. Sylvester (1814–1897), and many other leading mathematicians and physicists. Since the introduction of high-speed electronic computers, matrices have found increasing applications in biology, sociology, economics, engineering, physics, psychology, statistics, and many other applied areas of mathematics. Matrices are frequently used in linear programming, in Markov processes, in queuing (waiting-line) theory, in regression analysis, and in game theory and have interesting applications in business and management. In this chapter we discuss the basic arithmetic of matrices. Then we examine the solutions of linear equations using matrix theory.

3.2 WHAT IS A MATRIX?

Suppose that a contractor of two housing projects needs 70 units of steel, 90 units of wood, 50 units of glass, and 150 units of labor for project I and 85 units of steel, 75 units of wood, 100 units of glass, and 175 units of labor for project II. This information can be presented by means of rectangular array as follows:

$$
\begin{array}{c c c c c}
 & \text{Steel} & \text{Wood} & \text{Glass} & \text{Labor} \\
\text{Project I} & \begin{bmatrix} 70 & 90 & 50 & 150 \\ \text{Project II} & 85 & 75 & 100 & 175 \end{bmatrix}
\end{array}
$$

The numbers in the first row indicate the requirements for project I, and those in the second row pertain to the material and labor required for project II. This rectangular array, which is a simple example of a *matrix*, has two rows and four columns and is called a 2×4 matrix.

| DEFINITION 3.2.1 | A **matrix** is a rectangular array of numbers arranged in rows and columns. |

What does this definition mean? It just means that a matrix is a set of numbers arranged in a pattern suggesting the geometric form of a rectangle. Some simple examples are given below.

$$[2 \quad 1 \quad 3 \quad 4] \qquad \begin{bmatrix} 3 & 1 & 4 \\ 5 & 3 & 12 \end{bmatrix} \qquad \begin{bmatrix} 1 & 2 \\ 3 & 4 \\ 5 & 6 \end{bmatrix} \qquad \begin{bmatrix} -1 & 4 \\ 2 & -3 \end{bmatrix} \qquad \begin{bmatrix} 8 \\ 1 \\ 2 \\ 3 \\ -1 \end{bmatrix}$$

The first example is a 1×4 matrix, the second a 2×3 matrix, the third a 3×2 matrix, the fourth a square matrix with two rows and two columns, and the fifth a 5×1 matrix.

It is sometimes necessary to make a general reference to a matrix. In that case, we usually say that the matrix A consists of mn numbers arranged in m rows and n columns. The individual numbers in matrix A are denoted by a_{ij}, where i denotes the row in which the element is located and j refers to the column. For example, in matrix A, a_{12} is an element in the first row and second column,

$$A = \begin{bmatrix} a_{11} & a_{12} & a_{13} & \cdots & a_{1n} \\ a_{21} & a_{22} & a_{23} & \cdots & a_{2n} \\ a_{31} & a_{32} & a_{33} & \cdots & a_{3n} \\ \cdots\cdots\cdots\cdots\cdots\cdots\cdots\cdots \\ a_{m1} & a_{m2} & a_{m3} & \cdots & a_{mn} \end{bmatrix}$$

a_{21} is an element in the second row and first column, a_{33} is the element in the third row and third column, a_{32} is the element in the third row and second column, and so on.

| DEFINITION 3.2.2 | A matrix with m rows and n columns is called an **$m \times n$** matrix, read "m by n matrix." |

| DEFINITION 3.2.3 | The **dimensions of a matrix** are given in the form **$m \times n$**, where m represents the number of rows and n the number of columns. |

Thus, a matrix

$$A = \begin{bmatrix} 3 & 2 & 4 \\ 1 & 6 & 5 \end{bmatrix}$$

with two rows and three columns is of dimensions 2×3. Note that in stating the dimensions of a matrix, we always list the number of rows *first* and the number of columns *second*.

DEFINITION 3.2.4

A matrix of dimensions $1 \times n$ is called a **row vector.**

Examples of row vectors are

$$[4 \quad 5] \qquad [2 \quad -3 \quad 5] \qquad [1 \quad -2 \quad 3 \quad 6]$$

DEFINITION 3.2.5

A matrix of dimensions $m \times 1$ is called a **column vector.**

Some simple examples of column vectors are

$$\begin{bmatrix} 5 \\ 3 \end{bmatrix} \qquad \begin{bmatrix} 1 \\ 2 \\ 6 \end{bmatrix} \qquad \begin{bmatrix} 10 \\ 23 \\ 65 \\ 71 \end{bmatrix} \qquad \begin{bmatrix} 8 \\ -12 \\ 16 \\ 7 \\ 3 \end{bmatrix}$$

Thus, an $m \times n$ matrix is simply a set of m row vectors each with n elements placed one under the other, or a set of n column vectors each with m elements placed from left to right.

DEFINITION 3.2.6

A matrix A of dimensions $m \times n$ is called a **zero matrix** if all its elements are zero.

The matrix

$$A = \begin{bmatrix} 0 & 0 & 0 & 0 \\ 0 & 0 & 0 & 0 \\ 0 & 0 & 0 & 0 \end{bmatrix}$$

is a zero matrix of dimensions 3×4.

It is important that we distinguish a zero matrix from the real number 0. Generally, it is clear from the context whether the symbol 0 is to be interpreted as a *zero matrix* or an ordinary number. We will represent a zero matrix by the symbol 0.

DEFINITION 3.2.7

A matrix with n rows and n columns is called a **square matrix of order n.**

For example,

$$B = \begin{bmatrix} b_{11} & b_{12} & b_{13} \\ b_{21} & b_{22} & b_{23} \\ b_{31} & b_{32} & b_{33} \end{bmatrix}$$

is a square matrix of order 3. The elements b_{11}, b_{22}, and b_{33} are said to constitute the **main diagonal** of the matrix B.

DEFINITION 3.2.8

A square matrix of order n that has 1's along the main diagonal and 0's elsewhere is called the **identity matrix** and is denoted by I.

Thus,

$$I = \begin{bmatrix} 1 & 0 \\ 0 & 1 \end{bmatrix} \quad \text{is an identity matrix of order 2}$$

and

$$I = \begin{bmatrix} 1 & 0 & 0 \\ 0 & 1 & 0 \\ 0 & 0 & 1 \end{bmatrix} \quad \text{is an identity matrix of order 3}$$

In general,

$$I = \begin{bmatrix} 1 & 0 & 0 & \cdots & 0 \\ 0 & 1 & 0 & \cdots & 0 \\ 0 & 0 & 1 & \cdots & 0 \\ \cdots & \cdots & \cdots & \cdots & \cdots \\ 0 & 0 & 0 & \cdots & 1 \end{bmatrix}$$

is an identity matrix of order n.

DEFINITION 3.2.9

Two matrices A and B are **equal** if and only if they have the *same dimensions* and their corresponding elements are equal.

Thus, if

$$A = \begin{bmatrix} a_{11} & a_{12} & a_{13} \\ a_{21} & a_{22} & a_{23} \end{bmatrix} \qquad B = \begin{bmatrix} b_{11} & b_{12} & b_{13} \\ b_{21} & b_{22} & b_{23} \end{bmatrix}$$

then $A = B$ implies that $a_{11} = b_{11}, a_{12} = b_{12}, a_{13} = b_{13}$, and so on.

EXAMPLES 3.2

1. Given that $A = \begin{bmatrix} 6 & 4 & x & 5 \end{bmatrix}$ and $B = \begin{bmatrix} 6 & y & 1 & 5 \end{bmatrix}$, then $A = B$ if and only if $x = 1$ and $y = 4$. Note that A and B have the same dimensions.

2. Given that

$$\begin{bmatrix} x & 4 & 3 \\ 6 & y & 7 \end{bmatrix} = \begin{bmatrix} 7 & 4 & z \\ 6 & 0 & 7 \end{bmatrix}$$

then $x = 7$, $y = 0$, and $z = 3$.

3. Consider the matrices

$$A = \begin{bmatrix} 1 & 2 & -1 \\ 2 & 3 & 0 \end{bmatrix} \quad \text{and} \quad B = \begin{bmatrix} 1 & 2 & -1 \\ 2 & 3 & 4 \end{bmatrix}$$

Here $A \neq B$ because some of the corresponding elements are not equal. The matrices

$$C = \begin{bmatrix} 0 & 1 \\ 2 & 5 \end{bmatrix} \quad \text{and} \quad D = \begin{bmatrix} 0 & 1 & 6 \\ 2 & 5 & 4 \end{bmatrix}$$

are not equal because they are of different dimensions.

3.3 ADDITION AND SUBTRACTION OF MATRICES

DEFINITION 3.3.1

If A and B are two matrices each of dimensions $m \times n$, then their **sum (difference)** is another matrix C, also of dimensions $m \times n$, such that every element of the matrix C is the sum (difference) of the corresponding elements of the matrices A and B.

DEFINITION 3.3.2

Two matrices A and B are **conformable for addition (or subtraction)** if and only if they have the same dimensions.

EXAMPLES 3.3

1. (a) $[4 \quad 3 \quad 1] + [5 \quad 1 \quad 4] = [4 + 5 \quad 3 + 1 \quad 1 + 4] = [9 \quad 4 \quad 5]$

   (b) $[4 \quad 3 \quad 1] - [5 \quad 1 \quad 4] = [4 - 5 \quad 3 - 1 \quad 1 - 4] = [-1 \quad 2 \quad -3]$

   (c) $\begin{bmatrix} 3 \\ 1 \\ 5 \end{bmatrix} + \begin{bmatrix} -4 \\ 2 \\ 1 \end{bmatrix} = \begin{bmatrix} 3 - 4 \\ 1 + 2 \\ 5 + 1 \end{bmatrix} = \begin{bmatrix} -1 \\ 3 \\ 6 \end{bmatrix}$

   $\begin{bmatrix} 3 \\ 1 \\ 5 \end{bmatrix} - \begin{bmatrix} -4 \\ 2 \\ 1 \end{bmatrix} = \begin{bmatrix} 3 - (-4) \\ 1 - 2 \\ 5 - 1 \end{bmatrix} = \begin{bmatrix} 7 \\ -1 \\ 4 \end{bmatrix}$

2. If

$$A = \begin{bmatrix} 8 & 4 & 2 \\ 9 & 5 & 3 \end{bmatrix} \quad \text{and} \quad B = \begin{bmatrix} 5 & 1 & 0 \\ 3 & 2 & 4 \end{bmatrix}$$

then A and B are conformable for addition and subtraction. Clearly,

$$A + B = \begin{bmatrix} 8+5 & 4+1 & 2+0 \\ 9+3 & 5+2 & 3+4 \end{bmatrix} = \begin{bmatrix} 13 & 5 & 2 \\ 12 & 7 & 7 \end{bmatrix}$$

and

$$A - B = \begin{bmatrix} 8-5 & 4-1 & 2-0 \\ 9-3 & 5-2 & 3-4 \end{bmatrix} = \begin{bmatrix} 3 & 3 & 2 \\ 6 & 3 & -1 \end{bmatrix}$$

Thus, two matrices are conformable for addition (or subtraction) if and only if they have the same number of rows and the same number of columns. To understand this concept more clearly, consider the following example.

EXAMPLE 3.3.3

The stock report of automobile dealer A for the months of July and December 1979 is as follows:

	Granadas	Mavericks	Mustangs	Pintos
July	45	32	21	36
December	49	27	15	21

The inventory report of automobile dealer B for the same period is as follows:

	Granadas	Mavericks	Mustangs	Pintos
July	31	26	39	42
December	29	17	20	28

Over the period under study, the combined stock report for these dealers is the sum of the corresponding elements in the two matrices shown above. That is,

	Granadas	Mavericks	Mustangs	Pintos
July	45 + 31	32 + 26	21 + 39	36 + 42
December	49 + 29	27 + 17	15 + 20	21 + 28

$$= \begin{bmatrix} 76 & 58 & 60 & 78 \\ 78 & 44 & 35 & 49 \end{bmatrix}$$

THEOREM 3.3.1

If A, B, and C are three matrices conformable for addition, then

(1) $A + B = B + A$
(2) $A + 0 = 0 + A$
(3) $A + (B + C) = (A + B) + C$
(4) $A + C = B + C$ if and only if $A = B$

We advise the reader to write out in full the details for matrices of particular dimensions, to better understand the validity of Theorem 3.3.1.

3.4   SCALAR MULTIPLES OF MATRICES

DEFINITION
3.4.1

Let A be an $m \times n$ matrix and k be a scalar. Then kA is another $m \times n$ matrix B such that

$$b_{ij} = k \cdot a_{ij} \quad \text{for every } (i, j)$$

Thus, if

$$A = \begin{bmatrix} 2 & 3 \\ 1 & 5 \end{bmatrix}$$

then

$$A + A = \begin{bmatrix} 2 & 3 \\ 1 & 5 \end{bmatrix} + \begin{bmatrix} 2 & 3 \\ 1 & 5 \end{bmatrix} = \begin{bmatrix} 2+2 & 3+3 \\ 1+1 & 5+5 \end{bmatrix} = \begin{bmatrix} 4 & 6 \\ 2 & 10 \end{bmatrix} = 2A$$

$$A + A + A = \begin{bmatrix} 2 & 3 \\ 1 & 5 \end{bmatrix} + \begin{bmatrix} 2 & 3 \\ 1 & 5 \end{bmatrix} + \begin{bmatrix} 2 & 3 \\ 1 & 5 \end{bmatrix} = \begin{bmatrix} 2+2+2 & 3+3+3 \\ 1+1+1 & 5+5+5 \end{bmatrix}$$

$$= \begin{bmatrix} 6 & 9 \\ 3 & 15 \end{bmatrix} = 3A$$

Thus, if we add a matrix A to itself k times, we obtain another matrix each of whose elements are those of A multiplied by the number k.

EXAMPLES 3.4

1. (a) Let $A = \begin{bmatrix} 1 & 4 & -2 & 3 \end{bmatrix}$. Then $3A = \begin{bmatrix} 3 & 12 & -6 & 9 \end{bmatrix}$.

   (b) Let $A = \begin{bmatrix} 2 \\ -4 \\ 5 \end{bmatrix}$. Then $(-2)A = \begin{bmatrix} -4 \\ 8 \\ -10 \end{bmatrix}$.

2. If

$$A = \begin{bmatrix} 4 & -2 & 5 \\ -3 & 1 & 6 \end{bmatrix}$$

then

$$(-1)A = \begin{bmatrix} -4 & 2 & -5 \\ 3 & -1 & -6 \end{bmatrix}$$

and

$$4A = \begin{bmatrix} 16 & -8 & 20 \\ -12 & 4 & 24 \end{bmatrix}$$

In general, if k is a scalar, then

$$kA = \begin{bmatrix} 4k & -2k & 5k \\ -3k & k & 6k \end{bmatrix}$$

3. If

$$A = \begin{bmatrix} 3 & 2 \\ 4 & 1 \end{bmatrix} \quad \text{and} \quad B = \begin{bmatrix} -1 & 1 \\ 2 & -3 \end{bmatrix}$$

then

$$3A + 2B = \begin{bmatrix} 9 & 6 \\ 12 & 3 \end{bmatrix} + \begin{bmatrix} -2 & 2 \\ 4 & -6 \end{bmatrix} = \begin{bmatrix} 7 & 8 \\ 16 & -3 \end{bmatrix}$$

EXERCISES 3.4

Find x, y, and z in Exercises 1–3.

1. $[1 \quad 2 \quad 3] + [x \quad y \quad z] = [4 \quad 5 \quad 10]$

2. $[-1 \quad 2 \quad 3] + 3[x \quad y \quad z] = [8 \quad 17 \quad 6]$

3. $\begin{bmatrix} -1 \\ 2 \\ 3 \end{bmatrix} + 2\begin{bmatrix} x \\ y \\ z \end{bmatrix} = \begin{bmatrix} 4 \\ -3 \\ 5 \end{bmatrix}$

4. Given that

$$\begin{bmatrix} -2 & x & 4 & 0 \\ 3 & 2 & y & 3 \end{bmatrix} = \begin{bmatrix} -2 & 3 & z & 0 \\ 3 & 2 & 1 & u \end{bmatrix}$$

find the values of x, y, z, and u.

5. Given that

$$\begin{bmatrix} x+1 & 3 & -1 \\ 4 & -1 & 0 \\ 2 & z+3 & -3 \end{bmatrix} = \begin{bmatrix} 5 & 3 & -1 \\ 4 & y+2 & 0 \\ 2 & 0 & -3 \end{bmatrix}$$

find the values of x, y, and z.

Find $A + B$ and $A - B$ in Exercises 6–13.

6. $A = [4 \quad 2 \quad 1 \quad 6]$, $B = [5 \quad 4 \quad 3 \quad 2]$

7. $A = [2 \quad 3 \quad 4 \quad 1 \quad 7]$, $B = [-2 \quad -1 \quad 0 \quad 3 \quad -2]$

8. $A = \begin{bmatrix} 2 \\ 3 \\ 1 \\ -2 \end{bmatrix}$, $B = \begin{bmatrix} -1 \\ 1 \\ 4 \\ 3 \end{bmatrix}$

9. $A = \begin{bmatrix} 0 \\ 5 \\ 7 \end{bmatrix}$, $B = \begin{bmatrix} 4 \\ -1 \\ -3 \end{bmatrix}$

10. $A = \begin{bmatrix} 1 & 2 \\ 3 & 4 \end{bmatrix}$, $B = \begin{bmatrix} 2 & -1 \\ -1 & 3 \end{bmatrix}$

11. $A = \begin{bmatrix} 1 & 2 & 3 \\ 3 & 1 & 2 \end{bmatrix}$, $B = \begin{bmatrix} -2 & -1 & -4 \\ 1 & 3 & 2 \end{bmatrix}$

12. $A = \begin{bmatrix} 2 & -1 & 1 \\ 2 & 3 & 4 \\ 5 & -1 & 1 \end{bmatrix}$, $B = \begin{bmatrix} 4 & -2 & 3 \\ 5 & -1 & 2 \\ 2 & 1 & -3 \end{bmatrix}$

13. $A = \begin{bmatrix} 2 & 1 & 3 & 4 \\ 4 & -1 & 2 & 3 \end{bmatrix}$, $B = \begin{bmatrix} -1 & 2 & -1 & 3 \\ 3 & -1 & 2 & -5 \end{bmatrix}$

Given that

$$A = \begin{bmatrix} 5 & 8 \\ 7 & 9 \end{bmatrix} \qquad B = \begin{bmatrix} 1 & 2 \\ 3 & 4 \end{bmatrix} \qquad C = \begin{bmatrix} 7 & 5 \\ 6 & 9 \end{bmatrix}$$

perform the indicated operations in Exercises 14–17.

14. $3A + 2B$

15. $4A + 3B + 5C$

16. $A + (B + C)$

17. $(A + B) + C$

18. Given that

$$A = \begin{bmatrix} 2 & 1 & 3 \\ 4 & -1 & 2 \end{bmatrix} \qquad B = \begin{bmatrix} -3 & 2 & 0 \\ 1 & 5 & 2 \end{bmatrix} \qquad C = \begin{bmatrix} -2 & 5 & 4 \\ 2 & 1 & 3 \end{bmatrix}$$

verify that $A + (B + C) = (A + B) + C$.

3.5 MULTIPLICATION OF MATRICES

In Example 3.3.1 we saw how row (column) vectors having the same number of elements are added or subtracted. In Example 3.4.1 we examined the scalar multiples of a row (column) vector. It seems reasonable to ask why it is necessary to have two means of representing vectors, particularly when there is no real difference in the properties of row and column vectors. The answer lies in the product of two vectors, where two quantities are studied simultaneously and it becomes necessary to represent one as a row vector and the other as a column vector.

Suppose that the manager of Cinemas I, II, III, and IV reports that in a particular show 940 tickets were sold for Cinema I, 1005 for Cinema II, 1140 for Cinema III, and 685 for Cinema IV. These figures are represented by a row vector (a 1 × 4 matrix)

$$A = [940 \quad 1005 \quad 1140 \quad 685]$$

Suppose that a ticket costs \$3.50 in Cinema I, \$3.00 in Cinema II, \$2.75 in Cinema III, and \$2.50 in Cinema IV. This information can be represented in a column vector (a 4×1 matrix):

$$B = \begin{bmatrix} \$3.50 \\ \$3.00 \\ \$2.75 \\ \$2.50 \end{bmatrix}$$

What are total receipts for this show?

We must multiply the number of tickets sold by the corresponding price in each Cinema and then add these amounts to get the total receipts. Clearly, the multiplication should have the form

$$AB = \begin{bmatrix} 940 & 1005 & 1140 & 685 \end{bmatrix} \begin{bmatrix} \$3.50 \\ \$3.00 \\ \$2.75 \\ \$2.50 \end{bmatrix}$$

$$= (940)(\$3.50) + (1005)(\$3.00) + (1140)(\$2.75) + (685)(\$2.50)$$
$$= \$11{,}152.50$$

DEFINITION 3.5.1   Let $A = \begin{bmatrix} a_1 & a_2 & \cdots & a_n \end{bmatrix}$ be a row vector (a $1 \times n$ matrix) and

$$B = \begin{bmatrix} b_1 \\ b_2 \\ \vdots \\ b_n \end{bmatrix}$$

be a column vector (a $n \times 1$ matrix). Then

$$AB = \begin{bmatrix} a_1 & a_2 & \cdots & a_n \end{bmatrix} \begin{bmatrix} b_1 \\ b_2 \\ \vdots \\ b_n \end{bmatrix}$$

$$= a_1 b_1 + a_2 b_2 + \cdots + a_n b_n$$

Note that we write the row vector *first* and a column vector *second*; the vector multiplication is a row-by-column operation and is defined if and only if both row and column vectors have the same number of elements. Such a vector multiplication yields a real number.

EXAMPLE 3.5.1

Consider the following shopping list with unit prices given for each item.

8 cans of soup	$0.43
5 cans of orange juice	$0.89
2 dozen eggs	$0.99
2 gallons of milk	$1.79
4 pounds of hamburger meat	$1.43

Determine the total bill.

SOLUTION   The purchases can be represented by a row vector (1 × 5 matrix):

$$\begin{array}{ccccc} & \text{Orange} & \text{Eggs} & \text{Milk} & \text{Meat} \\ \text{Soup} & \text{juice} & \text{(dozen)} & \text{(gallon)} & \text{(pound)} \end{array}$$
$$A = \begin{bmatrix} 8 & 5 & 2 & 2 & 4 \end{bmatrix}$$

The prices of these items are represented by a column vector (a 5 × 1 matrix):

$$B = \begin{bmatrix} \$0.43 \\ \$0.89 \\ \$0.99 \\ \$1.79 \\ \$1.43 \end{bmatrix}$$

The multiplication that the cashier performs to obtain the correct bill is

$$AB = \begin{bmatrix} 8 & 5 & 2 & 2 & 4 \end{bmatrix} \begin{bmatrix} \$0.43 \\ \$0.89 \\ \$0.99 \\ \$1.79 \\ \$1.43 \end{bmatrix}$$

$$= (8)(\$0.43) + (5)(\$0.89) + (2)(\$0.99) + (2)(\$1.79) + (4)(\$1.43) = \$19.17$$

EXERCISES 3.5

Given that $A = \begin{bmatrix} 4 & 1 & 5 \end{bmatrix}$, $B = \begin{bmatrix} 3 & 4 & 2 \end{bmatrix}$, and $C = \begin{bmatrix} 6 \\ -2 \\ 5 \end{bmatrix}$, perform the indicated operations in Exercises 1–4.

1. AC 2. BC

3. $(A + B)C$ 4. $(A - B)C$

5. Solve for x.

(a) $[1 \quad 2 \quad 3 \quad x] \begin{bmatrix} 3 \\ -1 \\ 1 \\ 2 \end{bmatrix} = 10$

(b) $[-1 \quad x \quad 2 \quad 3] \begin{bmatrix} 1 \\ 5 \\ 3 \\ -4 \end{bmatrix} = 13$

6. Given that
$$A = [x \quad y] \qquad B = \begin{bmatrix} 2 \\ 3 \end{bmatrix} \qquad C = \begin{bmatrix} 3 \\ 2 \end{bmatrix}$$
if $AB = 13$ and $AC = 12$, find x and y.

7. Given that
$$A = [x \quad y] \qquad B = \begin{bmatrix} 1 \\ -2 \end{bmatrix} \qquad C = \begin{bmatrix} 5 \\ 3 \end{bmatrix}$$
if $AB = 0$ and $AC = 13$, find x and y.

8. James goes to a grocery store and purchases 6 apples, 8 oranges, and 10 pears. Write a row vector A expressing these purchases. If an apple costs $0.12, an orange costs $0.10, and a pear costs $0.08, express these prices as a vector B. What does AB represent?

9. Susan purchased 7 cans of soup, 4 cans of orange juice, 1 dozen eggs, 3 chickens, and 2 gallons of milk. Write a row vector A expressing these purchases. Suppose that a can of soup costs $0.37, a can of orange juice costs $0.89, 1 dozen eggs costs $0.99, a chicken costs $0.83, and a gallon of milk costs $1.69. Write a column vector B expressing these prices. Determine the total bill that Susan must pay at the checkout counter.

10. Roger goes to a package store and buys some liquor for a party he is giving to celebrate his graduation. He purchases 2 quarts of Scotch whiskey, 3 quarts of blended whiskey, 2 quarts of gin, 3 quarts of rum, 3 quarts of vodka, 1 keg of beer, and 10 bottles of wine. Write a row vector A expressing these purchases. Suppose that Scotch whiskey costs $9.25 per quart; blended whiskey, $7.30 per quart; gin, $6.50 per quart; rum, $5.50 per quart; vodka, $7.50 per quart; a keg of beer, $18.00; and a bottle of wine, $4.25. Express these prices as a column vector B. Determine the total amount that Roger must pay for these purchases.

11. Mr. Gates has purchased 100 shares of stock 1, 200 shares of stock 2, 300 shares of stock 3, and 500 shares of stock 4 through the New York Stock Exchange. Write a row vector A expressing these purchases. Suppose that

stock 1 costs $25 per share; stock 2, $34 per share; stock 3, $22 per share; and stock 4, $12.50 per share. Write a column vector B indicating these prices. Determine the total amount that Mr. Gates has invested in these stocks.

12. A business manager of the baseball team reports that 1400 persons bought box seats, 4100 bought reserved seats, and 5500 bought general admission tickets. Write a row vector A whose components represent this information. Suppose that a box seat costs $10; a reserved seat, $7.50; and a general admission ticket, $5.00. Write a column vector B representing these admission costs. Determine the total receipts for the game.

Now that we know how to multiply a $1 \times n$ row vector by an $n \times 1$ column vector, we shall see that Definition 3.5.1 can, in fact, be extended to multiplying matrices in general.

Suppose that a company has three stores located in different regions. The retail unit sales of the products in these stores is given by matrix A:

$$
\begin{array}{c}
\\
\\
\text{Store 1} \\
\text{Store 2} \\
\text{Store 3}
\end{array}
\begin{array}{c}
\text{Product} \\
\begin{array}{cccc}
\text{I} & \text{II} & \text{III} & \text{IV}
\end{array} \\
\begin{bmatrix}
63 & 50 & 35 & 43 \\
81 & 66 & 51 & 48 \\
85 & 60 & 48 & 32
\end{bmatrix}
\end{array}
$$

where the first row shows the number of products sold in store 1 and the second and third rows reflect the unit sales in store 2 and store 3, respectively. Assuming that the selling prices of the products were $13.00, $12.50, $12.00, and $10.75, respectively, we represent this information by a column vector (a 4×1 matrix):

$$
B = \begin{bmatrix}
\$13.00 \\
\$12.50 \\
\$12.00 \\
\$10.75
\end{bmatrix}
$$

The total revenue sales matrix is

$$
AB = \begin{bmatrix}
63 & 50 & 35 & 43 \\
81 & 66 & 51 & 48 \\
85 & 60 & 48 & 32
\end{bmatrix}
\begin{bmatrix}
\$13.00 \\
\$12.50 \\
\$12.00 \\
\$10.75
\end{bmatrix}
$$

The total sales revenue for store 1 is given by

$$63(\$13.00) + 50(\$12.50) + 35(\$12.00) + 43(\$10.75) = \$2326.25 \qquad (3.1)$$

The sales revenue for store 2 is

$$81(\$13.00) + 66(\$12.50) + 51(\$12.00) + 48(\$10.75) = \$3006.00 \qquad (3.2)$$

and the sales revenue for store 3 is

$$85(\$13.00) + 60(\$12.50) + 48(\$12.00) + 32(\$10.75) = \$2775.00 \qquad (3.3)$$

Thus,

$$AB = \begin{bmatrix} 63 & 50 & 35 & 43 \\ 81 & 66 & 51 & 48 \\ 85 & 60 & 48 & 32 \end{bmatrix} \begin{bmatrix} \$13.00 \\ \$12.50 \\ \$12.00 \\ \$10.75 \end{bmatrix} = \begin{bmatrix} \$2326.25 \\ \$3006.00 \\ \$2775.00 \end{bmatrix}$$

If we are interested in the sales revenue of all three stores, then the product matrix AB is to be premultiplied by the row vector $[1 \quad 1 \quad 1]$; that is,

$$[1 \quad 1 \quad 1] \begin{bmatrix} \$2326.25 \\ \$3006.00 \\ \$2775.00 \end{bmatrix} = \$2326.25 + \$3006.00 + \$2775.00 = \$8107.25$$

If the interest lies in the number of products sold in each store, then matrix A is premultiplied by the row vector $[1 \quad 1 \quad 1]$. Thus,

$$[1 \quad 1 \quad 1] \begin{bmatrix} 63 & 50 & 35 & 43 \\ 80 & 66 & 51 & 48 \\ 85 & 60 & 48 & 32 \end{bmatrix} = [228 \quad 176 \quad 134 \quad 123]$$

These examples *suggest* the following definition for multiplying matrices in general.

DEFINITION 3.5.2   Let A be an $m \times n$ matrix and B be an $n \times p$ matrix. The **product** AB is then defined to be the $m \times p$ matrix whose element in the ith row and jth column is the sum of the products of the elements in ith row of A and the corresponding elements of the jth column of B.

In other words,

$$\begin{bmatrix} a_{11} & a_{12} & \cdots & a_{1n} \\ a_{21} & a_{22} & \cdots & a_{2n} \\ \hdotsfor{4} \\ a_{i1} & a_{i2} & \cdots & a_{in} \\ \hdotsfor{4} \\ a_{m1} & a_{m2} & \cdots & a_{mn} \end{bmatrix} \begin{bmatrix} b_{11} & b_{12} & \cdots & b_{1j} & \cdots & b_{1p} \\ b_{21} & b_{22} & \cdots & b_{2j} & \cdots & b_{2p} \\ \hdotsfor{6} \\ \hdotsfor{6} \\ b_{n1} & b_{n2} & \cdots & b_{nj} & \cdots & b_{np} \end{bmatrix}$$

$$= \begin{bmatrix} c_{11} & c_{12} & \cdots & c_{1j} & \cdots & c_{1p} \\ c_{21} & c_{22} & \cdots & c_{2j} & \cdots & c_{2p} \\ \hdotsfor{6} \\ c_{i1} & c_{i2} & \cdots & c_{ij} & \cdots & c_{ip} \\ \hdotsfor{6} \\ c_{m1} & c_{m2} & \cdots & c_{mj} & \cdots & c_{mp} \end{bmatrix}$$

where
$$c_{ij} = \begin{bmatrix} a_{i1} & a_{i2} & \cdots & a_{in} \end{bmatrix} \begin{bmatrix} b_{1j} \\ b_{2j} \\ \vdots \\ b_{nj} \end{bmatrix}$$

$$= a_{i1}b_{1j} + a_{i2}b_{2j} + \cdots + a_{in}b_{nj}$$

Let $i = 1$ and $j = 1$. Then

$$c_{11} = a_{11}b_{11} + a_{12}b_{21} + \cdots + a_{1n}b_{n1}$$

Note that this is the direct result of multiplying the elements in the *first* row of matrix A by the corresponding elements in the *first* column of matrix B.

If $i = 2$ and $j = 3$, then

$$c_{23} = a_{21}b_{13} + a_{22}b_{23} + \cdots + a_{2n}b_{n3}$$

Again, this is the sum of the products of the elements in the *second* row of matrix A and the corresponding elements of the *third* column of matrix B.

It is important to observe that the product matrix AB is defined, or A is *conformable* to B for multiplication, if and only if the number of columns of matrix A equals the number of rows of matrix B. Further, the product matrix AB has the same number of rows as matrix A and the same number of columns as matrix B. Thus,

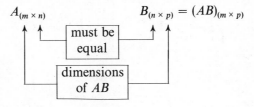

EXAMPLE 3.5.2

Let

$$A = \begin{bmatrix} 1 & 3 & 2 \\ -4 & 5 & 4 \end{bmatrix}_{(2 \times 3)} \quad \text{and} \quad B = \begin{bmatrix} 5 & 7 \\ 2 & 3 \\ 1 & -2 \end{bmatrix}_{(3 \times 2)}$$

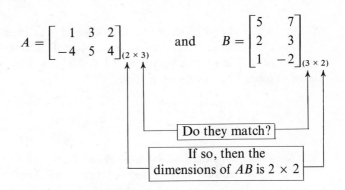

The matrix AB is defined, because the number of columns of matrix A is the same as the number of rows of matrix B. The product matrix $C = AB$ will have two rows and two columns. Thus

$$\begin{bmatrix} 1 & 3 & 2 \\ -4 & 5 & 4 \end{bmatrix} \begin{bmatrix} 5 & 7 \\ 2 & 3 \\ 1 & -2 \end{bmatrix} = \begin{bmatrix} c_{11} & c_{12} \\ c_{21} & c_{22} \end{bmatrix}$$

To obtain c_{11}, we multiply the elements in the *first* row of matrix A by the corresponding elements in the *first* column of matrix B and add. Thus,

$$c_{11} = \begin{bmatrix} 1 & 3 & 2 \end{bmatrix} \begin{bmatrix} 5 \\ 2 \\ 1 \end{bmatrix} = 1(5) + 3(2) + 2(1) = 13$$

Similarly,

$$c_{12} = \begin{bmatrix} 1 & 3 & 2 \end{bmatrix} \begin{bmatrix} 7 \\ 3 \\ -2 \end{bmatrix} = 1(7) + 3(3) + 2(-2) = 12$$

$$c_{21} = \begin{bmatrix} -4 & 5 & 4 \end{bmatrix} \begin{bmatrix} 5 \\ 2 \\ 1 \end{bmatrix} = (-4)(5) + 5(2) + 4(1) = -6$$

and $$c_{22} = \begin{bmatrix} -4 & 5 & 4 \end{bmatrix} \begin{bmatrix} 7 \\ 3 \\ -2 \end{bmatrix} = (-4)(7) + (5)(3) + (4)(-2) = -21$$

Thus,

$$C = AB = \begin{bmatrix} 13 & 12 \\ -6 & -21 \end{bmatrix}$$

3.6   PROPERTIES OF MATRIX MULTIPLICATION

We recall that the product of two matrices AB is defined, or A is conformable to B for multiplication, if and only if the number of columns of A equals the number of rows of B. The product matrix AB has the same number of rows as A and the same number of columns as B. Thus, if

$$A = \begin{bmatrix} 1 & 2 & -1 \\ 2 & 3 & 0 \end{bmatrix}_{(2 \times 3)} \quad \text{and} \quad B = \begin{bmatrix} 5 & 7 & 1 \\ 2 & 3 & 4 \\ 1 & -2 & 5 \end{bmatrix}_{(3 \times 3)}$$

then product AB is defined, because A has three columns and B has three rows. The resulting product matrix has two rows and three columns. In the product AB, we say that B is premultiplied by A and that A is postmultiplied by B.

The product BA is, however, not defined in the example above, because the number of columns in B is not equal to the number of rows in A. Even if the products AB and BA are defined, AB may not, in general, be equal to BA.

EXAMPLES 3.6

1. Let

$$A = \begin{bmatrix} 1 & 2 & 3 \\ 2 & 3 & 5 \end{bmatrix}_{(2 \times 3)} \qquad \text{and} \qquad B = \begin{bmatrix} 4 & 2 & 3 \\ 2 & 1 & 1 \\ 1 & 2 & 0 \end{bmatrix}_{(3 \times 3)}$$

Then

$$AB = \begin{bmatrix} 1 & 2 & 3 \\ 2 & 3 & 5 \end{bmatrix} \begin{bmatrix} 4 & 2 & 3 \\ 2 & 1 & 1 \\ 1 & 2 & 0 \end{bmatrix}$$

$$= \begin{bmatrix} 11 & 10 & 5 \\ 19 & 17 & 9 \end{bmatrix}$$

Note that the product BA is not defined. (Why?)

2. Let

$$A = \begin{bmatrix} 3 & 5 \\ 1 & -2 \\ 2 & 4 \end{bmatrix} \qquad \text{and} \qquad I = \begin{bmatrix} 1 & 0 \\ 0 & 1 \end{bmatrix}$$

Then

$$AI = \begin{bmatrix} 3 & 5 \\ 1 & -2 \\ 2 & 4 \end{bmatrix} \begin{bmatrix} 1 & 0 \\ 0 & 1 \end{bmatrix} = \begin{bmatrix} 3 & 5 \\ 1 & -2 \\ 2 & 4 \end{bmatrix}$$

Note that $AI = A$ but that IA is not defined.

3. Let

$$A = \begin{bmatrix} 1 & 2 & 4 \\ 2 & 3 & 0 \end{bmatrix} \qquad \text{and} \qquad B = \begin{bmatrix} 5 & 7 \\ 2 & 1 \\ 1 & 3 \end{bmatrix}$$

Then

$$AB = \begin{bmatrix} 1 & 2 & 4 \\ 2 & 3 & 0 \end{bmatrix} \begin{bmatrix} 5 & 7 \\ 2 & 1 \\ 1 & 3 \end{bmatrix} = \begin{bmatrix} 13 & 21 \\ 16 & 17 \end{bmatrix}$$

$$BA = \begin{bmatrix} 5 & 7 \\ 2 & 1 \\ 1 & 3 \end{bmatrix} \begin{bmatrix} 1 & 2 & 4 \\ 2 & 3 & 0 \end{bmatrix} = \begin{bmatrix} 19 & 31 & 20 \\ 4 & 7 & 8 \\ 7 & 11 & 4 \end{bmatrix}$$

Note that AB and BA are both defined, but AB is of dimensions 2×2 whereas BA is of dimensions 3×3.

4. Let

$$A = \begin{bmatrix} 3 & 4 \\ -1 & 2 \end{bmatrix} \quad \text{and} \quad B = \begin{bmatrix} 2 & 1 \\ 3 & 2 \end{bmatrix}$$

Then

$$AB = \begin{bmatrix} 3 & 4 \\ -1 & 2 \end{bmatrix}\begin{bmatrix} 2 & 1 \\ 3 & 2 \end{bmatrix} = \begin{bmatrix} 18 & 11 \\ 4 & 3 \end{bmatrix}$$

whereas

$$BA = \begin{bmatrix} 2 & 1 \\ 3 & 2 \end{bmatrix}\begin{bmatrix} 3 & 4 \\ -1 & 2 \end{bmatrix} = \begin{bmatrix} 5 & 10 \\ 7 & 16 \end{bmatrix}$$

Thus, AB and BA are both defined, but they are not equal.

5. Let

$$A = \begin{bmatrix} 2 & 3 & 4 \\ 1 & 5 & 2 \\ 0 & 6 & 1 \end{bmatrix} \quad \text{and} \quad I = \begin{bmatrix} 1 & 0 & 0 \\ 0 & 1 & 0 \\ 0 & 0 & 1 \end{bmatrix}$$

Then

$$AI = \begin{bmatrix} 2 & 3 & 4 \\ 1 & 5 & 2 \\ 0 & 6 & 1 \end{bmatrix}\begin{bmatrix} 1 & 0 & 0 \\ 0 & 1 & 0 \\ 0 & 0 & 1 \end{bmatrix} = \begin{bmatrix} 2 & 3 & 4 \\ 1 & 5 & 2 \\ 0 & 6 & 1 \end{bmatrix}$$

and

$$IA = \begin{bmatrix} 1 & 0 & 0 \\ 0 & 1 & 0 \\ 0 & 0 & 1 \end{bmatrix}\begin{bmatrix} 2 & 3 & 4 \\ 1 & 5 & 2 \\ 0 & 6 & 1 \end{bmatrix} = \begin{bmatrix} 2 & 3 & 4 \\ 1 & 5 & 2 \\ 0 & 6 & 1 \end{bmatrix}$$

Note that AI and IA are both defined and $AI = IA = A$.

We shall now show that the familiar rule in ordinary algebra that if $ab = ac$, then $b = c$ also breaks down in the multiplication of matrices. This is illustrated by the following example.

EXAMPLE 3.6.6

Let

$$A = \begin{bmatrix} 1 & 3 & 0 \\ 2 & 1 & 0 \\ -3 & 4 & 0 \end{bmatrix} \quad B = \begin{bmatrix} 1 & 3 & 2 \\ 2 & 1 & 3 \\ 1 & 1 & 1 \end{bmatrix} \quad C = \begin{bmatrix} 1 & 3 & 2 \\ 2 & 1 & 3 \\ 2 & 2 & 2 \end{bmatrix}$$

Then

$$AB = \begin{bmatrix} 7 & 6 & 11 \\ 4 & 7 & 7 \\ 5 & -5 & 6 \end{bmatrix} = AC$$

Thus, $AB = AC$, although $B \neq C$. In other words, the cancellation property in the real number system does not carry over to the algebra of matrices. However, the cancellation law holds in a special case which we shall explore later.

Another important rule in ordinary algebra, which states that if the product of two numbers a and b is zero, then at least one of them must be zero, also fails to hold for matrix multiplication. This is illustrated by the following example.

EXAMPLE 3.6.7

Form the product AB, where

$$A = \begin{bmatrix} 1 & 0 \\ 4 & 0 \end{bmatrix} \quad \text{and} \quad B = \begin{bmatrix} 0 & 0 \\ 2 & -1 \end{bmatrix}$$

SOLUTION

$$AB = \begin{bmatrix} 1 & 0 \\ 4 & 0 \end{bmatrix} \begin{bmatrix} 0 & 0 \\ 2 & -1 \end{bmatrix} = \begin{bmatrix} 0 & 0 \\ 0 & 0 \end{bmatrix}$$

The reader may observe that BA is not a zero matrix.

In summary, we wish to emphasize that three fundamental properties of ordinary multiplication do not extend themselves to matrix multiplication.

1. The commutative law, $AB = BA$, does *not*, in general, hold.
2. If $AB = AC$ or $BA = CA$, we cannot, in general, cancel A from both sides, even if A is not a zero matrix.
3. If $AB = 0$, it does not necessarily follow that at least one of the matrices is a zero matrix.

The fact that some fundamental properties of ordinary arithmetic do not, in general, hold for matrix arithmetic might make one believe that matrix multiplication is nearly a worthless operation. That is, of course, not true, because some of the vital properties, in particular the associative and distributive laws, remain to be verified in matrix multiplication.

We shall now state (without proof) the property of *associativity* in matrix multiplication. If A, B, and C are three matrices whose dimensions are properly related, then multiplication is associative; that is, $A(BC) = (AB)C$. Consider the following example.

EXAMPLE 3.6.8

Let

$$A = \begin{bmatrix} 2 & 1 \\ 4 & 3 \end{bmatrix} \qquad B = \begin{bmatrix} 3 & 5 \\ 1 & 7 \end{bmatrix} \qquad C = \begin{bmatrix} -1 \\ 2 \end{bmatrix}$$

Then

$$(AB)C = \begin{bmatrix} 7 & 17 \\ 15 & 41 \end{bmatrix} \begin{bmatrix} -1 \\ 2 \end{bmatrix} = \begin{bmatrix} 27 \\ 67 \end{bmatrix}$$

and

$$A(BC) = \begin{bmatrix} 2 & 1 \\ 4 & 3 \end{bmatrix} \begin{bmatrix} 7 \\ 13 \end{bmatrix} = \begin{bmatrix} 27 \\ 67 \end{bmatrix}$$

Thus,

$$A(BC) = (AB)C$$

The next property we state (without proof) is that of the **distributive law**. The property in the number system that $a(b + c) = ab + ac$ can be extended for matrices of proper dimension and

$$A(B + C) = AB + AC$$

The matrices B and C must be of the same dimensions for matrix addition. If B and C are two matrices both with n rows and p columns, then A must have n columns so that the products AB and AC can be formed. Under these conditions a distributive law in the real number system is carried over to arithmetic of matrices.

EXAMPLE 3.6.9

Let

$$A = \begin{bmatrix} 2 & 3 \\ 1 & 4 \end{bmatrix} \qquad B = \begin{bmatrix} -1 & -2 \\ 3 & 5 \end{bmatrix} \qquad C = \begin{bmatrix} 1 & 2 \\ 2 & 5 \end{bmatrix}$$

Then

$$B + C = \begin{bmatrix} 0 & 0 \\ 5 & 10 \end{bmatrix}$$

$$A(B + C) = \begin{bmatrix} 2 & 3 \\ 1 & 4 \end{bmatrix} \begin{bmatrix} 0 & 0 \\ 5 & 10 \end{bmatrix} = \begin{bmatrix} 15 & 30 \\ 20 & 40 \end{bmatrix}$$

$$AB = \begin{bmatrix} 2 & 3 \\ 1 & 4 \end{bmatrix} \begin{bmatrix} -1 & -2 \\ 3 & 5 \end{bmatrix} = \begin{bmatrix} 7 & 11 \\ 11 & 18 \end{bmatrix}$$

$$AC = \begin{bmatrix} 2 & 3 \\ 1 & 4 \end{bmatrix} \begin{bmatrix} 1 & 2 \\ 2 & 5 \end{bmatrix} = \begin{bmatrix} 8 & 19 \\ 9 & 22 \end{bmatrix}$$

$$AB + AC = \begin{bmatrix} 7 & 11 \\ 11 & 18 \end{bmatrix} + \begin{bmatrix} 8 & 19 \\ 9 & 22 \end{bmatrix} = \begin{bmatrix} 15 & 30 \\ 20 & 40 \end{bmatrix}$$

Thus,

$$A(B + C) = AB + AC$$

EXERCISES 3.6

For each of the matrices in Exercises 1–5, determine the dimensions of the product matrix AB.

1. A is a 3×3 matrix and B is a 3×5 matrix.

2. A is a 5×3 matrix and B is a 3×5 matrix.

3. A is a 3×5 matrix and B is a 5×3 matrix.

4. A is a 2×7 matrix and B is a 7×4 matrix.

5. A is a 3×2 matrix and B is a 2×4 matrix.

For the pairs of matrices in Exercises 6–11, determine whether it is possible to compute AB, BA, both, or neither.

6. A is a 3×5 matrix and B is a 5×2 matrix.

7. A is a 2×3 matrix and B is a 3×2 matrix.

8. A is a 4×4 matrix and B is a 4×4 matrix.

9. A is a 4×5 matrix and B is a 4×2 matrix.

10. A is a 5×7 matrix and B is a 7×5 matrix.

11. A is a 3×4 matrix and B is a 4×4 matrix.

In Exercises 12–15, find the products AB and BA.

12. $A = \begin{bmatrix} 3 & 4 \\ 2 & 3 \end{bmatrix}$, $B = \begin{bmatrix} 5 & 1 \\ 1 & 2 \end{bmatrix}$

13. $A = \begin{bmatrix} 6 & 1 \\ 2 & 5 \end{bmatrix}$, $B = \begin{bmatrix} 3 & 7 \\ 4 & -2 \end{bmatrix}$

14. $A = \begin{bmatrix} 1 & 2 \\ 4 & 5 \end{bmatrix}$, $B = \begin{bmatrix} 3 & 1 \\ 2 & 7 \end{bmatrix}$

15. $A = \begin{bmatrix} 1 & 2 & -1 \\ 2 & 3 & 0 \end{bmatrix}$, $B = \begin{bmatrix} 5 & 7 \\ 2 & 3 \\ 1 & -2 \end{bmatrix}$

Given that

$$A = \begin{bmatrix} 1 & 5 & 3 \\ 7 & 4 & 5 \end{bmatrix}, \quad B = \begin{bmatrix} 2 & -1 & 1 \\ 3 & 2 & 8 \\ 2 & 5 & 4 \end{bmatrix} \quad C = \begin{bmatrix} 3 & 1 & 2 \\ 1 & 2 & 3 \\ 2 & 3 & 1 \end{bmatrix}$$

perform the indicated operations in Exercises 16–19.

16. AB 17. AC

18. BC 19. CB

Given the matrices

$$A = \begin{bmatrix} 1 & 2 \\ 5 & 4 \end{bmatrix} \quad B = \begin{bmatrix} 2 & 7 \\ 3 & 9 \end{bmatrix} \quad C = \begin{bmatrix} 5 & -3 \\ 2 & 5 \end{bmatrix}$$

perform the indicated operations in Exercises 20–29.

20. AB 21. BA

22. AC 23. BC

24. CA 25. CB

26. $A(BC)$ 27. $(AB)C$

28. $A(B + C)$ 29. $(B + C)A$

30. Given that

$$A = \begin{bmatrix} 3 & 1 & 2 \\ 2 & 3 & 4 \end{bmatrix} \quad B = \begin{bmatrix} 1 & 3 \\ 2 & 1 \\ 4 & 5 \end{bmatrix} \quad C = \begin{bmatrix} 0 & 3 \\ 2 & 4 \\ 3 & 5 \end{bmatrix}$$

verify that $A(B + C) = AB + AC$.

31. Given that

$$A = \begin{bmatrix} 2 & 0 & -1 \\ -1 & -2 & 2 \end{bmatrix} \quad B = \begin{bmatrix} -1 & 0 & -1 \\ 2 & -3 & -1 \\ 3 & 0 & -1 \end{bmatrix} \quad C = \begin{bmatrix} 1 \\ 1 \\ -2 \end{bmatrix}$$

verify that $A(BC) = (AB)C$.

32. Given that

$$A = \begin{bmatrix} 1 & 2 & 3 \\ 2 & 0 & 5 \end{bmatrix} \quad B = \begin{bmatrix} 2 & 4 & 1 \\ 4 & 1 & 2 \\ 1 & 2 & 3 \end{bmatrix} \quad C = \begin{bmatrix} -1 & 4 \\ 3 & 2 \\ 5 & 1 \end{bmatrix}$$

verify that $A(BC) = (AB)C$.

33. Given that

$$A = \begin{bmatrix} 2 & 6 \\ 3 & 9 \end{bmatrix} \quad \text{and} \quad B = \begin{bmatrix} -6 & 15 \\ 2 & -5 \end{bmatrix}$$

Is AB a zero matrix? If so, can you conclude that if $AB = 0$, then either A or B or both are zero matrices?

34. Given that

$$A = \begin{bmatrix} 3 & 2 \\ 1 & 0 \end{bmatrix} \quad B = \begin{bmatrix} 2 & 4 \\ 1 & 2 \end{bmatrix} \quad C = \begin{bmatrix} 1 & 6 \\ 3 & -4 \end{bmatrix}$$

compare the products AB and CB. Is $AB = CB$? If so, does $AB = CB$ imply that $A = C$?

35. In a grocery store, Allan purchased 5 cans of soup, 2 cans of orange juice, 1 dozen eggs, 2 chickens, and 2 gallons of milk. Judy bought 7 cans of soup, 4 cans of orange juice, 2 dozen eggs, 1 chicken, 1 gallon of milk, and 1 gallon of ice cream. The respective costs of these purchases per unit are: a can of soup, $0.31; a can of orange juice, $0.79; a dozen eggs, $0.95; a chicken, $0.73; a gallon of milk, $1.89; and a gallon of ice cream, $1.65.
(a) Represent these purchases in a 2×6 matrix.
(b) Write the price vector of these purchases as a 6×1 matrix.
(c) Determine the amount Allan spent on his purchases. Compute the total amount Judy spent on her purchases.

36. An automobile manufacturer offers five pollution-free models A, B, C, D, and E and expects orders for 1 million, 3 million, 2 million, 1 million, and 4 million, respectively. The following matrix gives the amount of raw material needed for each model expressed in conveniently chosen units.

	Steel	Chromium	Glass	Rubber	Paint
A	8	3	5	7	6
B	5	2	6	6	5
C	4	2	7	5	5
D	5	4	6	8	7
E	3	4	5	6	4

The respective costs of material per unit are $12.00 for steel, $6.00 for chromium, $5.00 for glass, $3.50 for rubber, and $3.00 for paint.
(a) Express the expected demand for the various models as a 1×5 matrix. Using matrix multiplication, determine the total amount of raw material needed.
(b) Express the price vector as a 5×1 matrix. Using matrix multiplication, determine the total cost of material per car for each model.
(c) Determine the total investment as a 1×1 matrix.

3.7 SOLUTION OF LINEAR EQUATIONS—
GAUSS-JORDAN ELIMINATION METHOD

When a system of linear equations involves more than two equations in the same number of unknowns, solving the system by means of graphs or substitution methods is hardly a rewarding experience. We shall therefore present a *method of elimination* that not only is useful for machine calculations but also is fundamental to the understanding of the basic concepts involved.

Consider the following set of linear equations:

$$
\begin{aligned}
2x + 3y + 4z &= 12 \\
x + 2y + z &= 6 \\
3x - y + z &= 2
\end{aligned}
\qquad (3.4)
$$

Our objective is first to transform this set of linear equations into a triangular system:

$$
\begin{aligned}
x + b_{12}y + b_{13}z &= c_1 \\
y + b_{23}z &= c_2 \\
z &= c_3
\end{aligned}
$$

and then reduce it subsequently to the form

$$
\begin{aligned}
x + 0 \cdot y + 0 \cdot z &= c_1^* \\
0 \cdot x + y + 0 \cdot z &= c_2^* \\
0 \cdot x + 0 \cdot y + z &= c_3
\end{aligned}
$$

so that $x = c_1^*$, $y = c_2^*$, and $z = c_3$ is the obvious solution to the original set of linear equations (3.4). An interchange in the first two equations in (3.4) yields

$$
\begin{aligned}
x + 2y + z &= 6 \\
2x + 3y + 4z &= 12 \\
3x - y + z &= 2
\end{aligned}
\qquad (3.5)
$$

The first equation in (3.5) has 1 as its leading coefficient and can therefore be used to eliminate the variable x from the second and third equations in (3.5). Subtracting two times the first equation from the second and three times the first equation from the third equation, we obtain

$$
\begin{aligned}
x + 2y + z &= 6 \\
-y + 2z &= 0 \\
-7y - 2z &= -16
\end{aligned}
\qquad (3.6)
$$

Now we multiply the second equation in (3.6) by (-1) so as to have 1 as its leading coefficient. This yields the following system:

$$
\begin{aligned}
x + 2y + z &= 6 \\
y - 2z &= 0 \\
-7y - 2z &= -16
\end{aligned}
\qquad (3.7)
$$

Adding seven times the second equation in (3.7) to the third equation results in the elimination of the variable y from the third equation, and we have

$$
\begin{aligned}
x + 2y + \quad z &= \quad 6 \\
y - \quad 2z &= \quad 0 \\
-16z &= -16
\end{aligned}
\tag{3.8}
$$

We now divide the third equation in (3.8) by the coefficient of z. We have the following system in the triangular form:

$$
\begin{aligned}
x + 2y + \quad z &= 6 \\
y - 2z &= 0 \\
z &= 1
\end{aligned}
\tag{3.9}
$$

To eliminate z from the first two equations, we add two times the third equation to the corresponding elements of the second equation and (-1) times the third equation to the corresponding elements of the first equation. Thus, we obtain

$$
\begin{aligned}
x + 2y \quad &= 5 \\
y \quad &= 2 \\
z &= 1
\end{aligned}
\tag{3.10}
$$

Adding (-2) times the elements of the second equation in (3.10) to the corresponding elements of the first equation, we get

$$
\begin{aligned}
x \quad &= 1 \\
y \quad &= 2 \\
z &= 1
\end{aligned}
\tag{3.11}
$$

We leave it to the reader to verify by substitution in the original set of linear equations (3.4) that this is, in fact, a solution of three linear equations in three unknowns.

DEFINITION 3.7.1

Two systems of linear equations in n variables are said to be **equivalent** if and only if every solution of one system is also a solution of the other.

The elimination process of solving a system of equations reduces the given set of linear equations to an equivalent set of equations from which the existence of solutions can be easily read. Thus, (3.5) and (3.6) are equivalent. Similarly, the system of equations (3.6), (3.7), (3.8), and (3.9) are also equivalent, as are (3.10) and (3.11).

Observe that in finding the solution to the original set of linear equations (3.4), we used operations of the following form:

1. the interchange of any two equations of a system
2. the addition of an arbitrary multiple of one equation to another equation of the system
3. the multiplication of an equation of a system by a constant

It is easy to show that the foregoing operations preserve the solution of the set of linear equations.

In reducing the given system of linear equations to equivalent systems, we actually operate on the coefficients and the corresponding constants. The variables simply serve to keep their coefficients properly aligned in columns.

The system of linear equations (3.4) can be written in matrix form as

$$\begin{bmatrix} 2 & 3 & 4 \\ 1 & 2 & 1 \\ 3 & -1 & 1 \end{bmatrix}\begin{bmatrix} x \\ y \\ z \end{bmatrix} = \begin{bmatrix} 12 \\ 6 \\ 2 \end{bmatrix}$$

or more precisely, as

$$AX = B$$

where A is the *coefficient matrix* and B the *vector of constants*.

DEFINITION
3.7.2

The **augmented matrix** for the system $AX = B$ is the matrix $[A:B]$ found by adjoining the column vector B of constants to the right of the coefficient matrix A.

Thus, the augmented matrix of the system of equations (3.4) is

$$[A:B] = \begin{bmatrix} 2 & 3 & 4 & 12 \\ 1 & 2 & 1 & 6 \\ 3 & -1 & 1 & 2 \end{bmatrix}$$

Each row vector of the augmented matrix contains the coefficients of the unknowns and the constant term of the corresponding equation.

The technique of solving linear equations using matrices involves first obtaining a 1 in the a_{11}-position either by dividing the first row by its leading term or by interchanging the first row with another row having 1 as its leading term. Suitable multiples of the new first row are then added to the remaining rows to obtain 0's in the a_{21}- and a_{31}-positions. The process is repeated by selecting a new row among the altered rows and dividing it by its leading nonzero term, if necessary, to locate a 1 in the a_{22}-position. The new second row thus obtained is then used to obtain 0 in the a_{32}-position. The process is continued until we obtain an equivalent augmented matrix in triangular form:

$$[A^*:B^*] = \begin{bmatrix} 1 & b_{12} & b_{13} & c_1 \\ 0 & 1 & b_{23} & c_2 \\ 0 & 0 & 1 & c_3 \end{bmatrix}$$

Adding suitable multiples of the last row to the remaining rows yields 0's in b_{13}- and b_{23}-positions. Again, adding a suitable multiple of the second row to the first row places a 0 in the b_{12}-position. Thus, we have

$$\begin{bmatrix} 1 & 0 & 0 & c_1^* \\ 0 & 1 & 0 & c_2^* \\ 0 & 0 & 1 & c_3 \end{bmatrix}$$

so that $x = c_1^*$, $y = c_2^*$, and $z = c_3$ is a solution to the original set of equations (3.4).

The operations we performed on the system of linear equations (3.4) correspond exactly to those we perform now on the *rows* of the augmented matrix $[A:B]$. The augmented matrix is

$$\begin{array}{c} \text{Row} \\ R_1 \\ R_2 \\ R_3 \end{array} \begin{array}{cccc} x & y & z & \text{Constant} \\ \begin{bmatrix} 2 & 3 & 4 & 12 \\ 1 & 2 & 1 & 6 \\ 3 & -1 & 1 & 2 \end{bmatrix} \end{array} \qquad (3.4a)$$

Since 1 happens to be in the a_{21}-position, we begin by interchanging the first two rows of the preceding matrix so as to obtain a 1 in the a_{11}-position:

$$\begin{array}{c} \text{Row Operation} \\ R_1' = R_2 \\ R_2' = R_1 \\ R_3 \end{array} \begin{array}{cccc} x & y & z & \text{Constant} \\ \begin{bmatrix} 1 & 2 & 1 & 6 \\ 2 & 3 & 4 & 12 \\ 3 & -1 & 1 & 2 \end{bmatrix} \end{array} \qquad (3.5a)$$

Next, we need to obtain 0's in the a_{21}- and a_{31}-positions. This can be accomplished by multiplying the elements of the new first row, R_1', by (-2) and (-3) and then adding the results to the corresponding elements of the new second row, R_2', and R_3, respectively. The new matrix, together with the symbolic representation of the operations involved, is

$$\begin{array}{c} \text{Row Operation} \\ R_1' \\ R_2'' = R_2' + (-2)R_1' \\ R_3' = R_3 + (-3)R_1' \end{array} \begin{array}{cccc} x & y & z & \text{Constant} \\ \begin{bmatrix} 1 & 2 & 1 & 6 \\ 0 & -1 & 2 & 0 \\ 0 & -7 & -2 & -16 \end{bmatrix} \end{array} \qquad (3.6a)$$

Now we need to get a 1 in the a_{22}-position. This can be accomplished by simply multiplying the elements of the second row, R_2'', by (-1). Thus, we obtain

$$\begin{array}{c} \text{Row Operation} \\ R_1' \\ R_2''' = (-1)R_2'' \\ R_3' \end{array} \begin{array}{cccc} x & y & z & \text{Constant} \\ \begin{bmatrix} 1 & 2 & 1 & 6 \\ 0 & 1 & -2 & 0 \\ 0 & -7 & -2 & -16 \end{bmatrix} \end{array} \qquad (3.7a)$$

To obtain 0 in the a_{32}-position, we multiply the elements in the second row, R_2''', by seven and add the results to the corresponding elements of the third row, R_3'. The resulting matrix, together with the symbolic representation of the operations involved, is

$$\begin{array}{cc}
\text{Row Operation} & \begin{array}{cccc} x & y & z & \text{Constant} \end{array} \\
\begin{array}{c} R_1' \\ R_2''' \\ R_3'' = R_3' + 7R_2''' \end{array} & \left[\begin{array}{ccc|c} 1 & 2 & 1 & 6 \\ 0 & 1 & -2 & 0 \\ 0 & 0 & -16 & -16 \end{array}\right]
\end{array} \qquad (3.8a)$$

Dividing the elements of the third row, R_3'', by (-16), the matrix above becomes

$$\begin{array}{cc}
\text{Row Operation} & \begin{array}{cccc} x & y & z & \text{Constant} \end{array} \\
\begin{array}{c} R_1' \\ R_2''' \\ R_3''' = (-\frac{1}{16}) \end{array} & \left[\begin{array}{ccc|c} 1 & 2 & 1 & 6 \\ 0 & 1 & -2 & 0 \\ 0 & 0 & 1 & 1 \end{array}\right]
\end{array} \qquad (3.9a)$$

Now we use the last row, R_3''', to obtain 0's in the a_{13}- and a_{23}-positions. Adding two and (-1) times the elements of R_3''' to the corresponding elements of the first and second rows of the preceding matrix yields

$$\begin{array}{cc}
\text{Row Operation} & \begin{array}{ccc c} x & y & z & \text{Constant} \end{array} \\
\begin{array}{c} R_1'' = R_1' + (-1)R_3''' \\ R_2^{iv} = R_2''' + 2R_3''' \\ R_3''' \end{array} & \left[\begin{array}{ccc|c} 1 & 2 & 0 & 5 \\ 0 & 1 & 0 & 2 \\ 0 & 0 & 1 & 1 \end{array}\right]
\end{array} \qquad (3.10a)$$

Finally, we multiply the elements of the second row, R_2''', in the preceding matrix by (-2) and add the results to the corresponding elements of the first row, R_1''. The resulting matrix is

$$\begin{array}{cc}
\text{Row Operation} & \begin{array}{ccc c} x & y & z & \text{Constant} \end{array} \\
\begin{array}{c} R_1''' = R_1'' + (-2)R_2^{iv} \\ R_2^{iv} \\ R_3''' \end{array} & \left[\begin{array}{ccc|c} 1 & 0 & 0 & 1 \\ 0 & 1 & 0 & 2 \\ 0 & 0 & 1 & 1 \end{array}\right]
\end{array} \qquad (3.11a)$$

Thus,

$$x = 1 \qquad y = 2 \qquad z = 1$$

is the solution to the system of linear equations.

We wish to remark that the sequence of operations used to obtain an identity matrix on the left and a solution vector on the right is arranged in each case to take full advantage of any obvious circumstances and is by no means unique. Note carefully that each equation in the final matrix involves one unknown with a 1 as its coefficient so that the corresponding constant term is the solution for that unknown.

To apply these operations on the augmented matrix, we give next a definition for the corresponding operations on the rows of this matrix, since these rows correspond precisely to the set of simultaneous equations. These are referred to as *elementary operations*.

DEFINITION 3.7.3

The **elementary operations** on the rows of a matrix are of the following form:

Type 1 Multiplying the elements of any row, say R_i, by a constant k and then replacing it by R_i'. Symbolically, we may express this operation as

$$R_i' = k \cdot R_i$$

Type 2 Interchanging any two rows R_i and R_j. In symbols, we may write

$$R_j' = R_i$$

and

$$R_i' = R_j$$

Type 3 Multiplying the elements of any row, say R_i, by a scalar k and adding the results to the corresponding elements of another row, say R_j. This operation replaces the original row R_j by the new row R_j', where

$$R_j' = R_j + k \cdot R_i$$

To apply these operations to the augmented matrix $[A : B]$, we first check the first column to find a nonzero element that could be easily changed to 1. The row with the selected element is then interchanged with the first row by using an elementary operation of type 2. The new first row is divided, if necessary, by its first component to obtain a 1 in the a_{11}-position. This is elementary row operation of type 1. Then, using an elementary row operation of type 3, suitable multiples of this new first row are added to the other rows to obtain 0's in the rest of the first column. The process is then repeated on the second column, with the exception that the first row is no longer available for searching the nonzero element. Using an elementary operation of type 2, the row containing the selected element is then interchanged with the second row. The new second row is then divided by the element in the a_{22}-position so as to obtain a 1 in that place. Suitable multiples of the second row are then added to the remaining rows to obtain 0's in the rest of the second column, including the first row. This is an elementary row operation of type 3. The process is continued until we have an identity matrix on the left augmented by the solution vector on the right. This process is called the **Gauss-Jordan elimination method**.

The following examples will further illustrate these points.

EXAMPLE 3.7.1

Using the Gauss-Jordan elimination method, solve the following system of linear equations:

$$
\begin{aligned}
x + y + z &= 3 \\
2x + 4y - 3z &= -6 \\
4x + 3y - 5z &= -5
\end{aligned}
$$

SOLUTION   The augmented matrix is

$$\begin{array}{c} \\ R_1 \\ R_2 \\ R_3 \end{array} \begin{array}{cccc} x & y & z & \text{Constant} \\ \left[\begin{array}{ccc|c} 1 & 1 & 1 & 3 \\ 2 & 4 & -3 & -6 \\ 4 & 3 & -5 & -5 \end{array}\right] \end{array}$$

Since 1 happens to be in the a_{11}-position, we perform the following operations:

1. $R'_2 = R_2 + (-2)R_1$

2. $R'_3 = R_3 + (-4)R_1$

These operations transform the preceding matrix into

$$\begin{array}{c} \\ R_1 \\ R'_2 \\ R'_3 \end{array} \begin{array}{cccc} x & y & z & \text{Constant} \\ \left[\begin{array}{ccc|c} 1 & 1 & 1 & 3 \\ 0 & 2 & -5 & -12 \\ 0 & -1 & -9 & -17 \end{array}\right] \end{array}$$

Next, we multiply the elements of R'_3 by (-1) and then interchange the row so obtained with the second row, R'_2. Symbolically,

1. $R''_2 = (-1)R'_3$

2. $R''_3 = R'_2$

Thus, we obtain the matrix

$$\begin{array}{c} \\ R_1 \\ R''_2 \\ R''_3 \end{array} \begin{array}{cccc} x & y & z & \text{Constant} \\ \left[\begin{array}{ccc|c} 1 & 1 & 1 & 3 \\ 0 & 1 & 9 & 17 \\ 0 & 2 & -5 & -12 \end{array}\right] \end{array}$$

We now add the suitable multiples of R''_2 to the remaining rows to obtain 0's in the rest of the second column, including the first row. Thus, we perform the following operations:

1. $R'_1 = R_1 + (-1)R''_2$

2. $R'''_3 = R''_3 + (-2)R''_2$

which transform the preceding matrix into

$$\begin{array}{c} \\ R'_1 \\ R''_2 \\ R'''_3 \end{array} \begin{array}{cccc} x & y & z & \text{Constant} \\ \left[\begin{array}{ccc|c} 1 & 0 & -8 & -14 \\ 0 & 1 & 9 & 17 \\ 0 & 0 & -23 & -46 \end{array}\right] \end{array}$$

Since we need to have 1 in the a_{33}-position now, we divide the elements of R'''_3 by (-23). Symbolically, $R^{iv}_3 = (-\frac{1}{23})R'''_3$. Thus, we obtain

$$\begin{array}{c} \\ R'_1 \\ R''_2 \\ R^{iv}_3 \end{array} \begin{array}{cccc} x & y & z & \text{Constant} \\ \left[\begin{array}{ccc|c} 1 & 0 & -8 & -14 \\ 0 & 1 & 9 & 17 \\ 0 & 0 & 1 & 2 \end{array}\right] \end{array}$$

Finally, we perform the following operations in the order specified:

1. $R''_1 = R_1 + 8R^{iv}_3$

2. $R'''_2 = R_2 + (-9)R^{iv}_3$

and thus obtain the matrix

$$\begin{array}{c} \\ R''_1 \\ R'''_2 \\ R_3 \end{array} \begin{array}{cccc} x & y & z & \text{Constant} \\ \left[\begin{array}{ccc|c} 1 & 0 & 0 & 2 \\ 0 & 1 & 0 & -1 \\ 0 & 0 & 1 & 2 \end{array}\right] \end{array}$$

Thus,

$$x = 2 \qquad y = -1 \qquad z = 2$$

is the solution to the system of linear equations.

EXAMPLE 3.7.2

Solve the following sets of linear equations using the Gauss-Jordan elimination method.

(a) $\quad x + 2y + 3z = 5$ (b) $\quad x + 2y + 3z = 1$ (c) $\quad x + 2y + 3z = 21$
$\quad 3x - 5y + 4z = 16$ $\quad 3x - 5y + 4z = -8$ $\quad 3x - 5y + 4z = 67$
$\quad 5x + 8y + 11z = 19$ $\quad 5x + 8y + 11z = 3$ $\quad 5x + 8y + 11z = 81$

SOLUTION   Note that the three systems of linear equations have the same coefficient matrix and differ only in their constant values. Since the calculations on the coefficient matrix remain the same for all three systems, we shall solve three sets of simultaneous equations at the same time. This system has an augmented matrix

$$\begin{array}{c} \\ \\ R_1 \\ R_2 \\ R_3 \end{array} \begin{array}{c} \text{Constant Column} \\ x \quad y \quad z \qquad \text{Vectors} \\ \left[\begin{array}{ccc|ccc} 1 & 2 & 3 & 5 & 1 & 21 \\ 3 & -5 & 4 & 16 & -8 & 67 \\ 5 & 8 & 11 & 19 & 3 & 81 \end{array}\right] \end{array}$$

that can be simplified by the series of row operations shown in Table 3.1. We leave it for the reader to verify that $x = 1$, $y = -1$, $z = 2$ is the solution of the system of equations in (a); $x = -1$, $y = 1$, $z = 0$ is the solution of the system of equations in (b); and $x = 5$, $y = -4$, and $z = 8$ is the solution of the system of equations in (c).

TABLE 3.1

Row Operation	x	y	z	Constant Column Vectors (a)	(b)	(c)
R_1	1	2	3	5	1	21
R_2	3	−5	4	16	−8	67
R_3	5	8	11	19	3	81
R_1	1	2	3	5	1	21
$R_2' = R_2 + (-3)R_1$	0	−11	−5	1	−11	4
$R_3' = R_3 + (-5)R_1$	0	−2	−4	−6	−2	−24
R_1	1	2	3	5	1	21
$R_2'' = (-\frac{1}{2})R_3'$	0	1	2	3	1	12
$R_3'' = R_2'$	0	−11	−5	1	−11	4
R_1	1	2	3	5	1	21
R_2''	0	1	2	3	1	12
$R_3''' = R_3'' + 11R_2''$	0	0	17	34	0	136
R_1	1	2	3	5	1	21
R_2''	0	1	2	3	1	12
$R_3^{iv} = (\frac{1}{17})R_3'''$	0	0	1	2	0	8
$R_1' = R_1 + (-3)R_3^{iv}$	1	2	0	−1	1	−3
$R_2''' = R_2'' + (-2)R_3^{iv}$	0	1	0	−1	1	−4
R_3^{iv}	0	0	1	2	0	8
$R_1'' = R_1' + (-2)R_2'''$	1	0	0	1	−1	5
R_2'''	0	1	0	−1	1	−4
R_3^{iv}	0	0	1	2	0	8

EXERCISES 3.7

Solve the following systems of linear equations using the Gauss-Jordan elimination method.

1. $x + 2y = 5$
    $2x + 3y = 8$

2. $x + y = 5$
    $2x + 3y = 12$

3. $2x + 3y = 13$
    $x + 2y = 8$

4. $3x + 4y = 25$
    $-2x + 5y = 14$

5. $3x + 2y = 12$
    $2x - y = 1$

6. $x - 2y = 7$
    $4x - 3y = 18$

7. $\begin{aligned} x - 3y + z &= -6 \\ 2x + 3y + 3z &= 5 \\ 3x - y + z &= 0 \end{aligned}$

8. $\begin{aligned} x - y - z &= 2 \\ 2x + 4y - 3z &= -4 \\ -2x + 3y + 2z &= -5 \end{aligned}$

9. $\begin{aligned} x - 2y + z &= 11 \\ 2x - 4y - 3z &= 7 \\ 3x + 5y + 6z &= 9 \end{aligned}$

10. $\begin{aligned} 3x + 4y - z &= 8 \\ 2x - 5y + 3z &= 1 \\ x + y + z &= 6 \end{aligned}$

11. $\begin{aligned} x + 2y + z &= 5 \\ x - y + 2z &= 12 \\ 2x + 3y - z &= -1 \end{aligned}$

12. $\begin{aligned} 2x + 2y - z &= -1 \\ 3x + 2y - 2z &= -2 \\ x - y + z &= 6 \end{aligned}$

13. $\begin{aligned} x + y + 5z &= 8 \\ 2x + y - z &= 3 \\ 3x + 2y + 5z &= 12 \end{aligned}$

14. $\begin{aligned} 2x - 2y + z &= 17 \\ x - y - z &= -2 \\ x - 5y - z &= 6 \end{aligned}$

15. $\begin{aligned} x - 2y + 3z &= 9 \\ 3x - y - 2z &= 0 \\ 6x + 3y - z &= 1 \end{aligned}$

16. $\begin{aligned} x + 2y + 3z &= 14 \\ -x + y + 2z &= 7 \\ 3x - y + 5z &= 16 \end{aligned}$

17. $\begin{aligned} x + y - z &= 0 \\ 2x + y + 3z &= 9 \\ x + 3y + z &= 6 \end{aligned}$

18. $\begin{aligned} x + 2y + 3z &= 6 \\ 2x - 2y + 5z &= 5 \\ 4x - y - 3z &= 0 \end{aligned}$

Solve the following sets of linear equations.

19. (a) $\begin{aligned} x + 2y + 3z &= 2 \\ 3x - 5y + 4z &= -11 \\ 5x + 8y + 11z &= 10 \end{aligned}$

(b) $\begin{aligned} x + 2y + 3z &= 14 \\ 3x - 5y + 4z &= 5 \\ 5x + 8y + 11z &= 54 \end{aligned}$

20. (a) $\begin{aligned} x + 2y + z &= 4 \\ 2x + 3y + 4z &= 19 \\ 3x - y + z &= 9 \end{aligned}$

(b) $\begin{aligned} x + 2y + z &= 4 \\ 2x + 3y + 4z &= 4 \\ 3x - y + z &= 0 \end{aligned}$

(c) $\begin{aligned} x + 2y + z &= 8 \\ 2x + 3y + 4z &= 20 \\ 3x - y + z &= 4 \end{aligned}$

3.8   SOME SPECIAL CASES OF THE MATRIX SOLUTION

In Section 3.7 we investigated the Gauss-Jordan elimination method for solving n linear equations in n variables, in which the system of equations possesses a solution and the solution is unique. We consider next systems of linear equations in which it is not possible to obtain an identity matrix augmented by the unique solution vector. In some of these cases, we shall find that no solution of the system exists. In others, we shall find that there are infinitely many solutions to the system.

EXAMPLE 3.8.1

Solve the following set of linear equations.

$$x + 2y + 3z = 6$$
$$2x + 4y - z = 5$$
$$4x + 8y + 3z = 15$$

SOLUTION   The augmented matrix, in this case, is reduced by a series of elementary row operations as shown in Table 3.2. The last matrix in the table corresponds to the linear equations

$$x + 2y \quad = 3$$
$$z = 1$$

The first equation implies that $x = 3 - 2y$. Clearly, y can take any arbitrary value. Thus, if $y = c$, then

$$x = 3 - 2c \quad y = c \quad z = 1$$

We leave it for the reader to verify that by substituting $x = 3 - 2c$, $y = c$, and $z = 1$ in the original set of equations, this is, in fact, a solution of three linear equations in three unknowns. Note that for each value of c, we obtain a solution. Because c can assume any value, we conclude that the system has an *infinite* number of solutions.

TABLE 3.2

Row Operation	x	y	z	Constant Column Vector
R_1	1	2	3	6
R_2	2	4	-1	5
R_3	4	8	3	15
R_1	1	2	3	6
$R_2' = (-\frac{1}{7})[R_2 + (-2)R_1]$	0	0	1	1
$R_3' = (-\frac{1}{9})[R_3 + (-4)R_1]$	0	0	1	1
$R_1' = R_1 + (-3)R_2'$	1	2	0	3
R_2'	0	0	1	1
$R_3'' = R_3' + (-1)R_2'$	0	0	0	0

Another special case is illustrated by the following example.

EXAMPLE 3.8.2

Solve the following set of linear equations.

$$x - y + 2z = 2$$
$$-2x + 3y - 2z = -1$$
$$4x - 7y + 2z = -4$$

SOLUTION   The augmented matrix is simplified by a series of elementary row operations as shown in Table 3.3. The last row in the table determines the equation

$$0 \cdot x + 0 \cdot y + 0 \cdot z = -3$$

Because there are no values of x, y, and z for which this equation holds, we conclude that the system has no solution.

TABLE 3.3

Row Operation	x	y	z	Constant Column Vector
R_1	1	−1	2	2
R_2	−2	3	−2	−1
R_3	4	−7	2	−4
R_1	1	−1	2	2
$R_2' = R_2 + 2R_1$	0	1	2	3
$R_3' = R_3 + (-4)R_1$	0	−3	−6	−12
R_1	1	−1	2	2
R_2'	0	1	2	3
$R_3'' = R_3' + 3R_2'$	0	0	0	−3

EXERCISES 3.8

Find the solution or solutions, if they exist, for the following sets of linear equations.

1. $x + y = 3$
    $2x + 2y = 5$

2. $x + y = 4$
    $3x + 3y = 12$

3. $x - 3y = 6$
    $2x - 6y = 12$

4. $2x - y = 1$
    $3x + 2y = 12$

5. $\begin{aligned} x + y - z &= 4 \\ 2x + y + 3z &= 10 \\ 3x + 2y - 2z &= 10 \end{aligned}$

6. $\begin{aligned} x + y - z &= 5 \\ 2x + y + 3z &= 3 \\ 5x + 4y &= 18 \end{aligned}$

7. $\begin{aligned} x - y + z &= 5 \\ 3x + 5y - 13z &= -9 \\ 5x + 3y - 11z &= 1 \end{aligned}$

8. $\begin{aligned} x + 2y - 3z &= 7 \\ 2x - y + z &= 5 \\ 3x + y - 2z &= 12 \end{aligned}$

9. $\begin{aligned} x + 2y - z &= 6 \\ 3x - y + 2z &= 9 \\ 5x + 3y &= 21 \end{aligned}$

10. $\begin{aligned} -x + 2y + z &= 2 \\ 4x - y + 6z &= -4 \\ 2x + 3y + 8z &= 2 \end{aligned}$

11. $\begin{aligned} x - y + z &= 3 \\ 2x + 3y - z &= -5 \\ 4x + y + z &= 1 \end{aligned}$

12. $\begin{aligned} x + 2y - z &= 5 \\ 2x + 4y + 4z &= 22 \\ 4x + 8y + 3z &= 34 \end{aligned}$

13. $\begin{aligned} x - y - z &= 0 \\ 2x + y + 2z &= 11 \\ -x + 3y + z &= 2 \end{aligned}$

14. $\begin{aligned} x + 3y - 4z &= 5 \\ 2x - y + 5z &= 7 \\ 7x + 7y - 2z &= 31 \end{aligned}$

3.9   INVERSE OF A SQUARE MATRIX

In elementary algebra the equation

$$ax = b$$

can be solved for the unknown x by dividing both sides of the equation by a, thus obtaining

$$x = \frac{b}{a} \qquad \text{where } a \neq 0$$

Alternatively, we can multiply both sides of the equation by the multiplicative inverse of a, namely $1/a$ or a^{-1}. This yields

$$a^{-1}(ax) = a^{-1}b$$
$$(a^{-1}a)x = a^{-1}b$$

Since $a^{-1}a = 1$, we have

$$x = a^{-1}b$$

as the solution to the equation $ax = b$. This concept of multiplicative inverse can be extended to matrix multiplication, and then used effectively for solving a set of linear equations.

DEFINITION 3.9.1   If A is a square matrix of order n and B is another square matrix, also of order n, such that

$$AB = BA = I$$

then B is called an **inverse** of the matrix A.

EXAMPLES 3.9

1. Let

$$A = \begin{bmatrix} 9 & 7 \\ 5 & 4 \end{bmatrix} \quad \text{and} \quad B = \begin{bmatrix} 4 & -7 \\ -5 & 9 \end{bmatrix}$$

Then

$$AB = \begin{bmatrix} 9 & 7 \\ 5 & 4 \end{bmatrix}\begin{bmatrix} 4 & -7 \\ -5 & 9 \end{bmatrix} = \begin{bmatrix} 1 & 0 \\ 0 & 1 \end{bmatrix}$$

and

$$BA = \begin{bmatrix} 4 & -7 \\ -5 & 9 \end{bmatrix}\begin{bmatrix} 9 & 7 \\ 5 & 4 \end{bmatrix} = \begin{bmatrix} 1 & 0 \\ 0 & 1 \end{bmatrix}$$

Thus, B is an inverse of matrix A.

2. Let A and B be the following 3×3 matrices.

$$A = \begin{bmatrix} 2 & 0 & 1 \\ 1 & 1 & 2 \\ 1 & 2 & 0 \end{bmatrix} \quad \text{and} \quad B = -\frac{1}{7}\begin{bmatrix} -4 & 2 & -1 \\ 2 & -1 & -3 \\ 1 & -4 & 2 \end{bmatrix}$$

Then, we have

$$AB = \begin{bmatrix} 2 & 0 & 1 \\ 1 & 1 & 2 \\ 1 & 2 & 0 \end{bmatrix}\left(-\frac{1}{7}\right)\begin{bmatrix} -4 & 2 & -1 \\ 2 & -1 & -3 \\ 1 & -4 & 2 \end{bmatrix} = \begin{bmatrix} 1 & 0 & 0 \\ 0 & 1 & 0 \\ 0 & 0 & 1 \end{bmatrix}$$

and

$$BA = \left(-\frac{1}{7}\right)\begin{bmatrix} -4 & 2 & -1 \\ 2 & -1 & -3 \\ 1 & -4 & 2 \end{bmatrix}\begin{bmatrix} 2 & 0 & 1 \\ 1 & 1 & 2 \\ 1 & 2 & 0 \end{bmatrix} = \begin{bmatrix} 1 & 0 & 0 \\ 0 & 1 & 0 \\ 0 & 0 & 1 \end{bmatrix}$$

Consequently, matrix B is an inverse of matrix A and matrix A is an inverse of matrix B.

We do not wish to leave an impression that all nonzero square matrices have inverses. Consider the matrix

$$A = \begin{bmatrix} 1 & 0 \\ 0 & 0 \end{bmatrix}$$

and suppose that

$$B = \begin{bmatrix} a & b \\ c & d \end{bmatrix}$$

is an inverse of A. Then $AB = I$. That is,

$$\begin{bmatrix} 1 & 0 \\ 0 & 0 \end{bmatrix}\begin{bmatrix} a & b \\ c & d \end{bmatrix} = \begin{bmatrix} 1 & 0 \\ 0 & 1 \end{bmatrix}$$

$$\begin{bmatrix} a & b \\ 0 & \boxed{0} \end{bmatrix} = \begin{bmatrix} 1 & 0 \\ 0 & \boxed{1} \end{bmatrix}$$

Since the matrices are equal, we have, by Definition 3.2.9, that

$$0 = 1$$

which is false. Hence, we conclude that the matrix A does not have an inverse.

Later, we shall use the Gauss-Jordan elimination method to determine whether or not the inverse of a given square matrix exists.

THEOREM 3.9.1

The inverse of a square matrix, if it exists, is unique.

PROOF   Let A be a matrix with A^{-1} as its inverse. Suppose that B is another matrix, which is also an inverse of A. Then

$$AA^{-1} = A^{-1}A = I$$

and

$$AB = BA = I$$

Then, premultiplying both sides of $AB = I$ by A^{-1}, we have

$$A^{-1}(AB) = A^{-1}I = A^{-1}$$

Also,

$$A^{-1}(AB) = (A^{-1}A)B = IB = B$$

Hence,

$$B = A^{-1}$$

In view of this result, we shall henceforth denote the inverse of a matrix A by the symbol A^{-1}.

EXERCISES 3.9

1. Show that the inverse of

$$A = \begin{bmatrix} 4 & 7 \\ 1 & 2 \end{bmatrix} \quad \text{is} \quad A^{-1} = \begin{bmatrix} 2 & -7 \\ -1 & 4 \end{bmatrix}$$

by verifying that $AA^{-1} = A^{-1}A = I$.

2. Show that the inverse of

$$A = \begin{bmatrix} a & b \\ c & d \end{bmatrix} \quad \text{is} \quad A^{-1} = \frac{1}{ad - bc}\begin{bmatrix} d & -b \\ -c & a \end{bmatrix}$$

if and only if $ad - bc \neq 0$.

3. Show that the inverse of

$$A = \begin{bmatrix} 1 & 0 & 2 \\ 3 & 1 & 2 \\ 1 & -1 & 0 \end{bmatrix} \quad \text{is} \quad A^{-1} = \frac{1}{6}\begin{bmatrix} -2 & 2 & 2 \\ -2 & 2 & -4 \\ 4 & -1 & -1 \end{bmatrix}$$

by verifying that $AA^{-1} = A^{-1}A = I$.

4. Show that the inverse of

$$A = \begin{bmatrix} 1 & 1 & 1 \\ 3 & 4 & -1 \\ 2 & -5 & 3 \end{bmatrix} \quad \text{is} \quad A^{-1} = \frac{1}{27}\begin{bmatrix} -7 & 8 & 5 \\ 11 & -1 & -4 \\ 23 & -7 & -1 \end{bmatrix}$$

by verifying that $AA^{-1} = A^{-1}A = I$.

5. Show that the inverse of

$$A = \begin{bmatrix} 3 & 1 & 2 \\ 1 & -4 & 1 \\ 2 & 3 & 0 \end{bmatrix} \quad \text{is} \quad A^{-1} = \frac{1}{15}\begin{bmatrix} -3 & 6 & 9 \\ 2 & -4 & -1 \\ 11 & -7 & -13 \end{bmatrix}$$

6. Show that the inverse of

$$A = \begin{bmatrix} 3 & -1 & 2 \\ 1 & 2 & 1 \\ -2 & 1 & 3 \end{bmatrix} \quad \text{is} \quad A^{-1} = \frac{1}{30}\begin{bmatrix} 5 & 5 & -5 \\ -5 & 13 & -1 \\ 5 & -1 & 7 \end{bmatrix}$$

by verifying that $AA^{-1} = A^{-1}A = I$.

7. Prove that if A is a matrix with A^{-1} as its inverse and $AB = AC$, then $B = C$.

3.10 SOLUTION OF LINEAR EQUATIONS USING A^{-1}

In this section we shall attempt to illustrate how A^{-1} is used to solve the system of n linear equations in n unknowns. Consider the following example.

EXAMPLE 3.10.1

Solve the following system of linear equations.

$$\begin{aligned} x + 2y &= 4 \\ 2x + 3y &= 7 \end{aligned} \tag{3.12}$$

SOLUTION   Expressed in matrix notation, we have

$$\begin{bmatrix} 1 & 2 \\ 2 & 3 \end{bmatrix}\begin{bmatrix} x \\ y \end{bmatrix} = \begin{bmatrix} 4 \\ 7 \end{bmatrix}$$

$$AX = B \tag{3.13}$$

where $\qquad A = \begin{bmatrix} 1 & 2 \\ 2 & 3 \end{bmatrix} \qquad X = \begin{bmatrix} x \\ y \end{bmatrix} \qquad B = \begin{bmatrix} 4 \\ 7 \end{bmatrix}$

Assuming that we can find A^{-1} such that $AA^{-1} = A^{-1}A = I$, we multiply both sides of Equation (3.13) by A^{-1} and obtain

$$A^{-1}(AX) = A^{-1}B \tag{3.14}$$

Using the associative property in matrix multiplication, we get

$$(A^{-1}A)X = A^{-1}B \tag{3.15}$$

Since $A^{-1}A = I$, we obtain the solution

$$X = A^{-1}B \tag{3.16}$$

In the specific example,

$$A^{-1} = \begin{bmatrix} -3 & 2 \\ 2 & -1 \end{bmatrix}$$

since

$$A^{-1}A = \begin{bmatrix} -3 & 2 \\ 2 & -1 \end{bmatrix}\begin{bmatrix} 1 & 2 \\ 2 & 3 \end{bmatrix} = \begin{bmatrix} 1 & 0 \\ 0 & 1 \end{bmatrix}$$

$$AA^{-1} = \begin{bmatrix} 1 & 2 \\ 2 & 3 \end{bmatrix}\begin{bmatrix} -3 & 2 \\ 2 & -1 \end{bmatrix} = \begin{bmatrix} 1 & 0 \\ 0 & 1 \end{bmatrix}$$

Thus,

$$X = A^{-1}B = \begin{bmatrix} -3 & 2 \\ 2 & -1 \end{bmatrix}\begin{bmatrix} 4 \\ 7 \end{bmatrix} = \begin{bmatrix} 2 \\ 1 \end{bmatrix}$$

Hence, $x = 2$ and $y = 1$ are the solutions to the linear equations (3.12).

The problem that now remains is how to compute A^{-1}. The Gauss-Jordan elimination procedure that we have used in solving a system of linear equations can also be applied to compute the inverse of the coefficient matrix.

To compute A^{-1}, we set up the equations

$$AX = E_1, AX = E_2, \ldots, AX = E_n$$

where E_i is a column vector with a 1 in ith row and 0's elsewhere. Then the augmented matrix is

$$[A:E_1, E_2, \ldots, E_n] = [A:I_n]$$

If A^{-1} exists, we can reduce $[A:I_n]$ to $[I_n:A^{-1}]$ by repeatedly using elementary row operations. The following examples will illustrate these concepts.

EXAMPLE 3.10.2

Find the inverse of

$$A = \begin{bmatrix} 1 & 3 \\ 2 & 7 \end{bmatrix}$$

SOLUTION   The following elementary operations are performed in the order specified in Table 3.4.

TABLE 3.4

Row Operation	a_1	a_2	E_1	E_2
R_1	1	3	1	0
R_2	2	7	0	1
R_1	1	3	1	0
$R_2' = R_2 + (-2)R_1$	0	1	-2	1
$R_1' = R_1 + (-3)R_2'$	1	0	7	-3
R_2'	0	1	-2	1

The augmented matrix $[A:I_2]$ is reduced to $[I_2:A^{-1}]$, where

$$A^{-1} = \begin{bmatrix} 7 & -3 \\ -2 & 1 \end{bmatrix}$$

EXAMPLE 3.10.3

Find the inverse of the matrix

$$A = \begin{bmatrix} 1 & 1 & 1 \\ 3 & 4 & -1 \\ 2 & -5 & 3 \end{bmatrix}$$

and use A^{-1} to solve the following linear equations.

$$\begin{aligned} x + y + z &= 9 \\ 3x + 4y - z &= 13 \\ 2x - 5y + 3z &= 8 \end{aligned}$$

SOLUTION   The following elementary row operations are performed in the order specified in Table 3.5. So

$$A^{-1} = \frac{1}{27} \begin{bmatrix} -7 & 8 & 5 \\ 11 & -1 & -4 \\ 23 & -7 & -1 \end{bmatrix}$$

Thus,

$$\begin{bmatrix} x \\ y \\ z \end{bmatrix} = A^{-1} \begin{bmatrix} 9 \\ 13 \\ 8 \end{bmatrix} = \frac{1}{27} \begin{bmatrix} -7 & 8 & 5 \\ 11 & -1 & -4 \\ 23 & -7 & -1 \end{bmatrix} \begin{bmatrix} 9 \\ 13 \\ 8 \end{bmatrix} = \begin{bmatrix} 3 \\ 2 \\ 4 \end{bmatrix}$$

so that $x = 3$, $y = 2$, and $z = 4$ is the unique solution to this system of linear equations.

TABLE 3.5

Row Operation	a_1	a_2	a_3	E_1	E_2	E_3
R_1	1	1	1	1	0	0
R_2	3	4	-1	0	1	0
R_3	2	-5	3	0	0	1
R_1	1	1	1	1	0	0
$R'_2 = R_2 + (-3)R_1$	0	1	-4	-3	1	0
$R'_3 = R_3 + (-3)R_1$	0	-7	1	-2	0	1
$R'_1 = R_1 + (-1)R'_2$	1	0	5	4	-1	0
R'_2	0	1	-4	-3	1	0
$R''_3 = (-\frac{1}{27})(R'_3 + 7R'_2)$	0	0	1	$\frac{23}{27}$	$-\frac{7}{27}$	$-\frac{1}{27}$
$R''_1 = R'_1 + (-5)R''_3$	1	0	0	$-\frac{7}{27}$	$\frac{8}{27}$	$\frac{5}{27}$
$R''_2 = R'_2 + 4R''_3$	0	1	0	$\frac{11}{27}$	$-\frac{1}{27}$	$-\frac{4}{27}$
R''_3	0	0	1	$\frac{23}{27}$	$-\frac{7}{27}$	$-\frac{1}{27}$

We have observed earlier that not all nonzero square matrices have inverses. Consider, for instance, the square matrix

$$A = \begin{bmatrix} 1 & -1 & 2 \\ -2 & 3 & -2 \\ 4 & -7 & 2 \end{bmatrix}$$

Recall that for A^{-1} to exist, we must be able to reduce the augmented matrix

$$[A:E_1, E_2, E_3] = [A:I_3]$$

to the form $[I_3:A^{-1}]$ by repeatedly using elementary row operations. Let us perform the following elementary operations in the order specified, as shown in Table 3.6.

TABLE 3.6

Row Operation	a_1	a_2	a_3	E_1	E_2	E_3
R_1	1	-1	2	1	0	0
R_2	-2	3	-2	0	1	0
R_3	4	-7	2	0	0	1
R_1	1	-1	2	1	0	0
$R'_2 = R_2 + 2R_1$	0	1	2	2	1	0
$R'_3 = R_3 + (-4)R_1$	0	-3	-6	-4	0	1
$R'_1 = R_1 + R'_2$	1	0	4	3	1	0
R'_2	0	1	2	2	1	0
$R''_3 = R'_3 + 3R'_2$	0	0	0	2	3	1

Since we have a row of 0's in the first three columns, there is no way that we can reduce the coefficient matrix A to the identity matrix I. Hence, we conclude that A^{-1} does not exist. Further, note that it is impossible to solve the equations

$$\begin{bmatrix} 1 & -1 & 2 \\ -2 & 3 & -2 \\ 4 & -7 & 2 \end{bmatrix} \begin{bmatrix} x \\ y \\ z \end{bmatrix} = \begin{bmatrix} 1 \\ 0 \\ 0 \end{bmatrix}$$

since these equations are equivalent to the set of equations

$$\begin{bmatrix} 1 & 0 & 4 \\ 0 & 1 & 2 \\ 0 & 0 & 0 \end{bmatrix} \begin{bmatrix} x \\ y \\ z \end{bmatrix} = \begin{bmatrix} 3 \\ 2 \\ 2 \end{bmatrix}$$

and these equations are inconsistent. Specifically, the last row corresponds to the equation

$$0 \cdot x + 0 \cdot y + 0 \cdot z = 2$$

and there are no values of x, y, and z for which this equation holds.

Frequently, we wish to treat simultaneously the systems of linear equations

$$AX = B_1, AX = B_2, \ldots, AX = B_n$$

all of which have the same coefficient matrix A. We observed earlier that these systems can be solved simultaneously by using the Gauss-Jordan elimination method. Alternatively, we may observe that if the square matrix A has an inverse, then we can premultiply the augmented matrix

$$[A : B_1, B_2, \ldots, B_n]$$

by A^{-1} and obtain

$$[I_n : A^{-1}B_1, A^{-1}B_2, \ldots, A^{-1}B_n]$$

Note that $A^{-1}B_1$, $A^{-1}B_2$, ..., $A^{-1}B_n$ are the solutions to the systems of linear equations $AX = B_1$, $AX = B_2$, ..., $AX = B_n$, respectively.

EXAMPLE 3.10.4

Find the inverse of the matrix

$$A = \begin{bmatrix} 2 & 3 & -1 \\ 1 & 2 & 1 \\ -1 & -1 & 3 \end{bmatrix}$$

and use it to solve the following systems of linear equations.

(a)
$$\begin{aligned} 2x + 3y - z &= 4 \\ x + 2y + z &= 7 \\ -x - y + 3z &= 7 \end{aligned}$$

(b)
$$\begin{aligned} 2x + 3y - z &= 1 \\ x + 2y + z &= 4 \\ -x - y + 3z &= 5 \end{aligned}$$

(c)
$$\begin{aligned} 2x + 3y - z &= 26 \\ x + 2y + z &= 11 \\ -x - y + 3z &= -18 \end{aligned}$$

SOLUTION   We leave it for the reader to verify that

$$A^{-1} = \begin{bmatrix} 7 & -8 & 5 \\ -4 & 5 & -3 \\ 1 & -1 & 1 \end{bmatrix}$$

Note that the three systems of linear equations, which have the same coefficient matrix, differ only in the constant values. The augmented matrix associated with these systems is

Constant Column
Vectors

(a)   (b)   (c)

$$[A:B_1, B_2, B_3] = \begin{bmatrix} 2 & 3 & -1 & 4 & 1 & 26 \\ 1 & 2 & 1 & 7 & 4 & 11 \\ -1 & -1 & 3 & 7 & 5 & -18 \end{bmatrix}$$

Now we premultiply the augmented matrix $[A:B_1, B_2, B_3]$ by A^{-1} and obtain

$$\begin{bmatrix} 7 & -8 & 5 \\ -4 & 5 & -3 \\ 1 & -1 & 1 \end{bmatrix} \begin{bmatrix} 2 & 3 & -1 & 4 & 1 & 26 \\ 1 & 2 & 1 & 7 & 4 & 11 \\ -1 & -1 & 3 & 7 & 5 & -18 \end{bmatrix} = \begin{bmatrix} 1 & 0 & 0 & 7 & 0 & 4 \\ 0 & 1 & 0 & -2 & 1 & 5 \\ 0 & 0 & 1 & 4 & 2 & -3 \end{bmatrix}$$

Hence, $x = 7$, $y = -2$, $z = 4$ is the solution to the system of linear equations in (a); $x = 0$, $y = 1$, $z = 2$ is the solution to the system of linear equations in (b); and $x = 4$, $y = 5$, $z = -3$ is the solution to the system of linear equations in (c).

EXERCISES 3.10

Find, if possible, the inverse of the matrices in Exercises 1–4.

1. $\begin{bmatrix} 3 & 2 \\ 5 & 4 \end{bmatrix}$

2. $\begin{bmatrix} 4 & 1 \\ 5 & 2 \end{bmatrix}$

3. $\begin{bmatrix} -8 & 10 \\ 6 & -7 \end{bmatrix}$

4. $\begin{bmatrix} 1 & -1 \\ 4 & 5 \end{bmatrix}$

Compute the inverse of the matrices in Exercises 5–12.

5. $\begin{bmatrix} 2 & -1 & 2 \\ 4 & 1 & 2 \\ 8 & -1 & 1 \end{bmatrix}$

6. $\begin{bmatrix} 1 & 2 & 1 \\ 2 & 3 & 4 \\ 3 & -1 & 1 \end{bmatrix}$

7. $\begin{bmatrix} 1 & -3 & 1 \\ 2 & 3 & 3 \\ 3 & -1 & 1 \end{bmatrix}$

8. $\begin{bmatrix} 2 & 2 & -1 \\ 3 & 2 & -2 \\ 1 & -1 & 1 \end{bmatrix}$

9. $\begin{bmatrix} 1 & 2 & 1 \\ 1 & -1 & 2 \\ 2 & 3 & -1 \end{bmatrix}$ 10. $\begin{bmatrix} 3 & 1 & 0 \\ 1 & -1 & 2 \\ 1 & 1 & 1 \end{bmatrix}$

11. $\begin{bmatrix} 3 & 4 & -1 \\ 2 & -5 & 3 \\ 1 & 1 & 1 \end{bmatrix}$ 12. $\begin{bmatrix} 2 & 3 & -1 \\ 1 & 2 & 1 \\ -1 & -1 & 3 \end{bmatrix}$

13. Find the inverse of the matrix

$$A = \begin{bmatrix} 4 & -1 \\ 2 & 1 \end{bmatrix}$$

and use A^{-1} to solve the linear equations

$$4x - y = 17$$
$$2x + y = 7$$

14. Find the inverse of the matrix

$$A = \begin{bmatrix} 5 & 3 \\ 3 & 2 \end{bmatrix}$$

and use A^{-1} to solve the following systems of linear equations.

(a) $5x + 3y = 8$ (b) $5x + 3y = 2$
    $3x + 2y = 5$ $3x + 2y = 1$

(c) $5x + 3y = 13$ (d) $5x + 3y = 11$
    $3x + 2y = 8$ $3x + 2y = 7$

15. Find the inverse of the matrix

$$A = \begin{bmatrix} 3 & 2 \\ 4 & 1 \end{bmatrix}$$

and use A^{-1} to solve the following systems of linear equations.

(a) $3x + 2y = 12$ (b) $3x + 2y = 6$
    $4x + y = 11$ $4x + y = 13$

(c) $3x + 2y = 11$ (d) $3x + 2y = 1$
    $4x + y = -2$ $4x + y = 3$

16. Find the inverse of the matrix

$$A = \begin{bmatrix} 1 & 1 & 1 \\ 2 & 5 & -3 \\ 3 & 4 & -7 \end{bmatrix}$$

and use A^{-1} to solve the following systems of linear equations.

(a) $x + y + z = 3$ (b) $x + y + z = 6$
    $2x + 5y - 3z = 4$ $2x + 5y - 3z = 3$
    $3x + 4y - 7z = 0$ $3x + 4y - 7z = -10$

(c) $\begin{aligned} x + y + z &= -1 \\ 2x + 5y - 3z &= 13 \\ 3x + 4y - 7z &= 27 \end{aligned}$ (d) $\begin{aligned} x + y + z &= 6 \\ 2x + 5y - 3z &= 13 \\ 3x + 4y - 7z &= 10 \end{aligned}$

17. Find the inverse of the matrix

$$A = \begin{bmatrix} 1 & 1 & 5 \\ 2 & 1 & -1 \\ 3 & 2 & 5 \end{bmatrix}$$

and use A^{-1} to solve the following systems of linear equations.

(a) $\begin{aligned} x + y + 5z &= 0 \\ 2x + y - z &= 8 \\ 3x + 2y + 5z &= 7 \end{aligned}$ (b) $\begin{aligned} x + y + 5z &= 26 \\ 2x + y - z &= -1 \\ 3x + 2y + 5z &= 30 \end{aligned}$

(c) $\begin{aligned} x + y + 5z &= 41 \\ 2x + y - z &= 1 \\ 3x + 2y + 5z &= 49 \end{aligned}$ (d) $\begin{aligned} x + y + 5z &= 7 \\ 2x + y - z &= 1 \\ 3x + 2y + 5z &= 9 \end{aligned}$

Solve the systems of linear equations in Exercises 18–21 using the inverse of the coefficient matrix.

18. $\begin{aligned} x + 3y + 5z &= 5 \\ 3x + 2y + z &= 6 \\ 6x - 3y + 2z &= 5 \end{aligned}$ 19. $\begin{aligned} x + 3y + z &= 16 \\ 3x + 2y - z &= 1 \\ 6x - y + 3z &= -8 \end{aligned}$

20. $\begin{aligned} x + y + 2z &= 22 \\ 2x + y &= 11 \\ x + 2y + 2z &= 27 \end{aligned}$ 21. $\begin{aligned} 2x + 3y - z &= 4 \\ x + y + z &= 3 \\ 2y - z &= 1 \end{aligned}$

22. A department store has 90 men's suits of three different types which it must sell this spring. If it sells type I suit for $50, type II suit for $60, and type III suit for $75, the net sales amounts to $5800, but if these suits are sold at $45, $55, and $60, respectively, then the total revenue in sales is $4950. Using the inverse of the coefficient matrix, determine the number of suits of each type the store has in stock.

23. A shipping company has three different types of trucks, T_1, T_2, and T_3 to transport the number of station wagons, full-size cars, and intermediate-size cars as shown in the following matrix:

		Station Wagons	Full-size Cars	Intermediate-size Cars
Truck	T_1	2	6	9
	T_2	3	7	12
	T_3	6	6	8

Using the inverse of the matrix, determine the number of trucks of each type required to supply 58 station wagons, 75 full-size cars, and 62 intermediate-size cars to a dealer in Boston.

For Further Reading

Bronson, R. *Matrix Methods: An Introduction.* New York: Academic Press, Inc., 1969.

Campbell, H. G. *Matrices with Applications.* New York: Appleton-Century-Crofts, 1968.

Davis, P. J. *The Mathematics of Matrices.* New York: John Wiley & Sons, Inc., 1965.

Steinberg, D. I. *Computational Matrix Algebra.* New York: McGraw-Hill Book Company, 1974.

4

EQUIVALENCE AND THE NUMBER SYSTEMS

4.1 INTRODUCTION

When we first entered elementary school we learned how to say numbers and then how to write them. Progressing through the grades, we encountered increasingly ingenious ways to manipulate numbers and widely varied uses for them. High school brought with it numbers in disguise (algebra), numbers related to shapes (geometry and trigonometry), and perhaps even numbers linked to motion (calculus) or to chance (probability). And yet throughout this kaleidoscope of quantitative experience, one question remained generally unasked and almost certainly unanswered: What *is* a number? This gap in curiosity seems paradoxical in that the equipment needed to provide a mathematically satisfactory answer is well within the scope of what is often called "the new math," yet these easily accessible techniques are seldom focused on this fundamental question.

Historically, "numbers" and "number systems" of various sorts developed to meet the needs for specific kinds of computational aids, measurement techniques, and so forth. When the nineteenth-century trend in mathematics began to push for logical consistency and unity throughout all areas of the subject, mathematicians sought to unify arithmetic by developing all of the various number systems from a common simple base. One way to do this is to begin with the simple notion of a "set" or "collection" of things* as first explored by Cantor in the 1870s,†

---

*It suffices for our purposes to use the terms "set," "class," and "collection" informally and interchangeably, even though some distinctions are made among their meanings in formal abstract set theory.

† See Chapters 1 and 5 for further information about Cantor's work.

define something called an "equivalence relation," and build everything from these two primitive concepts; that is the course we shall pursue in this chapter. Little attention will be paid to operations and relations on the different sets of numbers; we shall focus on the question of what the numbers themselves are, pausing only briefly at each major step to indicate how their construction allows them to be manipulated in the ways we want.

"Wait a minute," we hear you say. "That sounds complicated. Can't we just define a number as an expression of quantity, or something like that?" We would agree that "quantity" expresses much the same idea as "number," but synonyms are not very informative. Even the common dictionary definitions of "number" merely list unenlightening paraphrases such as "one, or more than one" and "multitude." None of these purported definitions provides any way of distinguishing among specific numbers, like 3, 5, $\frac{1}{2}$, -7, and $\sqrt{2}$, or supplies a conceptual groundwork from which we can explain why the basic arithmetic operations work the way they do. Without such a foundation, arithmetic becomes an exercise in the rote learning of mystical chants, such as "one, two, three, . . . ," "two plus three is five," "five times six is thirty," and so on. But arithmetic is far too logical to be relegated to memory without understanding. Moreover, once the basic numerical ideas are understood there is much less to memorize, because the more complicated arithmetic facts are then obvious logical consequences of the simpler ones. Thus, what we really propose to do here is to take the magic out of arithmetic by showing that all numerical behavior is deducible by commonsense logic from a few basic principles.

EXERCISES 4.1

1. Look up "number" in a dictionary and comment on the utility of the definition you find.

2. Look up each of the following words in a dictionary and comment on whether or not the definition gives you enough information to work with the concept arithmetically. Also comment on how these definitions relate to the definition of "number" you found in Exercise 1.

   (a) one   (b) two   (c) three
   (d) five   (e) zero   (f) one-half (or half)
   (g) plus   (h) add   (i) multiply
   (j) subtract

4.2   EQUIVALENCE RELATIONS

The fundamental idea of a "relation" on a set is very simple; it is any rule or process by which elements of that set are paired up. If we remember (or believe) that $S \times S$ is the set of all ordered pairs of elements of a set S,* then the defini-

---

* In general, the **Cartesian product** of two sets A and B is $A \times B = \{(a, b) | a \in A \text{ and } b \in B\}$.

tion of a relation on a set can be stated easily and without the confusing vagueness of "rule or process":

DEFINITION
4.2.1

A **relation** on a set S is any subset of $S \times S$.

NOTATION: If $\mathscr{R}$ is a relation on S, then $\mathscr{R} \subseteq S \times S$. However, we usually write "$a \mathscr{R} b$" rather than "$(a, b) \in \mathscr{R}$" to denote a pair of elements in the relation.

EXAMPLES 4.2

1. The statement "x is less than y" describes a relation on any set of whole numbers. For instance, on the set $\{1, 2, 3\}$ this relation is $\{(1, 2), (1, 3), (2, 3)\}$, because 1 is less than both 2 and 3, and 2 is less than 3. We usually denote this relation by $<$, and hence would write $1 < 2$, $1 < 3$, and $2 < 3$.

2. Let B be a set of bricks of various solid colors. The statement "x is the same color as y" describes a relation on B.

3. Let C be the set of all cities in the United States. The statement "x is east of y" describes a relation on C. If we denote that relation by $\mathscr{E}$, then we could write Chicago $\mathscr{E}$ Denver (since Chicago is east of Denver), and New York $\mathscr{E}$ Chicago, but San Francisco $\not\mathscr{E}$ New York.*

4. Let $W = \{1, 2, 3, 4\}$, and define a relation $\mathscr{R}$ on W by "$x \mathscr{R} y$ if and only if either x and y are both even or they are both odd." Then $\mathscr{R} = \{(1, 1), (1, 3), (2, 2), (2, 4), (3, 1), (3, 3), (4, 2), (4, 4)\}$. We may write $1 \mathscr{R} 3$, $2 \not\mathscr{R} 3$, and so forth.

5. Note that the definition of relation is so general that *any* collection of ordered pairs of a set is a relation on that set; there need not be any obvious "rule" involved, except that those pairs have in fact been chosen to constitute the relation. For instance, if $S = \{*, o, \$, \#\}$, then $\mathscr{R} = \{(*, \$), (\#, o), (\#, *), (o, o), (\$, *)\}$ is a relation on S, and we would write $* \mathscr{R} \$$ because $(*, \$) \in \mathscr{R}$, but $o \not\mathscr{R} \#$ because $(o, \#) \notin \mathscr{R}$.

Among all the types of relations in mathematics, perhaps the most familiar one is exemplified by equality of whole numbers. We have worked with expressions such as $2 + 3 = 5$ since the first grade, yet the exact meaning of "$=$" may still be a bit vague. We know that it indicates sameness in some sense, but certainly not in every sense, for it is clear that the symbol 5 is quite different from the symbol $2 + 3$. Certain properties of this relation, seemingly too obvious to be worth mentioning, turn out to be the defining conditions for an important class of relations:

---

* If a relation does not hold between two elements, that is denoted by putting a slash through the relation symbol; recall $\neq$, $\not<$, and so on, from high school algebra.

DEFINITION
4.2.2

Let $\mathscr{R}$ be a relation on a set S such that

1. $x \mathscr{R} x$ for every element x in S (**reflexive** property).
2. If x and y are elements of S such that $x \mathscr{R} y$, then $y \mathscr{R} x$ (**symmetric** property).
3. If x, y, and z are elements of S (not necessarily distinct) such that $x \mathscr{R} y$ and $y \mathscr{R} z$, then $x \mathscr{R} z$ (**transitive** property).

Then we call $\mathscr{R}$ an **equivalence relation** on S, and we say that x **is equivalent to** y if $x \mathscr{R} y$.

EXAMPLES 4.2

6. As mentioned above, equality of numbers is a typical equivalence relation, since
   (a) If n is any number, $n = n$.
   (b) If m and n are numbers and $m = n$, then it must also be true that $n = m$.
   (c) If m, n, and p are numbers such that $m = n$ and $n = p$, it follows that $m = p$ also.

7. Congruence is an equivalence relation on any set of geometric figures.

8. Any two geographical positions may be considered equivalent if they have the same latitude.

9. Let P be the set of all cars in a large parking lot. If we say that two cars are related whenever they are the same make, we have an equivalence relation on P. Another (probably different) equivalence relation on P is defined by calling two cars related if their hoods are the same color.

Let us try to determine how an equivalence relation affects the set on which it is defined. Think for a moment about "equals" on any set of symbols representing numbers. It is easy to see that the set is broken down into subsets in a natural way, each subset containing symbols which stand for the same number. Thus, if

$$S = \{2, 5, 1 + 1, 3 - 2, 1, 7 - 5, 3 + 2\}$$

then S can be rewritten

$$S = \{2, 1 + 1, 7 - 5\} \cup \{5, 3 + 2\} \cup \{3 - 2, 1\}$$

and all symbols in each of these subsets are equal to each other. Moreover, any two subsets are disjoint.* This example suggests a justifiable conjecture about equivalence relations in general, but first let us give a name to such a division into subsets.

---

* Recall that sets A and B are **disjoint** if $A \cap B = \varnothing$; that is, A and B contain no elements in common.

DEFINITION
4.2.3

Let S be any set. If $\mathscr{C}$ is a collection of nonempty subsets of S such that

1. any two subsets in $\mathscr{C}$ are disjoint, and
2. the union of all the sets in $\mathscr{C}$ equals S,

then we say $\mathscr{C}$ is a **partition** of S, and any relation that gives rise to such a collection $\mathscr{C}$ is said to **partition** S.

THEOREM
4.2.1

Any equivalence relation partitions the set on which it is defined.

PROOF (optional)   Let $\mathscr{R}$ be an equivalence relation on a set S, and form a collection $\mathscr{C}$ of (nonempty) subsets by the following rule: For any $x, y \in S$, x and y are in the same subset if and only if $x \mathscr{R} y$. We must show that both conditions in the definition of partition are fulfilled.

1. Let A and B be any two distinct subsets from the collection $\mathscr{C}$, and suppose that A and B are not disjoint. Then there is some element $x \in A \cap B$. This implies that $x \in A$ and $x \in B$. Since A and B are distinct sets, either there is some element of A that is not in B, or there is some element of B not in A. The argument is the same in either case, so we may choose the latter without loss of generality. Let a be an element of A, and let b be an element of B that is not in A. Since $x \in A$, $x \mathscr{R} a$ by the way we formed the subsets. Similarly, $x \mathscr{R} b$, because $x \in B$. Now, since $\mathscr{R}$ is an equivalence relation, $x \mathscr{R} a$ implies that $a \mathscr{R} x$ by the symmetric property. Thus, we have $a \mathscr{R} x$ and $x \mathscr{R} b$, implying that $a \mathscr{R} b$ by the transitive property. This means that b must be in A, which is a contradiction. Therefore, the supposition that A and B have a nonempty intersection must be false, so A and B are disjoint. Since A and B stand for any distinct subsets in $\mathscr{C}$, it follows that $\mathscr{C}$ is a collection of pairwise disjoint subsets.
2. This part follows immediately from the reflexive property. Since every element is equivalent to itself, any element $s \in S$ that does not belong to any other subset by equivalence to other elements must then belong to the subset $\{s\}$.

The converse of this theorem is also true: Any partition of a set induces an equivalence relation on that set. (See Exercise 4.2.6.) Thus, the following terminology is reasonable:

DEFINITION
4.2.4

The subsets that form a partition are called **equivalence classes**.

EXAMPLES 4.2

10. In Example 4.2.9 each equivalence class determined by the first relation is the set of all cars of a particular make in that lot: all Dodges, all Hondas, all

Fords, and so on. The equivalence classes of the second relation may be represented by colors; they are all cars with red hoods, all cars with white hoods, and so forth.

11. Let $S = \{27, -3, 49, 1, -15, 13, -8, 36\}$, and define an equivalence relation on S by declaring $x \mathcal{R} y$ if and only if x and y have the same sign. (You should verify that $\mathcal{R}$ is actually an equivalence relation by checking to see that $\mathcal{R}$ has the reflexive, symmetric, and transitive properties.) Then the resulting partition is a collection consisting of two subsets, $\{27, 49, 1, 13, 36\}$ and $\{-3, -15, -8\}$.

Notice that part (1) of the proof of Theorem 4.2.1 establishes an important general fact about equivalence classes:

Two equivalence classes are equal if and only if they have an element in common.

Hence, to show equality between two equivalence classes it is always sufficient to show only that there is some element that belongs to both of them. This fact provides an efficient test for equality which will often be useful in later sections.

EXERCISES 4.2

1. Construct three relations (on any set you choose).

2. Which of the following statements define an equivalence relation on the set of all geographical positions? Why?
   (a) x is north of y.
   (b) x is less than 5 kilometers from y.
   (c) x has the same longitude as y.
   (d) x is closer to a pole than y.

3. Recall (or believe) that the set of all integers is the set $\{\ldots, -3, -2, -1, 0, 1, 2, 3, \ldots\}$ of all whole numbers and their negatives. Decide whether or not each of the following relations on this set is reflexive, symmetric, and/or transitive, giving counterexamples for the properties that do not hold.
   (a) $x \mathcal{R} y$ if and only if $x = y + 1$.
   (b) $x \mathcal{R} y$ if and only if 5 is a factor of $x - y$.
   (c) $x \mathcal{R} y$ if and only if x and y have opposite signs.
   (d) $x \mathcal{R} y$ if and only if 3 is a factor of both x and y.
   (e) $x \mathcal{R} y$ if and only if $x \neq y$.
   (f) $x \mathcal{R} y$ if and only if either x or y is 7.
   (g) $x \mathcal{R} y$ if and only if x is greater than y.
   (h) $x \mathcal{R} y$ if and only if x is not less than y.
   (i) $x \mathcal{R} y$ if and only if 5 is a factor of $x + y$.
   (j) $x \mathcal{R} y$ if and only if $x + y = 0$.

4. Exactly one of the relations in Exercise 4.2.3 is an equivalence relation. Describe the equivalence classes of its partition.

5. Which of the five relations in Examples 4.2.1 through 4.2.5 are equivalence relations? For each one that is, describe the resulting partition.

6. Prove that any partition of a set induces an equivalence relation on that set.

4.3   THE WHOLE NUMBERS

Up to now we have assumed an informal prior acquaintance with some types of numbers, just for the sake of having a few simple illustrative examples in the previous section. None of the theory of equivalence relations required the use of numbers, however, so it is correct at this point to consider ourselves at the very beginning of the logical development of the number systems. Thus, we shall no longer assume any prior knowledge of numbers, but instead shall build all the numerical concepts we need step by step, starting only with the fundamental notions of set and equivalence relation. In this way by the end of this unit we shall have a logically consistent development of all the various types of numbers. At each stage of the development we shall be able to *define* what a particular type of number is by choosing an appropriate set and defining a specific equivalence relation on that set; the "numbers" will be the equivalence classes we get! Thus, the entire structure of the number systems is unified by the single basic concept developed in the last section.

The most primitive type of number relates to counting. Historically, as soon as humans began to trade, comparing sizes of collections of things became a crucial issue. Let us transport ourselves back in time to witness an allegorical event in the prenumerical past.

Once upon a time, long, long ago, there lived an ambitious young man who owned a flock of sheep. One day, filled with a curiosity common to all budding mathematicians, he set out from his cave in the valley to investigate the far side of the nearest mountain. There he found another valley and another young man, who owned a herd of goats. Now, the first young man had never seen a goat before and was anxious to be the first person in his valley to own one, so he offered to exchange a sheep for a goat. After completing the transaction, the young shepherd brought the strange new animal back to his own village, and soon all his neighbors wanted goats of their own. The same reaction greeted the second young man and his sheep. Thus, the two men became sheep-and-goat traders for their respective valleys.

The process of exchange was slow and arduous, however, since each time the first valley wanted goats the second young man had to drive his entire herd to the top of the mountain, exchange a goat for a sheep until all the requests were filled, and then drive the entire herd back down the mountain. A similar process was required whenever the second valley wanted sheep. As you can well imagine, both men quickly began to tire of this routine, and indolence begot invention. One

afternoon, while resting atop the mountain, they resolved to trade by using twigs to represent goats and pebbles to represent sheep. Each trader would bring one of these objects for every animal required, they would exchange on a one-for-one basis, and then only the animals that were actually traded would have to be driven up the mountain. The method also proved useful in keeping track of orders, so the young men no longer had to build a separate corral for each customer's animals. News of their revolutionary system of barter and bookkeeping spread rapidly. As a result of their discovery the two young men acquired fame and fortune, became known as the first two mathematicians of all time, and lived happily ever after.

Thus may have been born the idea that collections of basically dissimilar objects may possess a common property that can be determined by a one-to-one matching process. This is the key idea underlying the counting numbers; it may be stated formally as follows:

DEFINITION 4.3.1

A matching between two sets A and B such that each element in either set corresponds to exactly one element in the other is called a **one-to-one correspondence** between A and B. If there exists a one-to-one correspondence between two sets A and B, then the sets are said to be **equivalent**; we write this $A \leftrightarrow B$.

EXAMPLE 4.3.1

There are a number of one-to-one correspondences between the sets $\{a, b, c, d\}$ and $\{*, \text{o}, \$, \#\}$; for instance,

$$\{a, b, c, d\} \qquad\qquad \{a, b, c, d\}$$
$$\mid \ \mid \ \mid \ \mid \qquad \text{and} \qquad \times \quad \times$$
$$\{*, \text{o}, \$, \#\} \qquad\qquad \{*, \text{o}, \$, \#\}$$

(There are 24 such correspondences; can you find all of them?) Thus, these two sets are equivalent. Neither set is equivalent to $\{x, y, z\}$, however, because it is not possible to set up a one-to-one correspondence between $\{a, b, c, d\}$ and $\{x, y, z\}$ or between $\{*, o, \$, \#\}$ and $\{x, y, z\}$.

This second and apparently ambiguous use of "equivalent" is really not a logical error. We are about to show that "$\leftrightarrow$" is an equivalence relation, so this use of the word is just a special case of the last.

THEOREM 4.3.1 $\leftrightarrow$ is an equivalence relation on the collection of all sets.

PROOF   Let $\mathscr{S}$ denote the collection of all sets. We must show that $\leftrightarrow$ is reflexive, symmetric, and transitive on $\mathscr{S}$.

1. Reflexive: If S is any set in $\mathscr{S}$, matching each element of S with itself is obviously a one-to-one correspondence between S and itself, so $S \leftrightarrow S$.
2. Symmetric: If S and T are sets in $\mathscr{S}$ such that $S \leftrightarrow T$, then each element of S is matched with exactly one element of T, and vice versa. Clearly, this matching is also a one-to-one correspondence between T and S; that is, $T \leftrightarrow S$.
3. Transitive: If S, T, and U are sets in $\mathscr{S}$ such that $S \leftrightarrow T$ and $T \leftrightarrow U$, then there is a natural one-to-one correspondence between S and U which can be built from the two given correspondences, as follows: For each $s \in S$ there is a corresponding $t \in T$ by the $S \leftrightarrow T$ matching; for this t, there is a corresponding $u \in U$ by the $T \leftrightarrow U$ matching; match s with that u. It is easy to check that this process yields a one-to-one correspondence between S and U, so $S \leftrightarrow U$, as required.

When considering equivalent sets we often refer to them as having the "same number" of elements, thereby recognizing the concept of a specific number as being a property common to two sets which can be put in one-to-one correspondence. But this "common property" notion is still too vague to be the basis for counting and arithmetic; for instance, just saying that "3" represents a property common to $\{a, b, c\}$ and $\{\#, *, !\}$ and "4" represents a property common to $\{w, x, y, z\}$ and $\{\$, 0, ?, \&\}$ does not provide much insight into how to multiply 3 times 4. We need a precise description of numbers in terms of things that can be manipulated. The key, of course, is to apply what we know about equivalence relations to the relation $\leftrightarrow$ on finite sets.

Specifically, since $\leftrightarrow$ is an equivalence relation, it partitions any collection of sets into equivalence classes. Now, if we apply $\leftrightarrow$ to the collection of all finite sets, each equivalence class will contain all sets that have the "same number" of elements as each other. For example, if we take the set $\{*, o, \square\}$, it will be in the class containing all those sets (and only those sets) which can be put in one-to-one correspondence with it; we can describe this in mathematical shorthand as $\{S \mid S \leftrightarrow \{*, o, \square\}\}$. But these are precisely the sets that share the common property we might call "threeness," and hence we can *define* "three" as being that collection of sets! In general, we are saying that the collection of all sets equivalent to a given set *is* the number of elements in that set. Formally,

DEFINITION
4.3.2

If A is any set, then the **number** of elements in A is the set of all sets that are equivalent to A. (This is also called the **cardinality** of A.*) Any number formed in this way using a finite set A is called a **whole number**, and the set A is sometimes called a **reference set** for that number.

NOTATION: If n denotes the number of elements in A, then we can write the definition above as $n = \{S \mid S \leftrightarrow A\}$. (Literally, this is read "n equals the set of all sets S such that S is equivalent to A.")

Thus, the whole numbers *are* the equivalence classes determined by the relation $\leftrightarrow$ on the collection of all finite sets, and each number can be represented by any set in that class. This is just a formal (but useful) way of saying, for example, that "three" is a concept that can be represented by $\{*, o, \square\}$ or $\{a, b, c\}$ or any set equivalent to them. Moreover, since different equivalence classes must be disjoint, we can say that two whole numbers are *equal* precisely when they have *equivalent* reference sets.

EXAMPLE 4.3.2

We know that $3 = \{S \mid S \leftrightarrow \{*, o, \square\}\}$. If you are told that a certain whole number n contains the set $\{x, y, z\}$, then you can conclude that n must be 3, because $\{x, y, z\}$ can be put in one-to-one correspondence with $\{*, o, \square\}$. (Note that the two reference sets are not equal in this example.)

Since each whole number is determined by some reference set, in order to specify all the whole numbers (as we "know" they ought to be) we must find a way of constructing larger and larger reference sets, without limit. The basic idea here is very simple: Each time we have a reference set, we can get a larger one by just "tacking on" one more element. Of course, we must be sure that the element tacked on is not already in the original set, for in that case we would not be changing the set at all. One convenient (and elegant) way to build these reference sets, and hence all the whole numbers, is as follows.

Starting with some single identifiable object, such as $\varnothing$,† make a reference set for 1; that is,

---

* You might find it interesting to compare this definition with the definition of cardinal number given in Section 5.8.

† Of course, you need not use the null set here, but within abstract set theory it is one of the few simple objects available without further definition.

$$1 = \{S \,|\, S \leftrightarrow \{\varnothing\}\}$$

Now observe that 1 is a well-defined object which clearly is not $\varnothing$ (why?), so $\{\varnothing, 1\}$ can be used to define 2; that is,

$$2 = \{S \,|\, S \leftrightarrow \{\varnothing, 1\}\}$$

By definition, 2 is different from both $\varnothing$ and 1, so $\{\varnothing, 1, 2\}$ can be used to define 3; that is,

$$3 = \{S \,|\, S \leftrightarrow \{\varnothing, 1, 2\}\}$$

Similarly,

$$4 = \{S \,|\, S \leftrightarrow \{\varnothing, 1, 2, 3\}\}$$
$$5 = \{S \,|\, S \leftrightarrow \{\varnothing, 1, 2, 3, 4\}\}$$

and so on. The collection of all numbers from 1 on which are obtained in this way is called the set of **natural numbers,** and is denoted by N. We also extend this pattern one step "backward," defining 0 as the class of all sets equivalent to the empty set, $\varnothing$. Of course, since there is only one such set, namely $\varnothing$ itself, we can say that $0 = \{\varnothing\}$. The set $\{0, 1, 2, 3, \ldots\}$ of all natural numbers and 0 is called the set of **whole numbers,** and is denoted by W.

Now that we know what whole numbers are, we can define the elementary arithmetic operations in a simple, straightforward way. Since the numbers are defined in terms of reference sets, the operations will also be defined in those terms. Let us look first at addition. A child in the first grade usually learns about addition by means of reference sets; for instance, $3 + 2$ is often explained by taking three pencils and two more pencils, gathering them together and asking the child to count them. Our definition of addition of whole numbers is nothing more than a formal description of this process.

DEFINITION 4.3.3

Let a and b be two whole numbers, with reference sets A in a and B in b chosen so that A and B have no elements in common. The **sum** of a and b is the whole number that contains $A \cup B$. We write this number as $a + b$.

EXAMPLE 4.3.3

To find $3 + 4$, consider $\{*, \circ, \#\} \in 3$ and $\{w, x, y, z\} \in 4$. Since $\{*, \circ, \#\} \cup \{w, x, y, z\} = \{*, \circ, \#, w, x, y, z\} \in 7$, we say that $3 + 4 = 7$.

Multiplication can also be defined in much the same way as it is treated in elementary school. For instance, $3 \cdot 4$ is often illustrated by taking three copies of a set of four objects and counting the total collection. We do the same thing.

**DEFINITION
4.3.4**

Let a and b be two whole numbers, with reference sets A in a and B in b.* The **product** of a and b is the whole number that contains the Cartesian product $A \times B$. We write this number as $a \cdot b$, or simply ab.

EXAMPLE 4.3.4

To find $2 \cdot 3$, consider $\{*, \#\} \in 2$ and $\{x, y, z\} \in 3$. Then $\{*, \#\} \times \{x, y, z\} = \{(*, x), (*, y), (*, z), (\#, x), (\#, y), (\#, z)\}$, which is a set in the class 6, so $2 \cdot 3 = 6$. Notice that the Cartesian product $\{*, \#\} \times \{x, y, z\}$ can be considered just a formal way of writing two distinct "copies" of the set $\{x, y, z\}$, one labeled by $*$ and the other labeled by $\#$.

We can also define the "less than" relationship between unequal whole numbers by looking at reference sets.

**DEFINITION
4.3.5**

Let a and b be two whole numbers. The number a is **less than** b (or b is **greater than** a) if a set from a is equivalent to a proper subset† of a set from b. We write this $a < b$ (or $b > a$).

NOTATION: If a is less than or equal to b, we write $a \leq b$.

EXAMPLE 4.3.5

To show that 2 is less than 3, consider the sets $\{*, \#\} \in 2$ and $\{x, y, z\} \in 3$. Since $\{*, \#\} \leftrightarrow \{x, y\}$ and $\{x, y\}$ is a proper subset of $\{x, y, z\}$, we can conclude that $2 < 3$.

With these definitions of $+$, $\cdot$, and $<$, all the familiar laws of whole number arithmetic can be *proved*; that is, the behavior of the whole number system is a direct logical consequence of a few elementary properties of sets. In particular, it is easy to prove that:

Addition and multiplication are associative: For all $a, b, c \in W$,

$$(a + b) + c = a + (b + c)$$

and

$$(a \cdot b) \cdot c = a \cdot (b \cdot c)$$

---

* A and B may have elements in common in this case.

† Recall that A is a **proper subset** of B if and only if A is a subset of B and A does not equal B.

Addition and multiplication are commutative: For all $a, b \in W$,

$$a + b = b + a$$

and

$$a \cdot b = b \cdot a$$

Multiplication is distributive over addition: For all $a, b, c \in W$,

$$a \cdot (b + c) = (a \cdot b) + (a \cdot c)$$

For all $a, b, c \in W$, if $a < b$, then $a + c < b + c$, and $a \cdot b < b \cdot c$ when $c \neq 0$. Although the proofs of these properties are easy, their inclusion here would only serve to distract us from the main purpose of this unit. Hence, we shall just examine one proof as an illustration of the way the definitions are used to produce the desired results.

THEOREM 4.3.2

Multiplication of whole numbers is commutative.

PROOF   We must show that, for all $a, b \in W$, $a \cdot b = b \cdot a$. Now, if a and b are any two whole numbers, we can choose a reference set from each, say A and B, respectively. By definition of multiplication, $a \cdot b$ can be represented by the set $A \times B$, and $b \cdot a$ can be represented by $B \times A$. As noted earlier, the numbers are equal if their reference sets are equivalent, so to complete the proof we only have to show $A \times B \leftrightarrow B \times A$. By the definition of Cartesian product,

$$A \times B = \{(a, b) | a \in A \text{ and } b \in B\}$$

$$B \times A = \{(b, a) | a \in A \text{ and } b \in B\}$$

It is easy to verify that matching each pair (a, b) in $A \times B$ with its reverse pair (b, a) in $B \times A$ establishes a one-to-one correspondence, so $A \times B \leftrightarrow B \times A$ and the proof is complete.

EXERCISES 4.3

1. (a) If two sets are equivalent, must they be equal? Why?
   (b) If two sets are equal, must they be equivalent? Why?

2. Give an example which illustrates the general fact that, for any two sets A and B, the sets $A \times B$ and $B \times A$ are equivalent.

3. Give five different one-to-one correspondences between the sets $\{a, b, c, d\}$ and $\{*, o, \$, \#\}$.

4. List five different sets that represent the whole number 6.

5. Give two different reference sets for each of the following numbers:
   (a) 2   (b) 10
   (c) 25   (d) 100

6. Use reference sets and the definitions of $+$ and $\cdot$ to compute:
   (a) $1 + 1$ (b) $4 + 7$
   (c) $2 + 0$ (d) $3 \cdot 5$
   (e) $6 \cdot 1$ (f) $9 \cdot 0$

7. Use the definition of $<$ to justify:
   (a) $5 < 8$ (b) $3 \leq 7$

8. Prove that $1 \cdot n = n$ for any $n \in W$.

9. Prove that addition of whole numbers is associative and commutative.

10. Prove that multiplication of whole numbers is distributive over addition.

11. Prove that $a < b$ implies $a + c < b + c$ for all whole numbers a, b, and c.

12. Show that Definition 4.3.3 of sum breaks down if the sets A and B are allowed to have elements in common.

13. For any whole number n, prove:
    (a) $n + 0 = n$ (b) $n \cdot 0 = 0$

Note: Our development of the whole numbers given here parallels the development of the transfinite cardinal numbers and their arithmetic in Sections 5.11 and 5.12. If both Chapters 4 and 5 are covered, an examination of this parallel might be worthwhile.

4.4   THE INTEGERS

Unlike addition and multiplication, subtraction can be defined on a set of numbers without direct reference to the construction of the numbers themselves. However, it requires that addition has already been defined, for subtraction is just addition "in reverse," so to speak.

DEFINITION 4.4.1

Let S be any set of numbers on which an addition operation $+$ has been defined. Then for any elements a and b in S, we say the **difference** $a - b$ equals a number c in S if and only if $b + c$ equals a. In symbols, $a - b = c$ iff $b + c = a$.

EXAMPLE 4.4.1

In the whole numbers, $5 - 3 = 2$ because $3 + 2 = 5$.

Of course, defining something does not automatically assure its existence. Just saying a hobbit* is a small, rotund, humanlike creature with furry feet does not guarantee that hobbits exist; similarly, our general definition for the difference of two numbers does not guarantee that such a difference always exists. In particular, the difference of two whole numbers does not necessarily exist (in that system). For instance, the definition of difference requires that $3 - 5$ be a whole number which when added to 5 yields 3, and there is no such number. (See Exercise 4.4.1.) This section seeks to remedy that shortcoming of the whole number system by extending it to a larger system, the "integers," in which differences always exist. As before, the key is to define an equivalence relation on an appropriate set.

Let us begin with the trivial observation that any "subtraction problem" involving two whole numbers is determined by the ordered pair of numbers themselves, whether or not the difference actually exists: $5 - 3$ is determined by the ordered pair $(5, 3)$, $2 - 9$ by $(2, 9)$, $87 - 64$ by $(87, 64)$, and so forth. Thus, the set of all possible difference questions of the form $x - y = ?$, where x and y are whole numbers can be represented by the set of all ordered pairs of whole numbers, $W \times W$. It seems reasonable to consider two of these difference questions "the same" with respect to the subtraction process if they yield the same answer. We use this idea to motivate the following definition, which allows us to extend that "sameness" to ordered pairs representing difference questions without whole number answers.

DEFINITION
4.4.2

Two ordered pairs (a, b) and (c, d) of whole numbers are **equivalent** iff $b + c = a + d$. In this case we write $(a, b) \sim (c, d)$.

EXAMPLES 4.4

2. $(5, 3) \sim (8, 6)$, because $3 + 8 = 5 + 6$. (Note that $5 - 3 = 2$ and $8 - 6 = 2$.)

3. $(2, 7) \sim (5, 10)$, because $7 + 5 = 2 + 10$. (Note that $2 - 7$ and $5 - 10$ do not exist in the whole numbers.)

4. $(7, 4)$ is not equivalent to $(14, 8)$, because $4 + 14 \neq 7 + 8$.

THEOREM
4.4.1

The relation $\sim$ defined in Definition 4.4.2 is an equivalence relation on the set $W \times W$.

---

*A complete, accurate definition may be found in J. R. R. Tolkein's *The Hobbit* (New York: Ballantine Books, Inc., 1966), p. 16. A fascinating but not compelling argument for their existence is made in this book.

PROOF (optional)

1. Reflexive: We must show that any pair (x, y) of whole numbers is related to itself. But $(x, y) \sim (x, y)$ iff $y + x = x + y$, which is true because addition of whole numbers is commutative.
2. Symmetric: If (w, x) and (y, z) are two pairs of whole numbers such that $(w, x) \sim (y, z)$, we must show that $(y, z) \sim (w, x)$. By definition of $\sim$, $(w, x) \sim (y, z)$ tells us that $x + y = w + z$. Again using commutativity of addition, as well as symmetry of $=$ in the whole number system, this equation can be rewritten as $z + w = y + x$, which is the required condition for $(y, z) \sim (w, x)$.
3. Transitive: If (u, v), (w, x), and (y, z) are three pairs of whole numbers such that $(u, v) \sim (w, x)$ and $(w, x) \sim (y, z)$, we must show that $(u, v) \sim (y, z)$. Now, $(u, v) \sim (w, x)$ means $v + w = u + x$, and $(w, x) \sim (y, z)$ means $x + y = w + z$. Adding these two equations, we get $(v + w) + (x + y) = (u + x) + (w + z)$, which can be rewritten $(v + y) + (w + x) = (u + z) + (w + x)$, using the commutativity and associativity of $+$ in W. But this last equation is true if and only if $v + y = u + z$, which is the required condition for $(u, v) \sim (y, z)$.

We have already shown that any equivalence relation partitions the set on which it is defined into equivalence classes. In particular, the relation $\sim$ partitions the set $W \times W$ into (disjoint) sets of ordered pairs of whole numbers; we call these equivalence classes "integers"!

DEFINITION 4.4.3

An **integer** is the set of all ordered pairs of whole numbers which are related to some particular pair—say (a, b)—(and hence to each other) by the equivalence relation $\sim$; we denote this set of pairs by $[a, b]$. In symbols, $[a, b] = \{(x, y) \mid (x, y) \sim (a, b)\}$.

NOTATION: The set of all integers will be denoted by I; that is,

$$I = \{[a, b] \mid a, b \in W\}.$$

EXAMPLE 4.4.5

The integer $[5, 3]$ contains the pairs $(5, 3)$, $(8, 6)$, $(75, 73)$, and infinitely many others. Among the pairs contained in $[2, 9]$ are $(6, 13)$, $(1, 8)$, and $(53, 60)$. Notice that the general fact about equivalence classes stated just before Exercises 4.2 tells us that $[5, 3] = [8, 6]$ because $(5, 3) \sim (8, 6)$ (i.e., $3 + 8 = 5 + 6$), but $[5, 3] \neq [2, 9]$ because $(5, 3) \not\sim (2, 9)$ (i.e., $3 + 2 \neq 5 + 9$).

A little thought about Definition 4.4.3 and Example 4.4.5 should produce a simple but useful observation: If $[5, 3]$ is intended to represent the solution to the subtraction problem $5 - 3$ (that is, 2), then $[5, 3]$ contains all other pairs whose difference is 2, such as $(6, 4)$, $(7, 5)$, $(8, 6)$, and so on. Notice that $(6, 4) = (5 + 1,$

$3 + 1), (7, 5) = (5 + 2, 3 + 2), (8, 6) = (5 + 3, 3 + 3)$, and so on. This suggests a general pattern: If we take any pair of whole numbers and add the same whole number to each number in that pair, then the original pair and the new pair both represent the same integer. More formally,

THEOREM 4.4.2

If a, b, and x are any whole numbers, then $[a + x, b + x] = [a, b]$.

PROOF   This one is easy. By the general fact about equivalence classes stated earlier, all we have to show is that $(a + x, b + x) \sim (a, b)$. But by Definition 4.4.2, this is true iff $(b + x) + a = (a + x) + b$, which is true because addition of whole numbers is both associative and commutative.

Using Theorem 4.4.2, it is easy to see that any integer can be represented by a pair containing at least one zero. For example, $[7, 3]$ can be represented by $(4, 0)$, because $7 = 4 + 3$ and $3 = 0 + 3$. Similarly, $[12, 5] = [7 + 5, 0 + 5] = [7, 0]$, $[17, 35] = [0 + 17, 18 + 17] = [0, 18]$, $[9, 9] = [0 + 9, 0 + 9] = [0, 0]$, and so forth. The general method here merely requires a representative pair of unequal numbers to be expressed in a form such that the larger number is the sum of the smaller number and something else; the theorem does the rest. [In the case of equal numbers in a representative pair, it should be clear that $(0, 0)$ is always an equivalent representative.] This means that any integer takes one of three forms: $[0, 0]$, $[n, 0]$, or $[0, n]$, where n can be any natural number. Since such a pair involves at most one nonzero whole number, it is a simple matter to abbreviate the way we write it; all we really need is that nonzero number and something to indicate whether it appears in the first place or in the second. There are many ways to do this (for instance, we might write $[5, 0]$ as 5_1 and $[0, 5]$ as 5_2), but the following notation carries the weight of tradition:

NOTATION: If n is any natural number, we write $[n, 0]$ as $+n$ and $[0, n]$ as $-n$. The integer $[0, 0]$ is simply denoted by 0. (The potential confusion between the whole number 0 and the integer 0 will generally be eliminated by the context in which the symbol appears.)

DEFINITION 4.4.4

Integers of the form $+n$ (for some $n \in N$) are called **positive** and those of the form $-n$ are called **negative**.

Now, if we look back to the initial motivation for using classes of pairs, this "new" notation should help to clarify what we have done. Take as an example the integer $[9, 5]$, which can also be written $[4, 0]$. This represents the solution to the subtraction problem $9 - 5$, and we write it as $+4$. Similarly, $[2, 8]$ represents the solution to $2 - 8$, and since $[2, 8] = [0, 6]$ we write that solution as -6; $[7, 7]$ represents the solution to $7 - 7$ and can be rewritten as $[0, 0]$, or simply 0. Since the difference $a - b$ of any two whole numbers can be represented by $[a, b]$, and this in turn can be represented in the form $[n, 0]$, $[0, n]$ or $[0, 0]$, for some $n \in N$, the solutions to all such subtraction problems can be written as $+n$, $-n$, or 0. Thus, we have constructed a *set* called the integers which looks familiar and appears to represent what we set out to find. However, before we can say that this collection of classes of pairs is a number system, we must define some arithmetic operations for them in such a way that they behave "properly." In particular, addition and multiplication of these new "numbers" should make sense and have nice properties such as commutativity, associativity, and distributivity. Moreover, since this new system is intended to be a worthwhile extension of the whole numbers, it should contain a copy of the whole number system and should also have a subtraction operation that always works. This may seem like a tall order, but the hard work is already done. A few careful but natural definitions allow all the remaining pieces of this puzzle to fall into place easily; the rest of this section is devoted to supplying these definitions and outlining their consequences.

As in the case of the whole numbers, the definitions of addition and multiplication for integers flow easily from a consideration of what integers "really" are and how we would like them to behave. For instance, what should it mean to add $[7, 3]$ and $[5, 2]$? Recalling that these two integers are equivalence classes of ordered pairs represented by $(7, 3)$ and $(5, 2)$, respectively, and that these pairs represent the (whole number) subtraction problems $7 - 3$ and $5 - 2$, it seems natural to expect that the sum of these two integers ought to be related to $(7 - 3) + (5 - 2)$. We ought to be able to write this last expression as $(7 + 5) - (3 + 2)$, which suggests that $[7, 3] + [5, 2]$ should come out to be $[7 + 5, 3 + 2]$. In thinking about this example, you should notice (among other things) that the symbol $+$ is being used in two different senses; in the expressions $7 + 5$ and $3 + 2$ it denotes the previously defined addition of whole numbers, but in $[7, 3] + [5, 2]$ it stands for the still-to-be-defined addition of integers. To preserve the distinction between what we are currently defining and what is already known, for the remainder of this section the new $+$ will be distinguished from the old by putting a circle around it. Thus, we may write an appropriate definition of addition of integers, $\oplus$, as follows:

DEFINITION 4.4.5   The **sum** of two integers $[a, b]$ and $[c, d]$ is the integer $[a + c, b + d]$. In symbols, $[a, b] \oplus [c, d] = [a + c, b + d]$.

EXAMPLES 4.4

6. As we saw above, $[7, 3] \oplus [5, 2] = [12, 5]$. Similarly,

$$[9, 1] \oplus [4, 8] = [9 + 4, 1 + 8] = [13, 9].$$

7. Using the abbreviated notation developed following Definition 4.4.4, we can see that this addition operation works just the way we would like. For instance,

$$(-2) \oplus (+3) = [0, 2] \oplus [3, 0] = [0 + 3, 2 + 0] = [3, 2] = [1, 0] = +1*$$
$$(+7) \oplus (-1) = [7, 0] \oplus [0, 1] = [7 + 0, 0 + 1] = [7, 1] = [6, 0] = +6$$
$$(-4) \oplus (-5) = [0, 4] \oplus [0, 5] = [0 + 0, 4 + 5] = [0, 9] = -9$$

Multiplication of integers appears to be a bit more complicated but really is not. Instead of motivating this by example, let us plunge right into the confusion by giving the formal definition first and then unraveling its origins. As before, the new operation will be distinguished from the old by a circle.

DEFINITION 4.4.6

The **product** of two integers $[a, b]$ and $[c, d]$ is the integer $[ac + bd, ad + bc]$. In symbols, $[a, b] \odot [c, d] = [ac + bd, ad + bc]$.

EXAMPLES 4.4

---

8. $[2, 5] \odot [7, 3] = [2 \cdot 7 + 5 \cdot 3, 2 \cdot 3 + 5 \cdot 7] = [29, 41]$. [This is just another way of writing $(-3) \odot (+4) = -12$; if that is not obvious, you should take a minute to convince yourself that the two statements are the same.]

9. $(+2) \odot (-5) = [2, 0] \odot [0, 5] = [2 \cdot 0 + 0 \cdot 5, 2 \cdot 5 + 0 \cdot 0] = [0, 10] = -10$.

10. $(-7) \odot (-1) = [0, 7] \odot [0, 1] = [0 \cdot 0 + 7 \cdot 1, 0 \cdot 1 + 7 \cdot 0] = [7, 0] = +7$.

"OK, I admit that multiplication works the way we want it to," you say grudgingly, "but where in the world did such a weird definition come from?" Well, it really is much more natural than it appears. Consider once again the integers as classes of ordered pairs that represent solutions to subtraction problems. Then $[a, b] \odot [c, d]$ is the solution to $(a - b) \cdot (c - d)$ (when such a whole number exists), and it is not hard to show by whole number arithmetic that $(a - b) \cdot (c - d) = (ac + bd) - (ad + bc)$, just as it was in high school algebra. And the integer generated by the right side of this equation is $[ac + bd, ad + bc]$!

A similar motivation can be used for the integer definition of "less than," whose symbol is again distinguished from the previous one by a circle.

DEFINITION 4.4.7

The integer $[a, b]$ is **less than** the integer $[c, d]$ if and only if $a + d$ is less than $b + c$. In symbols, $[a, b] \oslash [c, d]$ iff $a + d < b + c$. (If $[a, b]$ is less than $[c, d]$, we may also say $[c, d]$ is **greater than** $[a, b]$ and write $[c, d] \obslash [a, b]$.)

---

* Note here a third use of $+$, this time not to denote an operation but merely to indicate the position of a number in a pair. With all this ambiguity, it is no wonder that "simple" arithmetic can be confusing to children!

EXAMPLES 4.4

11. $[7, 5] \lessapprox [6, 3]$, because $7 + 3 < 5 + 6$.

12. $+2 \lessapprox +3$, because $+2 = [2, 0]$, $+3 = [3, 0]$, and $2 + 0 < 0 + 3$.

13. $-3 \lessapprox -2$, because $-3 = [0, 3]$, $-2 = [0, 2]$, and $0 + 2 < 3 + 0$.

Using these definitions of $\oplus$, $\odot$, and $\lessapprox$, it is not hard to prove the usual laws of integer arithmetic. In particular:

Addition and multiplication are associative: For all $x, y, z \in I$,

$$(x \oplus y) \oplus z = x + (y \oplus z)$$

and

$$(x \odot y) \odot z = x \odot (y \odot z)$$

Addition and multiplication are commutative: For all $x, y \in I$,

$$x \oplus y = y \oplus x$$

and

$$x \odot y = y \odot x$$

Multiplication is distributive over addition: For all $x, y, z \in I$,

$$x \odot (y \oplus z) = (x \odot y) \oplus (x \odot z)$$

If $x, y, z \in I$ and $x \lessapprox y$, then

$$x \oplus z \lessapprox y \oplus z$$

$$x \odot z \lessapprox y \odot z \qquad \text{if } z \text{ is positive}$$

$$x \odot z \gtrapprox y \odot z \qquad \text{if } z \text{ is negative}$$

As in Section 4.3, a proof of one of these properties should suffice to demonstrate the general approach; the rest are left as exercises.

THEOREM 4.4.3

Addition of integers is commutative.

PROOF We must show that, for all $x, y \in I$, $x \oplus y = y \oplus x$. Let $x = [a, b]$ and $y = [c, d]$, where $a, b, c, d \in W$. Then, by definition of $\oplus$,

$$x \oplus y = [a, b] \oplus [c, d] = [a + c, b + d]$$

and

$$y \oplus x = [c, d] \oplus [a, b] = [c + a, d + b]$$

Now, since addition of whole numbers is commutative, $a + c = c + a$ and $b + d = d + b$, making the representative pairs for $x \oplus y$ and $y \oplus x$ not only equivalent (which is all that is necessary) but actually equal. Hence, $x \oplus y = y \oplus x$, as required.

Besides the properties listed above, it is also possible to prove for the integers all the rules for signed numbers that you learned in elementary school. These rules, often regarded as mysterious and magical by schoolchildren (and some of their teachers!) are simple consequences of the way we define positive and negative numbers. We prove here one such rule which seems to be particularly bothersome in elementary school; other signed number rules appear as exercises.

THEOREM 4.4.4

The product of two negative integers is positive.

PROOF   Suppose that $-a$ and $-b$ are negative integers. Then $-a = [0, a]$ and $-b = [0, b]$, so

$$(-a) \odot (-b) = [0, a] \odot [0, b] = [0 \cdot 0 + a \cdot b, 0 \cdot b + a \cdot 0] = [ab, 0] = +ab$$

which is positive by definition.

It should come as no surprise that the whole number system appears within the integers under the guise of the positive integers and zero. In fact, if we identify each whole number a with its corresponding (nonnegative) integer $[a, 0]$, it is easy to see that integer addition and multiplication behave on these numbers exactly like the corresponding whole number operations: For any whole numbers a and b,

$$[a, 0] \oplus [b, 0] = [a + b, 0 + 0] = [a + b, 0] = +(a + b)$$

and $$[a, 0] \odot [b, 0] = [a \cdot b + 0 \cdot 0, a \cdot 0 + 0 \cdot b] = [a \cdot b, 0] = +(a \cdot b)$$

Thus, it is accurate to regard the system of integers as an extension of the whole number system.

Let us return now to the question of subtraction. We began this section by observing that the difference of two whole numbers does not always exist in W. However, it is now easy to show that the difference of two whole numbers always exists *in I*. Consider any two whole numbers a and b, and put them in integer form, $[a, 0]$ and $[b, 0]$, respectively. As you might have guessed from the way we defined integers, the difference $a - b$ is the integer $[a, b]$; that is, in accordance with Definition 4.4.1, when $[a, b]$ is added to b the result is a:

$$[b, 0] \oplus [a, b] = [b + a, 0 + b] = [a + b, 0 + b] = [a, 0] \qquad \text{(Why?)}$$

Thus, subtraction of whole numbers always makes sense in this expanded system.

But we can do even better than that with what we have developed. We can, in fact, show that the difference of any two *integers* always exists in I. There are several ways to establish this fact; perhaps the easiest begins by introducing the concept of "inverses" for addition.

DEFINITION 4.4.8

Let S be any set of numbers on which an addition operation $+$ has been defined, and let s be any element of S. A number t in S is called the **additive inverse** or **negative of** s if $s + t = 0$.

NOTATION: The additive inverse of s is usually written as $-s$.

EXAMPLES 4.4

14. In the integers, the additive inverse of $+3$ is -3, since $(+3) \oplus (-3) = [3, 0] \oplus [0, 3] = [3, 3] = [0, 0] = 0$. Notice that the apparently ambiguous use of the negative sign can be reconciled by observing the identification between the whole number 3 and the positive integer $+3$; that is, $-(+3) = -3$.

15. The negative of -5 is $+5$, because $(-5) \oplus (+5) = [0, 5] \oplus [5, 0] = [5, 5] = 0$. Notice that the negative *of* an integer may be a positive integer; in this case, $-(-5) = +5$.

It is easy to show that every integer has an additive inverse in I. In fact, if $[a, b]$ is any integer, its additive inverse is $[b, a]$, because

$$[a, b] \oplus [b, a] = [a + b, b + a] = [a + b, a + b] = [0, 0] = 0$$

That is, $-[a, b] = [b, a]$.

The importance of additive inverses for us lies in the fact that subtraction can be accomplished using them. Recall how subtraction of signed numbers was described in elementary school: When confronted by a problem such as $(+3) - (-2)$, for instance, we were instructed to "change the sign" of the second number and then add. Thus, we were shown examples like $(+3) - (-2) = (+3) + (+2) = 5$ and $(+2) - (+6) = (+2) + (-6) = -4$. But $+2$ is the negative of -2, and -6 is the negative of $+6$, so this change-sign-and-add process is nothing more than subtracting by adding the negative of the second number to the first one. Formally,

---

DEFINITION 4.4.9

The **difference** $x \ominus y$ of any two integers x and y is found by adding to x the negative of y. In symbols, $x \ominus y = x \oplus (-y)$.

EXAMPLES 4.4

16. $[2, 5] \ominus [3, 7] = [2, 5] \oplus (-[3, 7]) = [2, 5] \oplus [7, 3] = [9, 8]$.

17. $(+5) \ominus (+3) = [5, 0] \ominus [3, 0] = [5, 0] \oplus [0, 3] = [5, 3] = [2, 0] = +2$.

18. $(+1) \ominus (+8) = [1, 0] \ominus [8, 0] = [1, 0] \oplus [0, 8] = [1, 8] = [0, 7] = -7$.

19. $(+2) \ominus (-9) = [2, 0] \ominus [0, 9] = [2, 0] \oplus [9, 0] = [11, 0] = +11$.

(You should be able to give a reason for each step in these examples from the material in this section.)

If you have been following our development of the integers closely, you should be aware of a problem at this point. We have now defined the difference of two integers in two apparently different ways (pardon the pun!), once just above in Definition 4.4.9 and once back in Definition 4.4.1. That makes no sense, of course, unless the two definitions actually say the same thing. That is, to rescue our credibility we must show that the number we get from this additive inverse definition (4.4.9) is the number c required by Definition 4.4.1. In other words, $x \ominus y$ will satisfy the earlier definition of difference if and only if $y \oplus (x \ominus y) = x$.

THEOREM 4.4.5

For any two integers x and y: $y \oplus (x \ominus y) = x$.

PROOF   Let us write x and y in "pair-class" form as $x = [a, b]$ and $y = [c, d]$. Then

$$y \oplus (x \ominus y) = [c, d] \oplus ([a, b] \ominus [c, d])$$

$$= [c, d] \oplus ([a, b] \oplus (-[c, d])) \qquad \text{by definition of } \ominus$$

$$= [c, d] \oplus ([a, b] \oplus [d, c]) \qquad \text{by definition of additive inverse}$$

$$= [c, d] \oplus [a + d, b + c] \qquad \text{by definition of } \oplus$$

$$= [c + (a + d), d + (b + c)] \qquad \text{by definition of } \oplus$$

$$= [a + (c + d), b + (c + d)] \qquad \text{by commutativity and associativity of } + \text{ in } W$$

$$= [a, b] \qquad \text{by Theorem 4.4.2}$$

$$= x$$

as required.

This result (finally) wraps up the problem we began the section with. We have shown that every integer has a negative (Definition 4.4.8) and that the difference of two integers can be found by adding the negative of the second to the first (Definition 4.4.1 and Theorem 4.4.5). Hence, the integers form a number system containing the whole numbers in which subtraction always makes sense.

EXERCISES 4.4

1. Use the definition of $+$ for whole numbers to justify the discussion following Example 4.4.1 in which it is stated that there is no whole number x such that $5 + x = 3$.

2. Illustrate the fact that $\sim$ (as defined in Definition 4.4.2) is an equivalence relation on $W \times W$ by giving numerical examples of the three required properties.

3. List three different representative pairs for each of the following integers:
   (a) $+3$ (b) $+17$ (c) -2
   (d) -25 (e) 0

4. Use the definitions of $\oplus$, $\ominus$, and $\odot$ to verify the following:
   (a) $(+1) \oplus (+1) = +2$ (b) $(+7) \oplus (-3) = +4$
   (c) $(-6) \oplus (-8) = -14$ (d) $(+5) \oplus (-9) = -4$
   (e) $(+3) \ominus (+2) = +1$ (f) $(+3) \ominus (+10) = -7$
   (g) $(-2) \ominus (-2) = 0$ (h) $(+4) \ominus (-1) = +5$
   (i) $(+7) \odot (+8) = +56$ (j) $(-3) \odot (+10) = -30$
   (k) $(+17) \odot (-1) = -17$ (l) $(-4) \odot (-5) = +20$

5. (a) Use the definitions of 0, $\oplus$, and $\odot$ to verify that $(-5) \oplus 0 = -5$ and $0 \odot (+7) = 0$.
   (b) Prove that $x \oplus 0 = x$ for any integer x.
   (c) Prove that $0 \odot x = 0$ for any integer x.

6. Use the definition of $\odot$ to verify that
   (a) $+3 \odot +4$ (b) $-2 \odot -1$
   (c) $-6 \odot +5$ (d) $-7 \odot 0$

7. Let a and b be natural numbers. Show that $a < b$ if and only if $+a \odot +b$.

8. Prove:
   (a) Every negative integer is less than every positive integer.
   (b) Every negative integer is less than zero.
   (c) Zero is less than every positive integer.

9. Prove:
   (a) Addition of integers is associative.
   (b) Multiplication of integers is commutative.
   (c) Multiplication of integers is associative.
   (d) Multiplication of integers is distributive over addition.

10. Let x, y, and z be integers such that $x \odot y$. Prove:
    (a) $x \oplus z \odot y \oplus z$.
    (b) $x \odot z \odot y \odot z$ if z is positive.
    (c) $x \odot z \odot y \odot z$ if z is negative.

11. Let x and y be *integers*. (Hence, in this situation the sign $-$ denotes additive inverse.) Prove:
    (a) $(-x) \oplus (-y) = -(x \oplus y)$
    (b) $x \odot (-y) = (-x) \odot y = -(x \odot y)$
    (c) $-(-x) = x$

12. Show by example that subtraction of integers is neither commutative nor associative.

13. Using the material of this section, justify each step in the following computation:

$$(-4) \ominus (-7) = [0, 4] \ominus [0, 7]$$
$$= [0, 4] \oplus (-[0, 7])$$
$$= [0, 4] \oplus [7, 0]$$
$$= [7, 4]$$
$$= [3, 0]$$
$$= +3$$

4.5 THE RATIONAL NUMBERS

Before proceeding further, it will be useful to fix one notational inconvenience. In Section 4.4 we used circles around the signs $+$, $-$, $\cdot$, and $<$ to distinguish their integer usage from their whole number usage. However, as we saw in our discussion following Theorem 4.4.4, since the whole number system is actually contained within the integers there is no need to carry on this distinction any longer; the integer operations suffice for both systems. Thus, from now on, integer addition, subtraction, multiplication, and inequality will be denoted simply by $+$, $-$, $\cdot$, and $<$, respectively. We shall also write any natural number n and its corresponding positive integer $+n$ interchangeably.

One further preliminary comment is in order. This section and Section 4.4 complement each other. As you will see, the development of the rational numbers is virtually identical to that of the integers, with multiplication now playing the key role that addition played before. Thus, Section 4.4 should provide a step-by-step guide to this one, with the writing of the two sections deliberately emphasizing the parallel. It is also true, however, that because of the way elementary arithmetic is usually taught, the use of different ordered pairs to represent the same number should seem much more familiar and natural this time, and hence this section should provide some useful retrospective insight into the previous one.

Like subtraction, division can be defined on a set of numbers without direct reference to the construction of the numbers themselves. However, it requires that multiplication has already been defined, for division is just multiplication "in reverse," so to speak. (Does this look familiar?)

DEFINITION 4.5.1 Let S be any set of numbers on which a multiplication operation $\cdot$ has been defined. Then for any elements a and b in S, we say that the **quotient** $a \div b$ equals a number c in S if and only if $b \cdot c$ equals a. In symbols, $a \div b = c$ iff $b \cdot c = a$.

EXAMPLE 4.5.1

In the integers, $6 \div 3 = 2$, because $3 \cdot 2 = 6$.

Once again, as in the case of the hobbit, a definition does not assure existence. In particular, the quotient of two integers does not necessarily exist (in the integers). For instance, the definition of quotient requires that $1 \div 2$ be an integer which when multiplied by 2 yields 1, and there is no such integer. This section seeks to remedy that shortcoming of the system of integers by extending it to a larger system, the "rational numbers," in which quotients always exist. As before, the key is to define an equivalence relation on an appropriate set.

We begin by observing that any "division problem" involving two integers is determined by the ordered pair of integers themselves, whether or not the quotient actually exists: $6 \div 3$ is determined by the ordered pair $(6, 3)$, $1 \div 2$ by $(1, 2)$, $-14 \div 5$ by $(-14, 5)$, and so forth. Thus, the set of all possible quotient questions of the form $x \div y = ?$, where x and y are integers, can be represented by the set of all ordered pairs of integers. But some of these questions are not answerable, even under the best of conditions. For instance, $2 \div 0 = ?$ requires a number which when multiplied by 0 yields 2. However, we would expect 0 in any number system to yield 0 when multiplied by anything (as happens in I), so there is no hope of finding a suitable candidate for "?" in this case. In general, $x \div 0 = ?$ is not answerable for any nonzero integer x. Moreover, the question $0 \div 0 = ?$ is useless because there are too many possible answers, since $0 \cdot c = 0$ for any number c. Thus, we shall exclude from consideration all quotient questions whose second integer is 0. If we denote the set of all nonzero integers by I^*, then the set of quotient questions we want to consider can be represented by the set of ordered pairs $I \times I^*$.

It seems reasonable to consider two of these quotient questions "the same" with respect to the division process if they yield the same answer. We use this idea to motivate the following definition, which allows us to extend that "sameness" to ordered pairs representing quotient questions without integer answers.

DEFINITION 4.5.2

Two ordered pairs (w, x) and (y, z) of integers are **equivalent** iff $x \cdot y = w \cdot z$. In this case we write $(w, x) \approx (y, z)$.

EXAMPLES 4.5

2. $(6, 3) \approx (10, 5)$, because $3 \cdot 10 = 6 \cdot 5$. (Note that $6 \div 3 = 2$ and $10 \div 5 = 2$.)

3. $(1, 2) \approx (4, 8)$, because $2 \cdot 4 = 1 \cdot 8$. (Note that $1 \div 2$ and $4 \div 8$ do not exist in the integers.)

| 4. $(5, 3)$ is not equivalent to $(4, 2)$, because $3 \cdot 4 \neq 5 \cdot 2$.

THEOREM 4.5.1

The relation $\approx$ as defined in Definition 4.5.2 is an equivalence relation on the set $I \times I^*$.

PROOF (optional)

1. Reflexive: We must show that any pair (x, y) of integers (with $y \neq 0$) is related to itself. But $(x, y) \approx (x, y)$ iff $y \cdot x = x \cdot y$, which is true because multiplication of integers is commutative.

2. Symmetric: If (w, x) and (y, z) are two pairs of integers such that $(w, x) \approx (y, z)$, we must show that $(y, z) \approx (w, x)$. By definition of $\approx$, $(w, x) \approx (y, z)$ tells us that $x \cdot y = w \cdot z$. Again using commutativity of integer mulitplication, as well as symmetry of $=$, this equation can be rewritten as $z \cdot w = y \cdot x$, which is the required condition for $(y, z) \approx (w, x)$.

3. Transitive: If (u, v), (w, x), and (y, z) are three pairs of integers such that $(u, v) \approx (w, x)$ and $(w, x) \approx (y, z)$, we must show that $(u, v) \approx (y, z)$. Now, $(u, v) \approx (w, x)$ means $v \cdot w = u \cdot x$, and $(w, x) \approx (y, z)$ means $x \cdot y = w \cdot z$. Multiplying these two equations, we get

$$(v \cdot w) \cdot (x \cdot y) = (u \cdot x) \cdot (w \cdot z)$$

which can be rewritten as

$$(v \cdot y) \cdot (w \cdot x) = (u \cdot z) \cdot (w \cdot x)$$

using the commutativity and associativity of $\cdot$ in I. If $w \cdot x \neq 0$, this last equation is true if and only if $v \cdot y = u \cdot z$, which is the required condition for $(u, v) \approx (y, z)$. If $w \cdot x = 0$, it is not hard to verify the same fact, but the details are a bit tedious and so are omitted. (If you would like to complete the proof yourself, try to justify the steps in this argument: If $w \cdot x = 0$, then $w = 0$, so both $x \cdot y = 0$ and $u \cdot x = 0$. Hence, both u and y equal 0, implying that $v \cdot y = u \cdot z$.)

Since any equivalence relation partitions the set on which it is defined into equivalence classes, the relation $\approx$ partitions the set $I \times I^*$ into (disjoint) sets of ordered pairs of integers. We call these equivalence classes "rational numbers"!

DEFINITION 4.5.3

A **rational number** is the set of all ordered pairs of integers which are related to some particular pair—say (a, b)—(and hence to each other) by the equivalence relation $\approx$; we denote this set of pairs by $[a, b]$. In symbols, $[a, b] = \{(x, y) \mid (x, y) \approx (a, b)\}$.

NOTATION: The set of all rational numbers will be denoted by Q; that is,

$$Q = \{[a, b] \mid a, b \in I, b \neq 0\}$$

EXAMPLE 4.5.5

The rational number [6, 3] contains the pairs (6, 3), (10, 5), (−2, −1) and infinitely many others. Among the pairs contained in [−2, 3] are (−4, 6), (10, −15), and (−50, 75). Notice that the general fact about equivalence classes stated just before Exercises 4.2 tells us that [6, 3] = [10, 5] because (6, 3) ≈ (10, 5) (that is, 3 · 10 = 6 · 5), but [6, 3] ≠ [−2, 3] because (6, 3) ≉ (−2, 3) [that is, 3 · (−2) ≠ 6 · 3].

A little thought about Definition 4.5.3 and Example 4.5.5 should produce a simple but useful observation: If [6, 3] is intended to represent the solution to the division problem 6 ÷ 3 (that is, 2), then [6, 3] contains all other pairs whose quotient is 2, such as (14, 7), (20, 10), (−50, −25), and so on. Notice that (6, 3) = (2 · 3, 1 · 3), (14, 7) = (2 · 7, 1 · 7), (20, 10) = (2 · 10, 1 · 10), (−50, −25) = (2 · (−25), 1 · (−25)), and so on. This suggests a general pattern: If we take any pair of integers and multiply both of them by the same integer, then the original pair and the new pair both represent the same rational number. There is an obvious exception to this rule: If we multiply any pair of numbers by zero, the result will be (0, 0), so clearly that case will have to be excluded. However, in all other cases the pattern holds. More formally:

THEOREM 4.5.2   If a, b and x are any integers and $x \neq 0$, then $[a \cdot x, b \cdot x] = [a, b]$.

PROOF   By the general fact about equivalence classes stated earlier, all we have to show is that $(a \cdot x, b \cdot x) \approx (a, b)$. But by the definition of $\approx$ (Definition 4.5.2), this is true iff $(b \cdot x) \cdot a = (a \cdot x) \cdot b$, which is true because multiplication of integers is both associative and commutative.

Once again, as in the integers, we have a set of "numbers" in which each "number" is itself an infinite collection of ordered pairs. This time, however, the structure should not seem so strange, because you have been working with rational numbers this way, more or less, since elementary school. If you have not already seen the connection between our formal construction and your elementary arithmetic experience, one small notational adjustment should do the trick: Instead of writing (2, 3) to represent the solution to the division problem 2 ÷ 3, for example, write $\frac{2}{3}$. There! Doesn't that feel better? Both (2, 3) and $\frac{2}{3}$ are just pairs of integers, but we are used to thinking of the latter notation in terms of division. In elementary school we called these pairs "fractions" and referred to their first and second integers as "numerator" and "denominator," respectively. We were taught to check if two fractions are "equal" by "cross-multiplying"; that is, we were told that $\frac{2}{3} = \frac{4}{6}$ because 2 · 6 = 3 · 4. Moreover, we were told that "equal" fractions really represent the "same quantity"; that is, $\frac{2}{3}, \frac{4}{6}, \frac{6}{9}, \frac{8}{12}$, and so on, are all names for the same number because they are all "equal." We were also taught that we could multiply both the numerator and denominator of a fraction by the same (nonzero) number without changing its "value"; sometimes this process was known as "canceling" a common factor. But all of this is just what we have constructed here in our development of Q! The "fractions" are the ordered pairs, "cross-multiplying" is the basis for the definition of the equivalence relation $\approx$, and the representation of the same quantity by different fractions is nothing more than the assertion that a rational number $[a, b]$ is a class containing infinitely many equivalent pairs, any one of which can be used to represent that class. Finally, the "canceling" process is no more nor less than the statement of our last theorem. Thus, in elementary school notation, with all its attendant ambiguities, the rational number $\frac{2}{3}$ *is* the set of fractions $\ldots, \frac{2}{3}, \frac{4}{6}, \frac{6}{9}, \frac{8}{12}, \ldots$, and all the fractions in that set are equivalent. The only significant difference between the elementary school approach and ours is that we make a sharp distinction between the number $[2, 3]$, which is a class of pairs, and the pair $(2, 3)$ which represents that class, whereas no such explicit distinction is made in elementary school. Thus, a fifth-grader is expected to distinguish between the number $\frac{2}{3}$ and the representative fraction $\frac{2}{3}$ solely from context! ("What do you mean, $\frac{2}{3}$ and $\frac{4}{6}$ are the same? They sure look different!")

A further disadvantage of the less formal, seemingly easier elementary school approach to numbers should become clearer as you continue to study this section. As we mentioned earlier, the construction of the rationals exactly parallels that of the integers, with $\cdot$ substituting for $+$. Yet in elementary school this parallel is hidden, since integers are not developed using pairs, and thus much of the underlying unity of the number systems is obscured. Having said all this, we shall (grudgingly) begin to convert back to traditional fractional notation, simply because it is more familiar. Throughout the rest of this section examples will be given in both notations, in the hope that the contextual distinction between class and pair will become sufficiently clear by the time this section is complete. Then in the remaining sections of the chapter we shall use the traditional notation exclusively.

Now, if we look back to the initial motivation for using classes of ordered pairs, this "new" notation should help clarify what we have done. Take as an example the rational number $[6, 3]$, which can also be written $[2, 1]$. This represents the solution to the division problem $6 \div 3$, and we write it as $\frac{6}{3}$, or $\frac{2}{1}$. Similarly, $[4, 6]$ represents the solution to $4 \div 6$, and since $[4, 6] = [2, 3]$ we can write that solution as either $\frac{4}{6}$ or $\frac{2}{3}$. As you know, it is often convenient to represent a rational number in "lowest terms" by eliminating any factors common to both numerator and denominator (i.e., to both first and second integers in the representative pair); this practice is justified by Theorem 4.5.2. Moreover, it is convenient to restrict our attention to pairs whose denominator integer is positive. Since Theorem 4.5.2 provides that $(a, b) \approx (a \cdot (-1), b \cdot (-1))$ for any pair (a, b) in $I \times I^*$, we lose no equivalence classes if we restrict the set of pairs to $I \times I^+$ (where I^+ denotes the set of positive integers). We shall make that restriction from now on, and hence shall call a rational number "positive" or "negative" according to whether its numerator is positive or negative.

Thus, we have constructed a *set* called the rational numbers which looks familiar and appears to represent what we set out to find. However, before we can say that this collection of classes of pairs is a number system, we must define some arithmetic operations for them in such a way that they behave "properly." In particular, addition and multiplication of these new "numbers" should make sense and have nice properties such as commutativity, associativity, and distributivity. Moreover, since this new system is intended to be a worthwhile extension of the integers, it should contain a copy of the system of integers and should also have a division operation that always works. This may seem like a tall order, but the hard work is already done. Once again, a few careful but natural definitions allow all the pieces to fall into place easily; the rest of this section is devoted to supplying these definitions and outlining their consequences.

As in the case of the integers, the definitions of addition and multiplication for rational numbers flow from a consideration of what rational numbers "really" are and how we would like them to behave. In the interest of brevity, however, we shall give the formal definitions first, illustrating by example that they behave as we would expect. Once again, to preserve the distinction between what we are currently defining and what is already known, the new (rational number) operations will be distinguished from the old (integer) ones by putting circles around the symbols for the new ones.

DEFINITION 4.5.4

The **sum** of two rational numbers $[a, b]$ and $[c, d]$ is the rational number $[a \cdot d + b \cdot c, b \cdot d]$. In symbols,

$$[a, b] \oplus [c, d] = [a \cdot d + b \cdot c, b \cdot d]$$

or, in fractional notation

$$\frac{a}{b} \oplus \frac{c}{d} = \frac{a \cdot d + b \cdot c}{b \cdot d}$$

EXAMPLES 4.5

6. $[4, 2] \oplus [9, 3] = [4 \cdot 3 + 2 \cdot 9, 2 \cdot 3] = [30, 6]$. (Can you see that this "really" says $2 + 3 = 5$? Check that by working out the division problems represented by the rational number pairs.)

7. $[-1, 3] \oplus [1, 3] = [(-1) \cdot 3 + 3 \cdot 1, 3 \cdot 3] = [0, 9]$.

8. $[1, 2] \oplus [3, 4] = [1 \cdot 4 + 2 \cdot 3, 2 \cdot 4] = [10, 8] = [5, 4]$. In fractional notation, $\frac{1}{2} \oplus \frac{3}{4} = \frac{5}{4}$.

DEFINITION 4.5.5

The **product** of two rational numbers $[a, b]$ and $[c, d]$ is the rational number $[a \cdot c, b \cdot d]$. In symbols,

$$[a, b] \odot [c, d] = [a \cdot c, b \cdot d]$$

or, in fractional notation,

$$\frac{a}{b} \odot \frac{c}{d} = \frac{a \cdot c}{b \cdot d}$$

EXAMPLES 4.5

9. $[2, 1] \odot [3, 1] = [2 \cdot 3, 1 \cdot 1] = [6, 1]$. In fractional notation, $\frac{2}{1} \odot \frac{3}{1} = \frac{6}{1}$.

10. $[4, 9] \odot [3, 2] = [4 \cdot 3, 9 \cdot 2] = [12, 18] = [2, 3]$. (Can you justify each step of this computation?) In fractional notation, $\frac{4}{9} \odot \frac{3}{2} = \frac{12}{18} = \frac{2}{3}$.

11. $[-1, 2] \odot [-1, 2] = [(-1) \cdot (-1), 2 \cdot 2] = [1, 4]$; that is, $-\frac{1}{2} \odot -\frac{1}{2} = \frac{1}{4}$.

The relation "less than" can also be defined for rational numbers in a natural way, although the formal statement may seem artificial. Observe first that $6 \div 3$ is less than $9 \div 3$, and $6 \div 3$ is also less than $6 \div 2$. In general, the result of a division problem can be made larger either by increasing the first number or by decreasing the second. Thus, if we compare two division problems $a \div b$ and $c \div d$, we would expect $c \div d$ to be larger than $a \div b$ if c is larger than a and/or d is smaller than b. But this happens precisely when $a \cdot d$ is smaller than $b \cdot c$. Thus, we can define "less than" for rational numbers as follows, again distinguishing the new symbol from the old one by a circle.

DEFINITION 4.5.6
The rational number $[a, b]$ is **less than** the rational number $[c, d]$ if and only if $a \cdot d < b \cdot c$. In symbols, $[a, b] \lessdot [c, d]$ iff $a \cdot d < b \cdot c$. (If $[a, b]$ is less than $[c, d]$ we may also say $[c, d]$ is **greater than** $[a, b]$ and write $[c, d] \gtrdot [a, b]$.)

EXAMPLES 4.5

12. $[3, 2] \lessdot [8, 5]$, because $3 \cdot 5 < 2 \cdot 8$.

13. $\frac{5}{9} \lessdot \frac{7}{8}$, because $5 \cdot 8 < 9 \cdot 7$.

14. $-4/5 \lessdot 1/2$, because $(-4) \cdot 2 < 5 \cdot 1$.

Using these definitions of $\oplus$, $\odot$, and $\lessdot$, it is not hard to prove the usual laws of rational number arithmetic. In particular:

Addition and multiplication are associative: For all $r, s, t \in Q$,

$$(r \oplus s) \oplus t = r \oplus (s \oplus t)$$

and
$$(r \odot s) \odot t = r \odot (s \odot t)$$

Addition and multiplication are commutative: For all $r, s \in Q$,
$$r \oplus s = s \oplus r$$
and
$$r \odot s = s \odot r$$

Multiplication is distributive over addition: For all $r, s, t \in Q$,
$$r \odot (s \oplus t) = (r \odot s) + (r \odot t)$$

If $r, s, t \in Q$ and $r \lessdot s$, then
$$r \oplus t \lessdot s \oplus t$$
$$r \odot t \lessdot s \odot t \qquad \text{if } t \text{ is positive}$$
and
$$r \odot t \gtrdot s \odot t \qquad \text{if } t \text{ is negative}$$

Again, a proof of one of these properties should suffice to demonstrate the general approach; the rest are left as exercises.

THEOREM 4.5.3

Multiplication of rational numbers is commutative.

PROOF   We must show that, for all $r, s \in Q$, $r \odot s = s \odot r$. Let $r = [a, b]$ and $s = [c, d]$, where $a, b, c, d \in I$. Then, by definition of $\odot$,
$$r \odot s = [a, b] \odot [c, d] = [a \cdot c, b \cdot d]$$
and
$$s \odot r = [c, d] \odot [a, b] = [c \cdot a, d \cdot b]$$

Since multiplication of integers is commutative, $a \cdot c = c \cdot a$ and $b \cdot d = d \cdot b$, making the representative pairs for $r \odot s$ and $s \odot r$ not only equivalent (which is all that is necessary) but actually equal. Hence, $r \odot s = s \odot r$, as required.

All the other familiar rules of fractional arithmetic can also be proved from the foundation we have laid, but to investigate them further would take us too far from our main theme of using equivalence relations to build the number systems. Hence, we shall move on, leaving the pursuit of a few of these other rules to the exercises.

It should come as no surprise that the system of integers appears within the rationals under the guise of all rational numbers with denominator 1. That is reasonable from a consideration of division: Any integer a divided by 1 should result in a, so $[a, 1]$ ought to represent a itself. Once this identification is made, it is easy to see that rational number addition and multiplication behave on these numbers exactly like the corresponding integer operations: For any integers x and y,

$$x \oplus y = \frac{x}{1} \oplus \frac{y}{1} = [x, 1] \oplus [y, 1] = [x \cdot 1 + 1 \cdot y, 1 \cdot 1] = [x + y, 1] = \frac{x + y}{1} = x + y$$

and

$$x \odot y = \frac{x}{1} \odot \frac{y}{1} = [x, 1] \odot [y, 1] = [x \cdot y, 1 \cdot 1] = [x \cdot y, 1] = \frac{x \cdot y}{1} = x \cdot y$$

Thus, it is accurate to regard the rational number system as an extension of the system of integers.

Let us now return to the question of division. We began this section by observing that the quotient of two integers does not always exist in I. However, it is now easy to show that the quotient of two integers always exists *in* Q. Consider any two integers x and y, and put them in rational form, $[x, 1]$ and $[y, 1]$, respectively. As you might have guessed from the way we defined rational numbers, the quotient $x \div y$ is the rational number $[x, y]$; that is [in accordance with the definition of quotient (4.5.1)], when $[x, y]$ is multiplied by y, the result is x:

$$[y, 1] \odot [x, y] = [y \cdot x, 1 \cdot y] = [x \cdot y, 1 \cdot y] = [x, 1] \qquad \text{(Why?)}$$

Thus, division of integers always makes sense in this expanded system.

But we can do even better than that with what we have developed. We can, in fact, show that the quotient of any two *rational numbers* always exists in Q. As suggested by the parallel situation in I, to establish this fact we introduce the concept of "inverses," this time for multiplication. (Notice that the role of 1 under multiplication is the same as that of 0 under addition. Compare Exercise 4.4.5 and Exercise 4.5.5.)

DEFINITION 4.5.7

Let S be any set of numbers on which a multiplication operation $\cdot$ has been defined, and let s be any element of S. A number t in S is called the **multiplicative inverse** or **reciprocal** of s iff $s \cdot t = 1$.

NOTATION: The multiplicative inverse of s is usually written as s^{-1}.

EXAMPLES 4.5

15. In the rational numbers, the multiplicative inverse of 3 (or $\frac{3}{1}$) is $\frac{1}{3}$, because

$$3 \odot \tfrac{1}{3} = [3, 1] \odot [1, 3] = [3, 3] = [1, 1] = 1$$

16. The reciprocal of $\frac{2}{3}$ is $\frac{3}{2}$, because

$$\tfrac{2}{3} \odot \tfrac{3}{2} = [2, 3] \odot [3, 2] = [6, 6] = [1, 1] = 1$$

In this case, we write $\left(\frac{2}{3}\right)^{-1} = \frac{3}{2}$.

It is easy to show that every rational number has a multiplicative inverse in Q. In fact, if $[a, b]$ is any rational number, its multiplicative inverse is $[b, a]$, because

$$[a, b] \odot [b, a] = [a \cdot b, b \cdot a] = [a \cdot b, a \cdot b] = [1, 1] = 1$$

That is, $[a, b]^{-1} = [b, a]$.

The importance of multiplicative inverses for us lies in the fact that division can be accomplished using them. Recall how division of fractions was described in elementary school: When confronted by a problem such as $\frac{2}{3} \div \frac{5}{7}$, for instance, we were instructed to "invert" the second fraction and multiply. Thus, we were shown examples like

$$\frac{2}{3} \div \frac{5}{7} = \frac{2}{3} \cdot \frac{7}{5} = \frac{14}{15} \quad \text{and} \quad \frac{1}{2} \div 3 = \frac{1}{2} \cdot \frac{1}{3} = \frac{1}{6}$$

But $\frac{7}{5}$ is the reciprocal of $\frac{5}{7}$, and $\frac{1}{3}$ is the reciprocal of 3, so this invert-and-multiply process is nothing more than dividing by multiplying the reciprocal of the second number by the first one. Formally,

---

DEFINITION 4.5.8

The **quotient** $r \oslash s$ of any two rational numbers r and s is found by multiplying r by the reciprocal of s. In symbols, $r \oslash s = r \odot s^{-1}$.

EXAMPLES 4.5

17. $[2, 5] \oslash [3, 7] = [2, 5] \odot [3, 7]^{-1} = [2, 5] \odot [7, 3] = [14, 15]$.

18. $\frac{7}{8} \oslash \frac{2}{3} = [7, 8] \oslash [2, 3] = [7, 8] \odot [2, 3]^{-1} = [7, 8] \odot [3, 2] = [21, 16] = \frac{21}{16}$.

19. $\frac{10}{9} \oslash \frac{5}{3} = [10, 9] \oslash [5, 3] = [10, 9] \odot [5, 3]^{-1} = [10, 9] \odot [3, 5] = [30, 45] = [2, 3] = \frac{2}{3}$.

20. $\frac{12}{13} \oslash 4 = [12, 13] \oslash [4, 1] = [12, 13] \odot [4, 1]^{-1} = [12, 13] \odot [1, 4] = [12, 52] = [3, 13] = \frac{3}{13}$.

(You should be able to give a reason for each step in these examples from the material in this section.)

If you have been following our development of the rationals closely, you should be aware of a problem at this point. We have now defined the quotient of two rational numbers in two apparently different ways, once just above (Definition 4.5.8) and once earlier (Definition 4.5.1). That makes no sense, of course, unless the two definitions actually say the same thing. That is, to rescue our credibility (again) we must show that the number we get from this multiplicative inverse definition (4.5.8) is the number c required by Definition 4.5.1. In other words, $r \oslash s$ will satisfy the earlier definition of quotient if and only if $s \odot (r \oslash s) = r$.

THEOREM 4.5.4

For any two rational numbers r and s: $s \odot (r \oslash s) = r$.

PROOF   Let us write r and s in "pair-class" form as $r = [a, b]$ and $s = [c, d]$. Then

$$s \odot (r \oslash s) = [c, d] \odot ([a, b] \oslash [c, d])$$

$$= [c, d] \odot ([a, b] \odot [c, d]^{-1}) \qquad \text{by definition of } \oslash$$

$$= [c, d] \odot ([a, b] \odot [d, c]) \qquad \text{by definition of multiplicative inverse}$$

$$= [c, d] \odot [a \cdot d, b \cdot c] \qquad \text{by definition of } \odot$$

$$= [c \cdot (a \cdot d), d \cdot (b \cdot c)] \qquad \text{by definition of } \odot$$

$$= [a \cdot (c \cdot d), b \cdot (c \cdot d)] \qquad \text{by commutativity and associativity of } \cdot \text{ in } I$$

$$= [a, b] \qquad \text{by Theorem 4.5.2}$$

$$= r$$

as required.

This result wraps up the problem we began the section with. We have shown in the discussion following Example 4.5.16 that every rational number has a reciprocal and that the quotient of two rationals can be found by multiplying the reciprocal of the second by the first (Definition 4.5.1 and Theorem 4.5.4). Hence, the rationals form a number system containing the integers in which division always makes sense.

There are two further number system extensions, both of them mathematically important, but at this point our main goal has been achieved. We have shown how the number systems of elementary arithmetic can be placed on a firm, simple, logical foundation, starting with the concept of an equivalence relation on a set and using that concept at each major step. In the next two (optional) sections we shall sketch very briefly the two further number system extensions, mainly to indicate that they, too, can be derived from what we have already built.

EXERCISES 4.5

1. Rewrite in fractional notation Definition 4.5.8 of quotient, Examples 4.5.17 through 4.5.20, and Theorem 4.5.4.

Note: With only a few minor variations, Exercises 2–13 below are the multiplicative counterparts of the corresponding exercises in Section 4.4 on the integers; you might find it useful to look back and compare the two exercise sets and their results.

2. Illustrate the fact that $\approx$ (Definition 4.5.2) is an equivalence relation on $I \times I^*$ by giving numerical examples of the three required properties.

3. List three different representative pairs for each of the following rational numbers:
   (a) $\frac{5}{9}$ (b) $\frac{15}{6}$
   (c) $-\frac{27}{95}$ (d) 0
   (e) 1

4. Use the definitions of $\oplus$, $\odot$, and $\ominus$ to verify the following:
   (a) $1 \oplus 1 = 2$ (b) $\frac{3}{5} \oplus \frac{4}{7} = \frac{41}{35}$
   (c) $-\frac{3}{8} \oplus \frac{19}{8} = 2$ (d) $\frac{4}{3} \oplus -\frac{7}{4} = -\frac{5}{12}$
   (e) $\frac{2}{3} \odot \frac{5}{9} = \frac{10}{27}$ (f) $\frac{2}{5} \odot \frac{15}{8} = \frac{3}{4}$
   (g) $-\frac{7}{6} \odot 3 = -\frac{7}{2}$ (h) $\frac{9}{2} \odot \frac{4}{3} = 6$
   (i) $4 \oslash 7 = \frac{4}{7}$ (j) $\frac{3}{7} \oslash \frac{4}{5} = \frac{15}{28}$
   (k) $-\frac{9}{8} \oslash \frac{3}{4} = -\frac{3}{2}$ (l) $\frac{1}{3} \oslash \frac{1}{3} = 1$

5. (a) Use the definitions of 0, 1, $\oplus$, and $\odot$ to verify that $\frac{3}{7} \oplus 0 = \frac{3}{7}, \frac{3}{7} \odot 1 = \frac{3}{7}, 0 \oplus 1 = 1,$ and $1 \odot 0 = 0$.
   (b) Prove that $r \oplus 0 = r$ for any rational number r.
   (c) Prove that $r \odot 1 = r$ for any rational number r.
   (d) Prove that $r \odot 0 = 0$ for any rational number r.

6. Use the definition of $\leqslant$ to verify that
   (a) $\frac{2}{3} \leqslant \frac{3}{4}$ (b) $1 \leqslant \frac{10}{9}$
   (c) $-\frac{5}{2} \leqslant \frac{1}{3}$ (d) $-\frac{17}{22} \leqslant 0$

7. Let x and y be integers. Show that $x < y$ if and only if $x/1 \leqslant y/1$.

8. Prove:
   (a) Every negative rational number is less than every positive rational number.
   (b) Every negative rational number is less than zero.
   (c) Zero is less than every positive rational number.

9. Prove:
   (a) Addition of rational numbers is commutative.
   (b) Addition of rational numbers is associative.
   (c) Multiplication of rational numbers is associative.
   (d) Multiplication of rational numbers is distributive over addition.

10. Let r, s, and t be rational numbers such that $r \leqslant s$. Prove:
    (a) $r \oplus t \leqslant s \oplus t$.
    (b) $r \odot t \leqslant s \odot t$ if t is positive.
    (c) $r \odot t \geqslant s \odot t$ if t is negative.

11. Let r and s be rational numbers. Prove:
    (a) $r^{-1} \odot s^{-1} = (r \odot s)^{-1}$
    (b) $r \odot s^{-1} = (r^{-1} \odot s)^{-1}$
    (c) $(r^{-1})^{-1} = r$

12. Show by example that division of rational numbers is neither commutative nor associative.

13. Using the material in this section, justify each step in the following computation:

$$\tfrac{5}{2} \oplus \tfrac{10}{8} = [5, 2] \oplus [10, 8]$$
$$= [5, 2] \odot [10, 8]^{-1}$$
$$= [5, 2] \odot [8, 10]$$
$$= [40, 20]$$
$$= [2, 1]$$
$$= 2$$

ESSAY QUESTIONS

1. Although we have not discussed it explicitly, subtraction is also possible in the rational number system. Write a detailed development of that operation on Q, including
   (a) A general definition of difference (see Definition 4.4.1).
   (b) A suitable definition of additive inverses.
   (c) Verification that the additive inverse of a rational number $[a, b]$ is the number $[-a, b]$ (which always exists in Q).
   (d) A definition of difference in terms of additive inverses.
   (e) A proof that the definitions you gave for parts (a) and (d) agree.
   Be sure to include examples for each step in the outline.

2. Write a detailed outline of the development of the integers and of the rational numbers, emphasizing the parallel structure of the two systems. (A convenient way of organizing such an outline is to put it in two columns—integers on the left, rationals on the right.)

4.6   THE REAL NUMBERS (optional)

Once again we shall simplify the notation by dropping the distinction between the integer operations and those for rational numbers. Since the system of integers is contained within the rational number system (see the discussion following Theorem 4.5.3) the rational number operations suffice for both systems. Thus, from now on the rational numbers will be written in traditional fractional form and their arithmetic operations and inequality will be denoted simply by $+$, $-$, $\cdot$, $\div$, and $<$.

As we have seen, the rational number system is "rich" enough to accommodate all four operations of elementary arithmetic. However, it has some glaring deficiencies, both in terms of its own algebraic structure and in relation to its geometric usefulness. For example, although raising a number to any (whole number) power is just repeated multiplication and hence is always possible in Q, taking a root is not merely repeated division, and for the most part even square roots of rationals do not exist in Q. The major link between geometry and numbers depends on being able to give a numerical label to each point on a line, but this is not possible using only rational numbers. The latter problem motivates the next extension of the number system, which we outline in this section.

Our development is of necessity somewhat brief and superficial, since a detailed, rigorous construction of the real numbers would require much more space and sophistication than is appropriate for this chapter. Moreover, the main objective of our treatment is simply to show that, once again, the underlying structure is that of an equivalence relation on an appropriate set. We take advantage of the limited nature of this objective by beginning our construction with a few "definitions" whose imprecision should make any self-respecting mathematician cringe but which should be informative enough to establish a basis for further discussion.

**DEFINITION
4.6.1**

A **sequence** of numbers is an infinite succession of numbers that starts with a first number and proceeds step by step through a second, third, fourth, and so on. The individual elements of the sequence are called **terms**. It is sometimes possible to write a formula that determines the sequence term corresponding to each natural number n; this formula is called the **nth term** or **general term** of the sequence.

NOTATION: Since any sequence has a term for each natural number, we use the naturals as subscripts and write $s_1, s_2, s_3, \ldots, s_n, \ldots$ for the terms of the sequence, where s_n denotes the general term. The sequence as a whole is denoted by $\{s_1, s_2, s_3, \ldots, s_n, \ldots\}$ or simply $\{s_n\}$.

EXAMPLES 4.6

1. $\{2n\}$ is the sequence that begins $2, 4, 6, 8, \ldots$. Note that any specific term can be found by substituting its index number for n in the general term and computing the result. Thus, the 25th term of this sequence is $2 \cdot 25$, or 50.

2. $\{1/n\}$ is the sequence $1, \frac{1}{2}, \frac{1}{3}, \frac{1}{4}, \ldots$. Its 25th term is $\frac{1}{25}$.

3. $\{(-1)^n\}$ is the sequence $-1, 1, -1, 1, \ldots$. Its 25th term is $(-1)^{25} = -1$.

4. $\{(2n - 1)/2^n\}$ is the sequence $\frac{1}{2}, \frac{3}{4}, \frac{5}{8}, \frac{7}{16}, \ldots$. Its 10th term is $(2 \cdot 10 - 1)/2^{10} = \frac{19}{1024}$.

Now let us take up the problem of labeling all the points on a line with numbers. The ability to do this is the key to coordinate geometry, which in turn is essential to calculus and much of higher mathematics, so the problem is clearly important. The most natural way to proceed is to choose (arbitrarily) a unit of length and a point for 0, and then to attach each rational number to the point which is that (directed) distance from 0. The details of the process are unimportant for our purposes; it suffices to say that any rational number can be positioned like that. Now let us assume all the rational numbers have been placed on the line as described; have we labeled *all* the points on the line? No! In fact, most of the points are still without labels.* Hence, we need to extend Q to a larger system; that is, we need a system of numbers that contains the rationals and is large enough to have a number for each point on the line. Again we resort to an equivalence relation on an appropriate set.

The approach taken here is quite natural (despite the way it might appear to you on first reading). Suppose that we consider a point P whose distance from 0 cannot be expressed as a rational number. We can still use the rational numbers to distinguish P from any other point on the line, thanks to a particularly nice property of the rationals, which we shall use here without proof. The rational numbers are "dense" on the line; that is, between any two distinct points on the line, no matter how close together they are, there is always some rational number. Thus, to distinguish P from any other point A we simply find a rational number which is closer to P than A is. Now, if we take any decreasing chain of distances—$\frac{1}{2}, \frac{1}{3}, \frac{1}{4}, \frac{1}{5}, \ldots$, for instance—for each of these distances, there is always some rational number within that distance of P. As we choose a sequence of rationals getting closer and closer to P, these rationals are distinguishing P from more and more points on the line, and in this way the *entire* infinite sequence identifies P uniquely by distinguishing P from all other points on the line!

DEFINITION 4.6.2	We say that a sequence $\{s_n\}$ of rational numbers is **eventually** within some specific distance of a point P if there is some term of $\{s_n\}$ after which all the terms are less than the given distance from P. We say that a sequence $\{s_n\}$ **converges to** P if, for each natural number n, $\{s_n\}$ is eventually within distance $1/n$ of P. A sequence is called **convergent** if it converges to some point.

EXAMPLE 4.6.5

The sequence $\{1, \frac{1}{2}, \frac{1}{4}, \frac{1}{8}, \ldots, 1/2^{n-1}, \ldots\}$ is eventually within distance $\frac{1}{3}$ of 0, because every term after $\frac{1}{2}$ is less than distance $\frac{1}{3}$ from 0. This sequence is also eventually within distance $\frac{1}{3}$ of the point $\frac{1}{6}$, because every term after $\frac{1}{2}$ is less than distance

---

* There is a proof of this in Chapter 5. Section 5.6 shows that the set of points on a line cannot be put in one-to-one correspondence with Q, and Section 5.8 establishes that the points on the line form a (much) larger set.

$\frac{1}{3}$ from the point $\frac{1}{6}$. This sequence converges to 0, because, for any natural number n, all the terms after $1/2^{n-1}$ are less than the distance $1/n$ from 0. (For instance, if $n = 5$, all the numbers $\frac{1}{32}$, $\frac{1}{64}$, $\frac{1}{128}$, ... are within distance $\frac{1}{5}$ of 0.) However, this sequence does not converge to $\frac{1}{3}$, because, for example, if $n = 4$, all the terms $\frac{1}{16}$, $\frac{1}{32}$, $\frac{1}{64}$, ... are more than the distance $\frac{1}{4}$ from the point $\frac{1}{3}$.*

Put another way, a sequence that converges to P is just a succession of closer and closer rational approximations of the distance from 0 to P (whether or not that distance is itself rational). Since it is possible to get as close an approximation as we please, the sequence taken as a whole actually identifies P unambiguously. But many different sequences can converge to the same number. For instance, the sequences $\{\frac{1}{2}, \frac{2}{3}, \frac{3}{4}, \ldots, n/(n + 1), \ldots\}$, $\{\frac{1}{2}, \frac{3}{4}, \frac{7}{8}, \ldots, (2^n - 1)/2^n, \ldots\}$, and $\{\frac{3}{2}, \frac{5}{4}, \frac{9}{8}, \ldots,$ $(2^n + 1)/2^n, \ldots\}$ (and infinitely many others) all converge to 1. Thus, if we were to identify the points on the line with the rational sequences that converge to them, each point would have many different identities, a confusing situation at best.

It makes far more sense to consider two sequences "equivalent" for our purposes if they converge to the same point and to use the resulting equivalence classes as numerical labels for the points to which their sequences converge. That is, we must define an equivalence relation on the set of all convergent sequences and then turn the resulting collection of equivalence classes into a number system containing Q by defining operations on it in some appropriate way. (Despite the unfamiliarity of convergent sequences, you should see a familiar pattern of procedure here: set, equivalence relation, equivalence classes, operations on the classes, containment of the previous system, and finally verification that the extended system solves the motivating problem.)

DEFINITION 4.6.3

Two convergent rational number sequences $\{s_1, s_2, \ldots, s_n, \ldots\}$ and $\{t_1, t_2, \ldots, t_n, \ldots\}$ are **equivalent** if the sequence of differences of their corresponding terms $\{s_1 - t_1, s_2 - t_2, \ldots, s_n - t_n, \ldots\}$ converges to 0. In this case we write $\{s_n\} \simeq \{t_n\}$.

EXAMPLES 4.6

6. $\{\frac{3}{2}, \frac{5}{4}, \frac{9}{8}, \ldots, (2^n + 1)/2^n, \ldots\}$ and $\{\frac{1}{2}, \frac{3}{4}, \frac{7}{8}, \ldots, (2^n - 1)/2^n, \ldots\}$ are equivalent, because their difference sequence is $\{\frac{3}{2} - \frac{1}{2}, \frac{5}{4} - \frac{3}{4}, \frac{9}{8} - \frac{7}{8}, \ldots, (2^n + 1)/2^n - (2^n - 1)/2^n, \ldots\} = \{1, \frac{1}{2}, \frac{1}{4}, \ldots, 1/2^{n-1}, \ldots\}$, which clearly converges to 0.

7. $\{5, 5, 5, \ldots\} \simeq \{\frac{49}{10}, \frac{499}{100}, \frac{4999}{1000}, \ldots\}$, because the difference sequence $\{\frac{1}{10}, \frac{1}{100}, \frac{1}{1000}, \ldots, 1/10^n, \ldots\}$ converges to 0.

---

* To check this, observe that $\frac{1}{3} - \frac{1}{4} = \frac{1}{12}$, which is bigger than any term of the sequence from $\frac{1}{16}$ on.

8. $\{0, \frac{1}{2}, \frac{2}{3}, \frac{3}{4}, \ldots, (n-1)/n, \ldots\}$ and $\{1, \frac{1}{2}, \frac{1}{3}, \frac{1}{4}, \ldots, 1/n, \ldots\}$ are not equivalent, because their difference sequence $\{-1, 0, \frac{1}{3}, \frac{2}{4}, \ldots, (n-2)/n, \ldots\}$ does not converge to 0. (In fact, it converges to 1.)

The difference sequence device used in the definition of $\simeq$ may appear unnecessarily awkward, but in fact it is very useful because it provides a way of deciding if two sequences converge to the same point without having to know what their convergence points are. This is particularly important when the points have no rational number labels, and those points, after all, are the ones that concern us most.

The following theorem will be assumed without proof. (You might find it instructive to try to construct a proof for it.)

THEOREM 4.6.1

The relation $\simeq$ defined above is an equivalence relation on the set of all convergent sequences of rational numbers.

Since $\simeq$ is an equivalence relation on the set of convergent rational sequences, it partitions that set into disjoint sets of sequences; we call these classes of sequences "real numbers."

DEFINITION 4.6.4

A **real number** is the set of all convergent sequences of rational numbers that are related to some particular sequence—say $\{s_n\}$—(and hence to each other) by the equivalence relation $\simeq$; we denote this set of sequences by $[s_n]$. In symbols, $[s_n] = \{\{x_n\} | \{x_n\} \simeq \{s_n\}\}$.

NOTATION: The set of all real numbers is denoted by R.

EXAMPLES 4.6

9. $[1/2^n]$ is the real number that also contains the sequences $\{1/10^n\}$, $\{1/n\}$, $\{-1/2n\}$, and so forth. (Can you verify this?) It should be easy to see that this sequence class actually represents 0.

10. An example of a sequence class in R that does not represent a rational number is a bit more cumbersome to construct at this stage, but correspondingly more interesting. We need a representative sequence that converges to some point which does not have a rational number label. To make one such sequence, consider the following process. Start with the set of all rational numbers between 1 and 2, which we shall call the "interval" $\langle 1, 2 \rangle$, and take its midpoint, $\frac{3}{2}$. If the square of that midpoint is less than 2, the midpoint of the interval becomes the first term of the sequence and we repeat the process on the interval $\langle \frac{3}{2}, 2 \rangle$ to find the next sequence term. If $(\frac{3}{2})^2$ is not less than 2 (as is the

case, in fact), then 1 becomes the first term of the sequence and we repeat the process on the interval $\langle 1, \frac{3}{2} \rangle$ to find the second term. Thus:

1 is the first term, because $(\frac{3}{2})^2 = \frac{9}{4} > 2$; now consider $\langle 1, \frac{3}{2} \rangle$.

$\frac{5}{4}$ is the second term, because $\frac{5}{4}$ is the new midpoint and $(\frac{5}{4})^2 = \frac{25}{16} < 2$; now consider $\langle \frac{5}{4}, \frac{3}{2} \rangle$.

$\frac{11}{8}$ is the third term, because $\frac{11}{8}$ is the new midpoint and $(\frac{11}{8})^2 = \frac{121}{64} < 2$; now consider $\langle \frac{11}{8}, \frac{3}{2} \rangle$.

And so forth.

There is no convenient formula for the general term of this sequence but the construction described above defines it unambiguously. Moreover, since the maximum distance between successive terms is halved at each step, it is not hard to believe that the sequence converges to something and thus represents a class in R. In fact, the "something" turns out to be the point whose distance from 0 is commonly referred to as $\sqrt{2}$, which is *not* a rational number. [For a proof that $\sqrt{2}$ cannot be expressed as a fraction, see Exercise 4.6.8(a).] Another sequence in this class is $\{1, 1.4, 1.41, 1.414, 1.4145, \ldots\}$, the successive decimal approximations of $\sqrt{2}$.

As you might expect from working through the previous sections of this chapter, the equivalence classes that are the real numbers have standard representative sequences. Moreover, much like the case of the integers, these standard representatives are usually written in a shorthand form which disguises the fact that they are sequences and obscures the equivalence classes they represent. The real numbers are most commonly represented as "infinite decimals," such as 3.2555 . . . , .1010010001 . . . , 47.2000 . . . , and so on. But an infinite decimal is just an abbreviated way of writing a sequence of finite decimal approximations. For example, 3.2555 . . . actually represents the rational number sequence 3, 3.2, 3.25, 3.255, 3.2555, ...* The important fact for us (which we shall not prove here) is that *every* real number contains one such decimal sequence (and occasionally two), so the infinite decimal sequences can be used as representatives for *all* the real numbers. When this is done, the infinite decimal is considered to "be" the number (that is, to be the entire class), much as the integer -3 "is" an entire class of ordered pairs. Thus, 3.2555 . . . = [3, 3.2, 3.25, 3.255, 3.2555, . . .].

To make R into a number system we need arithmetic operations, and these can be defined in terms of representative sequences. However, the details of their development are beyond the scope of this section, so we shall merely assume the existence of the operations $+$, $-$, $\cdot$, and $\div$ and the relation $<$ on R, with all their familiar properties.† Finally, we observe that the real number system actually

---

* It should be clear that these finite decimals are just rational numbers whose denominators are powers of 10; that is, $3.2 = \frac{32}{10}$, $3.25 = \frac{325}{100}$, and so forth.

† A description of these operations may be found in Patrick Shanahan, *Introductory College Mathematics* (Englewood Cliffs, N.J.: Prentice-Hall, Inc., 1963), Chap. 7.

contains the rationals. If we use the standard decimal representatives for the real numbers, those that represent rational numbers have a distinctive form: each one is an infinite decimal in which, from some specific digit on, a finite string of digits repeats again and again in the same order, without interruption. Such decimals are often called **repeating decimals**, for the sake of brevity. A proof of this surprisingly easy characterization of Q within R can be found in Section 5.5 (which can be read independently of the rest of that chapter). Hence, the real number system is an extension of the rational number system large enough to provide a numerical label for every point on a line.

EXERCISES 4.6

1. Write the first four terms of the sequence whose general term is
   (a) $3n$
   (b) $\dfrac{1}{n+2}$
   (c) $\dfrac{n}{n+1}$
   (d) 4
   (e) $\dfrac{1+(-1)^n}{n}$
   (f) $\dfrac{n^2}{2^n}$

2. For each sequence below, find a general term that fits the first five terms given.
   (a) $3, 4, 5, 6, 7, \ldots$
   (b) $1, 3, 5, 7, 9, \ldots$
   (c) $3, \frac{3}{10}, \frac{3}{100}, \frac{3}{1000}, \frac{3}{10000}, \ldots$
   (d) $1, \frac{1}{4}, \frac{1}{9}, \frac{1}{16}, \frac{1}{25}, \ldots$
   (e) $2, 5, 10, 17, 26, \ldots$

3. In reference to the discussion following Example 4.6.4 concerning identifying a point P by a sequence of rational numbers, why can't we just use the sequence $\{P + \frac{1}{2}, P + \frac{1}{3}, P + \frac{1}{4}, \ldots\}$ to identify P?

4. Find two sequences that are eventually within distance $\frac{1}{10}$ of 4 but do not converge to 4. Justify your answer.

5. Write three different sequences which you think converge to 5 and three more which you think converge to -1.

6. Illustrate the fact that $\simeq$ (Definition 4.6.3) is an equivalence relation on the set of all convergent sequences of rational numbers by giving particular examples of the three required properties.

7. Write three different sequences in the class
   (a) $[2/n]$
   (b) $[-1 + 1/n]$
   (c) $[n/(n+1)]$

8. (a) Fill in the details of the following proof that $\sqrt{2}$ cannot be expressed as a fraction.

Suppose that $\sqrt{2} = a/b$, where a/b is in lowest terms. Then $2b^2 = a^2$, so 2 is a factor of a^2 and hence also of a. Thus, we can write $a = 2c$ for some integer c. Then $2b^2 = (2c)^2$, so $b^2 = 2c^2$. This implies that 2 is a factor of b^2 and hence also of b, contradicting the premise that a/b is in lowest terms. Since any fraction can be put in lowest terms, this contradiction implies that $\sqrt{2}$ cannot be written as a fraction.

(b) Mimic the argument of part (a) to prove that $\sqrt{3}$ cannot be written as a fraction.

9. Refer to Example 4.6.10.
   (a) Find the fourth and fifth terms of the sequence described there, $\{1, \frac{5}{4}, \frac{11}{8}, \ldots\}$.
   (b) Mimic the construction of this example to describe a sequence that represents $\sqrt{3}$.

10. (a) Give the first four terms of the decimal sequence represented by $2.51466\ldots$.
    (b) Use the definition of equivalent sequences to verify that $0.999\ldots$ and $1.000\ldots$ are both standard representatives of the same real number. (That is, show that $0.999\ldots = 1$.)

4.7   THE COMPLEX NUMBERS (optional)

We conclude this chapter with a brief description of the final extension of the number system. Despite the fact that an equivalence relation does not play a role here, the construction of the complex numbers is similar in many ways to that of the integers and of the rationals. Moreover, the complex numbers actually complete the number system in an algebraic sense which will be made more precise later, so they provide a natural conclusion for any discussion of the various types of numbers.

The motivating problem for this last extension is one left over from the previous section. After the rationals were built, we raised two problems. The real numbers take care of naming all the points on a line, and they also give us a way of handling square roots of some numbers, namely the positive ones. However, as you may remember from high school, the square of any positive or negative real number must be positive, so square roots of negative numbers cannot exist in the reals. The complex numbers are required to remedy this deficiency.

Once again, if the question is put in the proper form, that form itself becomes the recipe for the answer. Let us consider the problem of how to treat $\sqrt{-p}$, where p is a positive real number. Since the square of any nonzero real number must be positive and since $(\sqrt{-p})^2 = -p$ by definition of square root, it is clear that $\sqrt{-p}$ cannot be a real number, and hence any attempt to treat it as a number at all requires a larger system. An examination of the square roots of positive numbers provides a clue as to how this extension may be made. We know that, if a and b are positive real numbers, $\sqrt{a \cdot b} = \sqrt{a} \cdot \sqrt{b}$. If we suppose that numbers in the extended system could be treated in the same way, then $\sqrt{-p}$ could be factored into $\sqrt{p} \cdot \sqrt{-1}$.* This seemingly trivial observation is actually a giant stride forward, since $\sqrt{p}$ is real for any p, and thus, instead of having to consider the behavior of infinitely many square roots of negatives, we need only consider that of one, $\sqrt{-1}$. If the number system can be enlarged to contain $\sqrt{-1}$ as well as the reals, it must also contain all sums and products involving both real numbers and $\sqrt{-1}$; that is, it must contain all quantities of the form $a + b\sqrt{-1}$, where a and b are real numbers. The notation is usually simplified by using i for $\sqrt{-1}$, and thus we might say our extended number system is the set of all objects of the form $a + bi$, where $a, b \in R$, with the arithmetic operations the same as they were for the reals.

At this point you should notice a strong odor of deceit. A second look at the preceding paragraph reveals a carefully obscured circular argument. The proposed extended system was constructed by using $\sqrt{-1}$ and hence its existence depends upon $\sqrt{-1}$, but this "number" was derived only by presupposing the existence of the extended number system. That is, each of these things depends upon the other, and nowhere have we established the existence of either! We must now show that such a system exists by constructing it using real numbers only, just as each previous extension of the number system was constructed from the one before it. Although the foregoing ideas cannot be part of the formal construction, they are valuable aids in the choice of appropriate definitions and methods of procedure. In particular, since things of the form $a + bi$ are determined by the choice of the numbers a and b, it makes sense to begin with ordered pairs of real numbers.

DEFINITION 4.7.1

A **complex number** is an ordered pair of real numbers. The set of all complex numbers is denoted by C; that is, $C = R \times R$.

EXAMPLE 4.7.1

$(2, 3)$ is a complex number and can be thought of as $2 + 3\sqrt{-1}$, or $2 + \sqrt{-9}$. $(0, 5)$ is a complex number; it can be regarded as $\sqrt{-25}$.

Two complex numbers are equal only if they are identical pairs of real numbers; there is no need to form equivalence classes. The presence of the mythical $\sqrt{-1}$ within the system is captured in the definitions of the arithmetic operations. (See Exercise 4.7.5.) As we have seen before, the only operations we need to define are addition and multiplication, since the general definitions of subtraction and division given in earlier sections can then be applied here. Once again, circles distinguish new operations from old.

---

* Actually, this process cannot be applied indiscriminately to roots of numbers. See, for example, Exercise 4.7.6.

DEFINITION 4.7.2	The **sum** of two complex numbers (a, b) and (c, d) is $(a + c, b + d)$. In symbols, $(a, b) \oplus (c, d) = (a + c, b + d)$.

DEFINITION 4.7.3	The **product** of two complex numbers (a, b) and (c, d) is $(a \cdot c - b \cdot d, a \cdot d + b \cdot c)$. In symbols, $(a, b) \odot (c, d) = (a \cdot c - b \cdot d, a \cdot d + b \cdot c)$.

EXAMPLE 4.7.2

$$(1, 2) \oplus (3, 4) = (1 + 3, 2 + 4) = (4, 6)$$

$$(1, 2) \odot (3, 4) = (1 \cdot 3 - 2 \cdot 4, 1 \cdot 4 + 2 \cdot 3) = (-5, 10)$$

Both of these operations are commutative and associative, and multiplication is distributive over addition. $(0, 0)$ plays the role of zero for addition and $(1, 0)$ plays the role of 1 for multiplication (see Exercise 4.7.2). There are negatives for all complex numbers and reciprocals for all nonzero complex numbers, so both subtraction and division can be defined, as we have seen earlier. The only unusual feature of this system in comparison with earlier systems is that there is no useful way to define a "less than" relation for all complex numbers.

Taking our cue from the $a + bi$ motivation, if we identify each real number r with the pair $(r, 0)$, then the real number system is contained within the complex number system, since

$$r + s = (r, 0) \oplus (s, 0) = (r + s, 0 + 0) = (r + s, 0)$$

and $$r \cdot s = (r, 0) \odot (s, 0) = (r \cdot s - 0 \cdot 0, r \cdot 0 + 0 \cdot s) = (r \cdot s, 0)$$

Finally, the elusive $\sqrt{-1}$ emerges as the complex number $(0, 1)$, since

$$(0, 1)^2 = (0, 1) \odot (0, 1) = (0 \cdot 0 - 1 \cdot 1, 0 \cdot 1 + 1 \cdot 0) = (-1, 0) = -1$$

and the existence of this number allows us to take square roots of all negative real numbers by the way we suggested earlier in the section.

In fact, the complex numbers do much more than that. The particular algebraic problems that gave rise to each extension of the number system are all related to the process of solving equations, and the resolution of them in the complex number system is conveniently summarized in the following theorem, which is assumed here without proof.*

---

*For a proof of this theorem, see Richard Courant and Herbert Robbins, *What Is Mathematics?* (Oxford: Oxford University Press, 1941), pp. 101–103.

THE FUNDAMENTAL THEOREM OF ALGEBRA
Every equation of the form

$$a_n x^n + a_{n-1} x^{n-1} + \cdots + a_1 x + a_0 = 0$$

where n is a positive integer and all the a's are complex numbers, has a solution in the complex number system.

In the language of high school algebra, this says that all polynomial equations in one variable with complex coefficients can be solved in the complex number system. Thus, the theorem assures us that, regardless of the size or intricacy of a polynomial equation, no further extension of the number system will ever be necessary to solve it. In other words, from the point of view of arithmetic, the number system is complete.

EXERCISES 4.7

1. Compute the following sums and products of complex numbers.
   (a) $(2, 3) \oplus (1, 5)$ (b) $(-4, 7) \oplus (1, -9)$
   (c) $(6, 0) \oplus (\frac{1}{3}, 2)$ (d) $(2, 5) \odot (3, 8)$
   (e) $(4, -7) \odot (0, 1)$ (f) $(3, \frac{2}{5}) \odot (0, 0)$

2. (a) Use the definitions of $\oplus$ and $\odot$ to verify that $(5, 2) \oplus (0, 0) = (5, 2)$ and $(5, 2) \odot (1, 0) = (5, 2)$.
   (b) Prove that $(a, b) \oplus (0, 0) = (a, b)$ for any complex number (a, b).
   (c) Prove that $(a, b) \odot (1, 0) = (a, b)$ for any complex number (a, b).
   (d) Prove that $(a, b) \odot (0, 0) = (0, 0)$ for any complex number (a, b).

3. Prove:
   (a) Addition of complex numbers is commutative.
   (b) Addition of complex numbers is associative.
   (c) Multiplication of complex numbers is commutative.
   (d) Multiplication of complex numbers is distributive over addition.

4. If i is defined as the complex number $(0, 1)$, then the previously vague "number" $a + bi$ becomes

$$(a, 0) \oplus ((b, 0) \odot (0, 1))$$

by using the matching between the real numbers a and b and their complex counterparts $(a, 0)$ and $(b, 0)$. Show by computation that this expression is equal to (a, b), as you should expect.

5. Let a, b, c, and d be real numbers and let $i = \sqrt{-1}$. Using the usual rules for addition and multiplication of real numbers and the fact that $i^2 = -1$, compute $(a + bi) + (c + di)$ and $(a + bi) \cdot (c + di)$, and verify that your answers

correspond to the definitions of the sum and product of two complex numbers in the same way that $a + bi$ corresponds to (a, b).

6. Comment on the following "proof" that $1 = -1$:

$$1 = \sqrt{1} = \sqrt{(-1) \cdot (-1)} = \sqrt{-1} \cdot \sqrt{-1} = (\sqrt{-1})^2 = -1$$

For Further Reading

Berlinghoff, William. *Mathematics: The Art of Reason.* Lexington, Mass. : D. C. Heath and Company, 1968, Chap. V.

Courant, Richard, and Herbert Robbins. *What Is Mathematics?* Oxford : Oxford University Press, 1941.

Dantzig, Tobias. *Number, the Language of Science, 4th ed.* New York : Macmillan Publishing Co., Inc., 1954.

Dubisch, Roy. *The Nature of Number.* New York : The Ronald Press Company, 1952.

Shanahan, Patrick. *Introductory College Mathematics.* Englewood Cliffs, N.J.: Prentice-Hall, Inc., 1963.

5

SETS AND INFINITY

5.1 INTRODUCTION

Mathematics is many things to many people. To some it is just a routine tool, to others a convenient language, and to still others it is a strict science. And to a surprising number of people, mathematics is predominantly an art, pursued for its own sake out of curiosity and with an appreciation of abstract beauty, in much the same way as a chess grandmaster seeks an elegant checkmate of a respected opponent. This chapter treats some mathematics which was developed from that last viewpoint. It introduces the theory of sizes of infinite sets, an elegant, abstract topic with no immediate application to (and apparently no direct connection with) the physical world. Nevertheless, as soon as it was introduced by the German mathematician Georg Cantor in 1872, this theory caused intense controversy in mathematical, philosophical, and theological circles, a controversy that contributed to Cantor's eventual mental breakdown, caused severe divisions of opinion among European theologians, and resulted in a continuing three-way philosophical split regarding the foundations of mathematics. Besides these negative results, Cantor's work affected mathematics in a decidedly positive way. His basic set theory provided a simple unifying approach to many diverse areas of mathematics. Moreover, the strange paradoxes encountered in some early extensions of his work impelled mathematicians to put their logical house in order, as it were, resulting in a careful examination of the logical foundations of mathematics, leading to many new results in that area, and paving the way for even more abstract unifying ideas, such as category theory. It is no exaggeration to say that Cantor's curious speculation has profoundly affected both the form and the content of all of modern mathematics.

The key to the universality of Cantor's work is the simplicity of its starting point. This fact is especially convenient for us, because it allows us to get a good look at some of his most important results without requiring much preliminary material. In fact, we shall assume only a few very elementary ideas from your prior mathematical experience. Specifically, we assume you know that two points determine a unique line, and that a line can be "coordinatized" so that each of its points corresponds to a unique real number, which can be expressed as an infinite decimal. Moreover, we assume that you have some experience with the number system and are familiar with the elementary language of sets, including subset, union, intersection, and Cartesian product.

5.2   SIZES OF SETS

Given two collections of things (i.e., sets), what does it mean to say they are the same size? For example, if $A = \{a, b, c, d\}$ and $B = \{*, \$, \#, \&\}$, what are we saying when we observe that A and B are the same size? We might begin to explain by saying that A and B have the same number of elements, but what does "same number" mean? In particular, must we have numbers before we can decide whether or not A and B are "the same size"? Of course, we could use numbers to count A and B, and say that "same size" means that we ended with the same number both times we counted. But counting is nothing more than a way of matching things up with part of a known set (the numbers $1, 2, 3, \ldots$), and if we are going to match up sets, there is no need to use the numbers at all; we can simply match directly the sets we want to compare. For instance, we can pair off the elements of A and B as in

$$A = \{a, b, c, \ d\}$$
$$| \quad | \quad | \quad |$$
$$B = \{*, \$, \#, \&\},$$

and, observing that the pairing exhausts both sets, conclude that A and B are the same size. This process may sound artificial, since we have been using counting for most of our lives, but it is a very natural way of comparing sizes of collections of things. (In fact, young children who have not yet learned to count often use this method to find out whether they get their fair shares when candy is distributed!) Since we are preparing to investigate infinite sets, where counting all the elements is a hopeless task, it is useful to have a formal definition of this simpler comparison process.

DEFINITION
5.2.1

A **one-to-one correspondence** between two sets A and B is any rule or process by which each element of A is associated with a unique element of B and each element of B is associated with a unique element of A. (*Note:* "Unique" means "one and only one.")

**DEFINITION
5.2.2**

Two sets A and B are called **equivalent** if there exists a one-to-one correspondence between them. We shall write this as $A \leftrightarrow B$. (*Note:* "Equivalent" is the formal word for saying that two sets are the same size.)

EXAMPLES 5.2

1. If $A = \{a, b, c, d\}$ and $B = \{*, \$, \#, \&\}$, the diagram preceding Definition 5.2.1 illustrates a one-to-one correspondence between A and B. Another such correspondence is

$$A = \{a, b, c, \ d\}$$

$$B = \{*, \$, \#, \&\}$$

2. The matching

$$A = \{a, b, c, \ d\}$$

$$B = \{*, \$, \#, \&\}$$

is not a one-to-one correspondence. Even though each element of A corresponds to a unique element of B, the element $\$$ in B is not matched with anything in A and $\#$ is matched with two elements of A.

3.
$$\{1, \quad 2, \quad 3, \quad 4, \dots, \quad n, \dots\}$$

$$\{-1, -2, -3, -4, \dots, -n, \dots\}$$

illustrates a simple one-to-one correspondence between the set of all positive integers and the set of all negative integers. Since the two sets are infinite, we cannot list every element in the correspondence, but it is easy to see that the rule given allows us to determine specifically the correspondence for each element in either set. For instance, the rule clearly indicates that the positive integer 37 corresponds to the negative integer -37.

EXERCISES 5.2

1. Find three more one-to-one correspondences between the sets A and B of Example 5.2.1.

2. Give an example of two sets that cannot be put into one-to-one correspondence. Justify your answer.

3. (a) If two sets are equivalent, must they be equal? Why?
   (b) If two sets are equal, must they be equivalent? Why?

Phrased in terms of the previous definition, the first fundamental question of
this chapter is: Are all infinite sets equivalent? That is, can any two infinite sets be
put in one-to-one correspondence? Before attempting to answer that in general,
let us examine a few infinite sets that should be familiar from prior mathematical
experience. Listed here for reference are some infinite sets together with their
names and notation, which we shall discuss throughout the chapter.

$N = \{1, 2, 3, \ldots, n, \ldots\}$, the **natural numbers.**

$N^{\text{even}} = \{2, 4, 6, \ldots, 2n, \ldots\}$, the **even natural numbers**.

$N^{\text{odd}} = \{1, 3, 5, \ldots, 2n - 1, \ldots\}$, the **odd natural numbers**.

$I = \{\ldots, -3, -2, -1, 0, 1, 2, 3, \ldots\}$, the **integers**.

$Q = \{\frac{p}{q} | p \text{ and } q \text{ are integers}, q \neq 0\}$, the **rational numbers**.

R denotes the set of all infinite decimals, the **real numbers,** and can be thought of
  as representing the set of all points on a single line ("the number line").
$[0, 1] = \{x | x \in R \text{ and } 0 \leq x \leq 1\}$, the set of all real numbers between 0 and 1
  inclusive, called the **unit interval**.
For any two real numbers a and b, $[a, b] = \{x | x \in R \text{ and } a \leq x \leq b\}$, the set of
  all real numbers between a and b inclusive, called the **interval a, b**.
$R \times R = \{(x, y) | x, y \in R\}$, the set of all ordered pairs of real numbers, which
  represents the set of all points on a plane.
$R \times R \times R = \{(x, y, z) | x, y, z \in R\}$, the set of all ordered triples of real
  numbers, which represents the set of all points in three-dimensional space.

There are, of course, many other infinite sets within the realm of mathematics,
but the ones listed above should be somewhat familiar to you and provide sufficient
variety for exploring our basic question about different sizes. Specifically, let us
examine the sets on this list to see if we can find examples of different sizes of
infinity. Intuitive reactions (that is, "gut feelings") about that problem are nearly
as varied as the people who think about it. Many feel that all infinite collections are
the same size; others see essential size differences between almost any two of the
sets listed, and still others take intermediate positions scattered between these two
extremes. The only dependable conclusion which can be drawn from this spectrum
of reactions is that intuition unsupported by logic is unreliable in an area as un-
familiar as infinity. We human beings have never encountered actually infinite
collections of things in our material experience, so all our attempts to deal with
them must involve extending the understanding of our finite experience into an
area where its applicability is unknown. Hence, we must fall back on logical
reasoning to guarantee the validity of any statements we make about infinity. In
particular, we must apply our definition of equivalent sets carefully and then be

prepared to accept the outcome of our reasoning, regardless of whether or not it conforms to our intuitive feelings.

5.3   INFINITE SETS

An appropriate way to start a formal discussion of infinity is to define exactly what we mean by "infinite set:"

DEFINITION
5.3.1

A set A is **finite** if it is empty or if there is a natural number n such that A is equivalent to $\{1, 2, \ldots, n\}$. A set is **infinite** if it is not finite.

This definition simply says that a set is infinite if it cannot be counted, a very natural way of describing infinite sets. However, this definition is sometimes difficult to apply, since it presupposes a familiarity with the natural numbers, and since proving that a set is infinite according to this definition requires a proof that *no* natural number will determine an equivalent set. Hence, it would be convenient to have a definition of infinite set that applies directly and simply to any set proposed. The following characteristic property of infinite sets is often useful.

DEFINITION
5.3.2

A set is **infinite** if it is equivalent to a proper subset of itself. A set is **finite** if it is not infinite. (Recall that A is a **proper subset** of B if A is a subset of B that does not contain all the elements of B.)

It can be shown that both of these definitions of infinite set yield the same results; that is, a set that satisfies either definition of infinite will also satisfy the other. Hence, they may be used interchangeably, according to which seems more convenient in a given situation.

EXAMPLES 5.3

1. We can easily show that the set N of natural numbers is infinite, using Definition 5.3.2: N^{even} is clearly a proper subset of N, and the one-to-one correspondence

$$N = \{1, 2, 3, 4, \ldots, \; n, \ldots\}$$

$$N^{\text{even}} = \{2, 4, 6, 8, \ldots, 2n, \ldots\}$$

establishes the equivalence of N and N^{even}.

2. The easiest way to show that $\{a, b, c, d, e\}$ is finite is to count it, that is, to apply Definition 5.3.1 and observe the one-to-one correspondence

$$\{a, b, c, d, e\}$$

$$\{1, 2, 3, 4, 5\}$$

EXERCISES 5.3

1. Use Definition 5.3.2 to prove that the set N^{even} is infinite.
2. Prove that the set $\{*, \$, \#, \&\}$ is finite.
3. Use Definition 5.3.2 to prove that the set I of integers is infinite.
4. Use Definition 5.3.2 to prove that the set $\{a, b\}$ is finite.

5.4   THE SIZE OF N

One way to paraphrase Definition 5.3.2 is to say that a set is infinite if we can throw away some of it and still be left with a set of the same size. In particular, Example 5.3.1 shows that N and N^{even} are the same size, even though N^{even} is obtained by throwing away "half" of N (that is, throwing away all the odd natural numbers). It is also easy to see that N^{odd} is equivalent to N (see Exercise 5.4.1). These two correspondences can be used to extend one step further our comparison of sizes of the sets listed in Section 5.2. If we ignore the usual ordering of the integers, we can show that I is equivalent to N:

$$N = \{1, 2, \quad 3, 4, \quad 5, 6, \quad 7, 8, \quad 9, \ldots\}$$

$$I = \{0, 1, \ -1, 2, \ -2, 3, \ -3, 4, \ -4, \ldots\}$$

It should be easy to see that the pattern established by the first few terms of this correspondence can be continued as far as we please, matching the even natural numbers with the positive integers and the odd natural numbers with zero and the negative integers. For instance, the positive integer 10 corresponds to the natural number 20, the natural number 51 corresponds to the negative integer -25, and so forth. We can even write a convenient general description of the matching process: Any even natural number n will correspond to $n/2$ and any odd natural number m will correspond to $(1 - m)/2$. Thus, we have shown that N and I are the same size.

EXERCISES 5.4

1. Set up a one-to-one correspondence which proves that N^{odd} is equivalent to N.

2. Set up a one-to-one correspondence which proves that N^{odd} is equivalent to N^{even}.

3. Refer to the correspondence between N and I given above.
   (a) What integer is matched with the natural number 34?
   (b) What natural number is matched with the integer -30?
   (c) What integer is matched with the natural number 75?
   (d) What natural number is matched with the integer 100?
   (e) If p denotes a positive integer, what is its corresponding natural number?

4. Set up a one-to-one correspondence between N and the set $\{30, 60, 90, 120, 150, \ldots\}$.

5. Use your answers to Exercises 1 and 4 above to set up a one-to-one correspondence between N^{odd} and $\{30, 60, 90, 120, 150, \ldots\}$.

Exercise 5.4.5 illustrates a simple but useful observation: If a set A is equivalent to a set B and if that set B is equivalent to some set C, then A and C are also equivalent. (The required one-to-one correspondence between A and C may be obtained by "patching together" the A-to-B correspondence and the B-to-C correspondence.) In particular, then, our work so far has shown that the first four infinite sets in the list of Section 5.2 are all the same size. However, because there are infinitely many rational numbers between any two consecutive integers, it appears that the next set on the list, Q, contains far too many elements to be equivalent to N. But this is not the case. (Here we have another example of the unreliability of intuition in dealing with infinity.) By ignoring the usual ordering of the numbers once again, Cantor devised a method for putting the positive integers and the

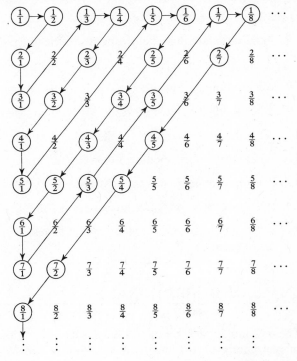

positive rationals in one-to-one correspondence. He arranged all the positive rationals in an infinite rectangular array, listing all fractions with numerator 1 in the first row, all fractions with numerator 2 in the second row, all fractions with numerator 3 in the third row, and so forth. He arranged the fractions in each row according to the natural order of their denominators. Then, starting in the upper left corner and omitting all fractions not in lowest terms, he proceeded to take the positive rationals in a zigzag diagonal order (as shown by the arrows in the accompanying figure), matching them with successive positive integers. Thus, the correspondence begins

$$I^+ = \{1, 2, 3, 4, 5, 6, 7, 8, 9, 10, 11, 12, 13, \ldots\}$$

$$Q^+ = \{1, \tfrac{1}{2}, 2, 3, \tfrac{1}{3}, \tfrac{1}{4}, \tfrac{2}{3}, \tfrac{3}{2}, 4, 5, \tfrac{1}{5}, \tfrac{1}{6}, \tfrac{2}{5}, \ldots\}$$

Hence, we have a one-to-one correspondence between the positive integers and the positive rationals. Moreover, it should be clear that this same process provides a one-to-one correspondence between the negative integers and all the negative rationals. The one-to-one correspondence between I and Q may then be completed by assigning the integer 0 to the rational number 0, and thus the sets I and Q are the same size!

EXERCISES 5.4

Questions 6–14 refer to the correspondence between I and Q described above.

6. What rational number corresponds to the integer 15?

7. What rational number corresponds to the integer 24?

8. What rational number corresponds to the integer 30?

9. What rational number corresponds to the integer -15?

10. What integer corresponds to the rational number $\tfrac{5}{2}$?

11. What integer corresponds to the rational number 8?

12. What integer corresponds to the rational number $\tfrac{4}{7}$?

13. What integer corresponds to the rational number $-\tfrac{5}{3}$?

14. Why were the fractions that are not in lowest terms skipped in the correspondence process?

5.5   RATIONALS AND IRRATIONALS

The next infinite sets to be considered, according to the list in Section 5.2, are sets of real numbers. Although real numbers are treated at least implicitly in high school mathematics courses, most elementary numerical examples in those courses usually involve rational numbers, and thus the distinction between rationals and reals often becomes blurred. Hence, before proceeding further with the discussion of infinite sizes, it is appropriate to pause and clarify this distinction.

The set R of real numbers may be described as the set of all infinite decimals. Now, every fraction can easily be converted to a decimal by dividing denominator into numerator. Hence, if we agree that any finite decimal can be made infinite simply by adding on an infinite "tail" of zeros, every rational number can be represented by an infinite decimal. For example, $\frac{2}{3} = 0.6666\ldots$, $\frac{3}{4} = 0.75000$ $\ldots$, and so forth. In other words, Q is a subset of R.

However, not every infinite decimal represents a rational number. To distinguish those decimals which represent fractions from those which do not, consider first an example of the division process mentioned above.

EXAMPLE 5.5.1

When we convert $\frac{47}{22}$ to a decimal, we divide 22 into 47 as follows:

$$
\begin{array}{r}
2.136 \\
22\overline{)47.0000\ldots} \\
44 \\
\hline
30 \\
22 \\
\hline
80 \\
66 \\
\hline
140 \\
132 \\
\hline
80 \\
\end{array}
$$

Since the remainder at the fourth step is the same as the one at the second, it is easy to see that the third and fourth digits of the quotient will repeat again and again from there on, without interruption. Thus, $\frac{47}{22} = 2.1363636\ldots.$

Must such a repetition of remainder happen with any division example, or is this just a carefully chosen special case? The answer is "yes" to both parts of that question. A repetition must always occur eventually, but the fact that it happens so conveniently soon in this example is a result of careful choice.

In general, when one integer is divided by another by the usual long-division method, the remainder at any step must be smaller than the divisor. Hence, if we are converting the fraction p/q to a decimal, the largest possible remainder is $q - 1$; that is, the number of different nonzero remainders in the division problem $q\overline{)p}$ can be no larger than $q - 1$. This means that after at most q steps in this division problem, the remainder *must* be a repetition of an earlier remainder, and hence the digits in the quotient that occur in the steps between the two equal remainders will repeat over and over from there on without interruption. Sometimes this repetition occurs early, as in Example 5.5.1; sometimes all possible nonzero remainders are used, as in Example 5.5.2.

EXAMPLE 5.5.2

$$
\begin{array}{r}
.571428 \\
7\overline{)4.00000000\ldots} \\
3\,5 \\
\overline{50} \\
49 \\
\overline{10} \\
7 \\
\overline{30} \\
28 \\
\overline{20} \\
14 \\
\overline{60} \\
56 \\
\overline{40}
\end{array}
$$

Thus, $\frac{4}{7} = 0.571428571428571428 \ldots$

The general argument given above establishes the following basic fact:

Every rational number can be expressed as an infinite decimal in which, from some specific digit on, a finite sequence of digits repeats again and again in the same order, without interruption.

For the sake of brevity, such decimals are simply called **repeating decimals**. Thus, we say every rational number can be expressed as a repeating decimal. We also use a notational convenience: To denote the repeating sequence of digits without having to write them several times over, we just put a bar over the sequence. For instance, the results of the two examples above are written $\frac{47}{22} = 2.1\overline{36}$ and $\frac{4}{7} = 0.\overline{571428}$.

Two very natural questions arise at this point:

1. Does *every* repeating decimal represent a rational number?

2. Are there really any (or many) such things as infinite nonrepeating decimals?

It is surprisingly easy to show that the answer to the first question is yes. To do this, we give a simple method for converting any repeating decimal to fractional form:

Suppose that d is a repeating decimal and that there are n digits in its repeating sequence. If we multiply d by 10^n and then subtract d from $10^n d$, we will get a *finite* decimal that equals $(10^n - 1)d$. Then d is just the fraction obtained by dividing that finite decimal by $10^n - 1$.

Perhaps the general argument just given reads more like a magic incantation than a logical procedure. It was stated in general terms simply to indicate that there is a procedure that works all the time. Consider the following examples, and then reread the general method; you should find it much clearer.

EXAMPLES 5.5

3. Suppose that $d = 0.\overline{35}$. Since there are two digits in the repeating sequence, we multiply d by 100 and subtract as follows:

$$
\begin{array}{rl}
100d = & 35.3535\ldots \\
- \quad d = & 0.3535\ldots \\
\hline
99d = & 35.0000\ldots \\
d = & 35/99
\end{array}
$$

4. Suppose that $d = 2.8\overline{473}$. Since there are three digits in the repeating sequence, we multiply d by 1000 and then subtract d:

$$
\begin{array}{rl}
1000d = & 2847.3473473\ldots \\
- \quad d = & 2.8473473\ldots \\
\hline
999d = & 2844.5000000\ldots \\
d = & 2844.5/999
\end{array}
$$

Strictly speaking, a fraction is the quotient of two integers, so to put the answer in proper form we should eliminate the decimal point from the numerator. This is easily accomplished by multiplying both numerator and denominator by 10, so that we have $d = \frac{28445}{9990}$.

Since every fraction can be expressed as a repeating decimal, and vice versa, we have characterized all those real numbers (in decimal form) which are rational; they are just the repeating decimals. The other real numbers are called **irrational** numbers; they are the infinite nonrepeating decimals. But are there any such numbers? Certainly—here is one example:

$$0.101001000100001\ldots$$

where the "$\ldots$" implies that the continuing pattern requires one more 0 after each 1 than was used after the previous 1. Clearly, there is no finite sequence of digits that repeats again and again in the same order without interruption, so this is not a repeating decimal. This example should suggest how you can construct many other irrational numbers, but the question "How many are there?" still persists. Are there more or fewer irrationals than rationals, or are those two sets the same size? A complete answer to that question will emerge from the material in the next several sections.

EXERCISES 5.5

1. Convert to a repeating decimal. (Give the entire repeating sequence of digits.)
   (a) $\frac{1}{9}$ (b) $\frac{2}{5}$
   (c) $\frac{13}{6}$ (d) $\frac{15}{52}$
   (e) $\frac{83}{44}$ (f) $\frac{93}{14}$

2. Convert to fractional form.
   (a) $1.\overline{4}$ (b) $0.0\overline{2}$
   (c) $0.\overline{9}$ (d) $0.\overline{02}$
   (e) $62.1\overline{75}$ (f) $3.2\overline{148}$

3. Write five examples of irrational numbers in infinite decimal form.

5.6 A DIFFERENT SIZE

The equivalence of such apparently different infinite sets as I and Q leads easily to the conjecture that any two infinite sets must be equivalent. Let us assume for the moment that this is true and see where that assumption leads us. In particular, let us suppose that N is equivalent to [0, 1], the set of all real numbers between 0 and 1. Recall that the real numbers between 0 and 1 can be described as all infinite decimals with 0 to the left of the decimal point. Then any one-to-one correspondence between N and [0, 1] will simply be a sequential listing of these infinite decimals, one for each natural number. For example, such a correspondence might begin as follows:

N		[0, 1]
1	———	. 3 0 1 2 5 9 4 . . .
2	———	. 1 6 6 5 2 1 8 . . .
3	———	. 4 1 1 2 1 0 7 . . .
4	———	. 2 0 5 0 9 6 3 . . .
5	———	. 0 0 0 1 1 1 1 . . .
6	———	. 8 5 7 3 0 9 9 . . .
⋮		⋮

Now, suppose that we have a proposed one-to-one correspondence between N and [0, 1], beginning, let us say, as in the listing above. That means there is an infinite decimal corresponding to each natural number. Let us use this correspondence to define a particular infinite decimal, according to the following general rule: For each natural number n, we look at the nth place of its corresponding decimal; if that digit is 1, we put a 2 in the nth place of our new decimal, and otherwise we put a 1 in that place. By looking at the list above, we see that our new decimal would begin 0.112121 . . . . Notice that this new number differs from the first number of [0, 1] in at least the first decimal place, differs from the second number in at least the second place, differs from the third number in at least the third place, and so forth. Because this new number differs from each number of [0, 1] in the list, it does not correspond to any natural number, yet it is clearly a real number between 0 and

1. In other words, we have found a number in [0, 1] which was left out at the alleged one-to-one correspondence between all of N and all of [0, 1]! Therefore, the proposed correspondence does *not* verify the equivalence of N and [0, 1]. But *any* one-to-one correspondence between N and [0, 1] must be of the same form as the one described above, and, despite the specific choice of the first few decimals to illustrate the correspondence, the method we gave for constructing the "new" decimal number is general enough to apply to any proposed listing. Thus, we have shown that *any* proposed one-to-one correspondence between N and [0, 1] must necessarily leave out at least one element of [0, 1], and hence N and [0, 1] cannot be equivalent! This means that the infinite sets N and [0, 1] are not the same size, so there are at least two different sizes of infinity!

([*Optional*] If you are bothered by the apparent use of a particular example to prove a general conclusion, the procedure described above can easily be written in a more formal and completely general way. We can represent any proposed one-to-one correspondence between N and [0, 1] by

N		[0, 1]					
1	——	. a_{11}	a_{12}	a_{13}	$\cdots$	a_{1i}	$\cdots$
2	——	. a_{21}	a_{22}	a_{23}	$\cdots$	a_{2i}	$\cdots$
3	——	. a_{31}	a_{32}	a_{33}	$\cdots$	a_{3i}	$\cdots$
$\vdots$							
n	——	. a_{n1}	a_{n2}	a_{n3}	$\cdots$	a_{ni}	$\cdots$
$\vdots$							

where the a_{ni}'s represent the single digits of the infinite decimals in the listing. The omitted infinite decimal . $b_1 \, b_2 \, b_3 \cdots b_n \cdots$ is then constructed according to the rule: For each n, $b_n = 2$ if $a_{nn} = 1$, and $b_n = 1$ if $a_{nn} \neq 1$. This description applies to any proposed correspondence, and hence justifies the general claim that *every* proposed correspondence is faulty. This procedure is called **Cantor's diagonalization process**.)

EXERCISES 5.6

1. Suppose that an alleged one-to-one correspondence between N and [0, 1] begins

N		[0, 1]
1	——	. 0 4 0 4 0 4 ...
2	——	. 1 1 1 1 1 1 ...
3	——	. 2 5 7 3 4 6 ...
4	——	. 1 0 1 0 0 1 ...
5	——	. 7 6 3 9 8 8 ...
6	——	. 1 2 1 1 2 1 ...
$\vdots$		$\vdots$

(a) Construct the first six decimal places of a number in [0, 1] that cannot be matched with any natural number in this correspondence.

(b) Construct three more such numbers, all different from each other, and all different from the number you found in part (a).

(c) Using Cantor's diagonalization process, how many different such numbers can be constructed?

2. Are I and [0, 1] equivalent? Are Q and [0, 1] equivalent? Why?

5.7   THE SIZE OF [0, 1]

Having discovered that there are different sizes of infinite sets by seeing that the unit interval [0, 1] on the number line is not equivalent to N, it seems natural to ask whether the length of an interval relates to its size as a set of points. For example, can we get an infinite set that is not equivalent to [0, 1] by choosing a longer interval, such as [0, 5] or [−36, 247], or a shorter interval, such as [$\frac{1}{2}$, $\frac{3}{4}$] or [0.00001, 0.00002]? It would seem reasonable to assume that an interval of length 5000 should contain too many points to be equivalent to an interval of length 1, but again intuition is unreliable. A surprisingly simple geometric argument proves that any interval [a, b] of any length can be put in one-to-one correspondence with [0, 1]: Place the interval [0, 1] perpendicular to [a, b], with point 0 lying on point a. Draw the line determined by the two points 1 and b, and choose a point P on that line which is somewhere on the extension of that line from b past the point 1, as shown in the accompanying diagram. Now, if we choose any point x in [0, 1], the

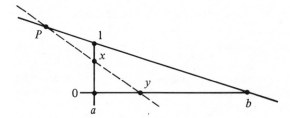

line through P and x necessarily intersects [a, b] at some point y; similarly, if we choose a point y in [a, b], the line through P and y necessarily intersects [0, 1] at some point x. This process gives us a matching between the points of [0, 1] and the points of [a, b]. Moreover, different points of [0, 1] determine different lines through P, so these lines must in turn intersect [a, b] at different points, and vice versa (since two points determine one and only one line); hence, the correspondence is one-to-one, so [0, 1] and [a, b] are equivalent sets. This means that *all* intervals, regardless of length, are equivalent to [0, 1], and hence to each other; thus, the length of an interval does not relate to its size as a set of points!

A similar argument shows that the interval [0, 1] is equivalent to the entire number line R. To do this, it is convenient to begin by observing that the size of

[0, 1] is unchanged if we delete its endpoints (see Exercise 5.7.2). We shall denote the unit interval without its endpoints by the symbol (0, 1) (sometimes called the "open" unit interval), and give an argument which shows that (0, 1) is equivalent to R. Place (0, 1) perpendicular to R so that the point $\frac{1}{2}$ in (0, 1) coincides with the point 0 in R, as shown in the accompanying diagram, and consider the lines ℓ_0 and

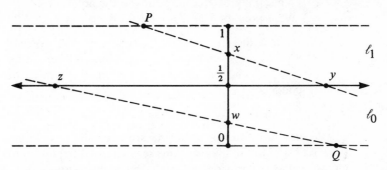

ℓ_1 parallel to R through the interval endpoints 0 and 1, respectively. Pick a point P on ℓ_1 somewhere to the left of 1. Using this point as we did in the interval argument before, we can match any point x between $\frac{1}{2}$ and 1 in (0, 1) with a unique positive real number y in R, and vice versa. Similarly, the choice of some point Q on ℓ_0 somewhere to the right of 0 enables us to match any point w between 0 and $\frac{1}{2}$ in (0, 1) with a unique negative real number z in R, and vice versa. The one-to-one correspondence between (0, 1) and R is completed by matching the interval point $\frac{1}{2}$ with the line point 0. Thus, we have shown that (0, 1) and R are equivalent; hence (by Exercise 5.7.2), as a set of points, R is the same size as [0, 1] or any other interval!

EXERCISES 5.7

1. Using a ruler, draw a 2-inch line segment T perpendicular to a 5-inch line segment F, with an endpoint of one coinciding with an endpoint of the other.

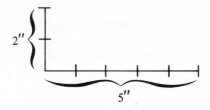

Then, mimicking the argument in this section, answer the following questions as precisely as you can by measuring.
(a) What point of F corresponds to the point $1''$ of T?
(b) What point of F corresponds to the point $\frac{1}{2}''$ of T?

(c) What point of F corresponds to the point $1\frac{1}{2}''$ of T?

(d) What point of T corresponds to the point $1''$ of F?

(e) What point of T corresponds to the point $2''$ of F?

(f) What point of T corresponds to the point $3''$ of F?

(g) What point of T corresponds to the point $4''$ of F?

(h) What point of T corresponds to the point $5''$ of F?

(i) What point of T corresponds to the point $0''$ of F?

2. Prove that the unit interval $[0, 1]$ is equivalent to the "open" unit interval $(0, 1) = \{x \mid x \in R \text{ and } 0 < x < 1\}$. (*Hint:* Observe that the set $\{a, b, 1, 2, 3, \ldots, n, \ldots\}$ and the set $\{1, 2, 3, \ldots, n, \ldots\}$ can be put in one-to-one correspondence.)

Because any two line segments are equivalent, we cannot get infinite sets of different sizes just by choosing line segments of different lengths. Perhaps, however, if we change dimension we might get a different-size set. Intuitively, it seems plausible that the set of all points in a plane, for example, is far too extensive to be equivalent to the set of points on a single line. But, here again, it can be shown that this is not the case. It is possible to set up a one-to-one correspondence between the set R of all real numbers and the set $R \times R$ of all ordered pairs of real numbers. (The details of the argument are beyond the scope of this course, so we simply state this fact without proof.) But R can represent all points on a line, as we have seen, and by using a rectangular coordinate system as in high school we can represent all points on a plane by $R \times R$. Hence, the set of all points in a plane, which is a two-dimensional geometric object, is the same size as the set of all points on a line, a one-dimensional geometric object!

Moreover, this correspondence can be extended to provide a matching between the set $R \times R$ of all points in a plane and the set $R \times R \times R$ of all points in three-dimensional space. Since we can describe the plane by all ordered pairs of real numbers and three-dimensional space by all ordered triples of real numbers, the matching between R and $R \times R$ can be used to establish a one-to-one correspondence between the first elements (or "first coordinates") of the ordered pairs and the first two coordinates of the ordered triples. Matching the last coordinate of each pair with the same last coordinate of each triple completes the correspondence. It is not hard to see that this matching is a one-to-one correspondence, so we have established that the set of all points in a single plane and the set of all points everywhere in three-dimensional space are the same size!

EXERCISES 5.7

3. Are the unit interval and the set of all points on a plane equivalent? Why?

4. Is the set of all points on a line equivalent to the set of all points in three-dimensional space? Why?

5. Is the unit interval equivalent to the set of all points in three-dimensional space? Why?

6. Is the set of rational numbers equivalent to the set of all points in three-dimensional space? Why?

5.8 CARDINAL NUMBERS

We have now examined all the infinite sets listed in Section 5.2 and have found that they can be separated into two distinct sizes, those which are equivalent to N and those which are equivalent to R. It is convenient to have names for these sizes, so we shall adopt for them the names that Cantor used.

DEFINITION 5.8.1

1. The collection of all sets that are equivalent to N is denoted by $\aleph_0$. ($\aleph_0$ is read "aleph null." $\aleph$ is the first letter of the Hebrew alphabet; the significance of the subscript 0 will appear later.)
2. The collection of all sets that are equivalent to R is denoted by **c**. (**c** is called the **cardinality of the continuum**.)

The sizes $\aleph_0$ and **c** are actually numbers, very much like the usual counting numbers 1, 2, 3, ... The number 3, for example, is simply an abbreviated way of referring to the common property possessed by all sets that can be put in one-to-one correspondence with $\{a, b, c\}$. But the notion of "common property" (or "three-ness," if you prefer) is not precise enough to work with, so instead we consider 3 to be the collection of all sets that are equivalent to $\{a, b, c\}$. In general:

DEFINITION 5.8.2

A **cardinal number** is the collection of all sets that are equivalent to a particular set. The particular set used to define a cardinal number is sometimes called a **reference set** for that number.*

Thus, as we have just seen, 3 is a cardinal number and $\{a, b, c\}$ is a reference set for it. Of course, we could use $\{*, \#, \$\}$ or any other three-element set as a reference set for 3. Similarly, the definitions given for $\aleph_0$ and **c** indicate that they are also cardinal numbers and supply the most commonly used reference sets for these numbers—N for $\aleph_0$ and R for **c**. Hence, we have two kinds of cardinal numbers: the **finite** cardinal numbers, whose reference sets are finite sets, and the **transfinite** cardinal numbers, whose reference sets are infinite. The whole numbers 0, 1, 2, 3, . . . , n, . . . are the finite cardinals and $\aleph_0$ and **c** are transfinite cardinals.

---

* You might find it interesting to compare this definition with the definition of "whole number" given in Section 4.3.

EXERCISES 5.8

1. Define the cardinal number 5, then give three different sets that could be used as reference sets for 5.

2. (a) Is -3 a cardinal number? Why?
   (b) Is $\frac{1}{2}$ a cardinal number? Why?
   (c) What is a reference set for the cardinal number 0? Is there more than one? Why?

3. (a) What is the cardinal number whose reference set is the set Q of all rational numbers? Why?
   (b) What is the cardinal number whose reference set is the unit interval $[0, 1]$? Why?

Several natural questions arise at this point. One that fairly cries out for an answer is: Are there any other transfinite cardinal numbers? Another is: Does it make sense to ask which of the numbers $\aleph_0$ and c is bigger? We might also wonder whether transfinite numbers can be added, subtracted, multiplied, and so forth, as we can do with the finite numbers. This last question will be postponed for awhile. Our immediate goal is to answer the first question, but to do that with some elegance we must begin with the second question.

Our intuitive understanding about sizes of collections of things is grounded almost exclusively on finite experience, and we have already had ample evidence that this intuition is often unreliable when applied to infinite situations. However, a careful analysis of the ways we handle finite cardinal numbers usually can provide the key to a similar treatment of the transfinite numbers. For instance, what does it mean to say that 5 is larger than 3? Since we are dealing with cardinal numbers, any description of their behavior should relate somehow to their definitions in terms of reference sets. Intuitively, 5 is larger than 3 because any attempted one-to-one correspondence between a reference set for 5 and a reference set for 3 results in having some elements of 5's reference set "left over." This observation can be generalized to provide a definition of "larger than" which applies to any cardinal numbers:

DEFINITION 5.8.3

Let $\mathscr{A}$ and $\mathscr{B}$ be cardinal numbers, with reference sets A and B, respectively. We say the number $\mathscr{A}$ **is larger than** the number $\mathscr{B}$ (or $\mathscr{B}$ **is smaller than** $\mathscr{A}$) if

1. There exists a one-to-one correspondence between all of set B and a proper subset of set A, and
2. There does not exist a one-to-one correspondence between B and all of A.

We denote "$\mathscr{A}$ is larger than $\mathscr{B}$" by writing $\mathscr{A} > \mathscr{B}$ (or $\mathscr{B} < \mathscr{A}$).

Condition (2) might seem redundant at first, and in fact for finite numbers it is, but the following example shows that it is necessary in order for the definition to apply to transfinite numbers. Consider two different reference sets for $\aleph_0$, namely N and N^{even}. Since there is obviously a one-to-one correspondence between N^{even} and a proper subset of N (namely, N^{even} itself), condition (1) of the definition is satisfied. Thus, if we were to use that condition alone as the definition of "larger than," we would find ourselves in the ridiculous position of having to say $\aleph_0$ is larger than itself. Condition (2) allows us to avoid this absurdity, however, because there is a one-to-one correspondence between N^{even} and all of N, so the proper definition of "larger than" does not apply to $\aleph_0$ and itself.

Now that we have a workable definition for "larger than," it is easy to prove that $\mathbf{c}$ is larger than $\aleph_0$ by using the results of our previous work. Choose N as the reference set for $\aleph_0$ and $[0, 1]$ as the reference set for $\mathbf{c}$. We have already proved that there is no one-to-one correspondence between N and all of $[0, 1]$, so condition (2) of the definition is satisfied. To verify condition (1), observe that $\{1, \frac{1}{2}, \frac{1}{3}, \frac{1}{4}, \ldots, 1/n, \ldots\}$ is a proper subset of $[0, 1]$ which can be put in one-to-one correspondence with N by matching each fraction of the form $1/n$ with its denominator, the natural number n. Thus, we have shown that $\aleph_0 < \mathbf{c}$.

EXERCISES 5.8

4. Prove that 7 is larger than 4.

5. Prove that $\aleph_0$ is larger than 3.

6. Prove that $\aleph_0$ is larger than any finite cardinal number.

7. Prove that $\mathbf{c}$ is larger than 5.

8. Prove that $\mathbf{c}$ is larger than any finite cardinal number.

9. Prove: For any three cardinal numbers $\mathscr{A}$, $\mathscr{B}$, and $\mathscr{C}$, if $\mathscr{A} < \mathscr{B}$ and $\mathscr{B} < \mathscr{C}$, then $\mathscr{A} < \mathscr{C}$. (*Hint:* You may assume without proof the following fact: If S and T are sets such that S is equivalent to a proper subset of T and T is equivalent to a proper subset of S, then S and T are themselves equivalent. This seemingly obvious fact of set theory is surprisingly difficult to prove; it is called the Cantor-Schröder-Bernstein Theorem.)

5.9 CANTOR'S THEOREM

Up to now, our investigation of particular infinite sets has led us to classify all the infinite sets we have seen into two sizes: the "discretely" infinite size $\aleph_0$ for sets like the natural numbers, and the "continuously" infinite size $\mathbf{c}$, for sets like the real number line. It is tempting, especially after discovering that even a change of dimension does not produce a set of larger size than **c**, to guess that there are no other infinite sizes. But once again, our expectations are deceptive. In the following theorem, Cantor shows that no matter what size set we have, there is a way to find a set of a larger size, thus proving that there are *infinitely* many different sizes of infinity!

**CANTOR'S
THEOREM**

Let $\mathscr{A}$ be any cardinal number, with reference set A, and let S be the set of all subsets of A. If $\mathscr{S}$ is the cardinal number that represents the size of S, then $\mathscr{S}$ is larger than $\mathscr{A}$. (Informally, Cantor's Theorem simply states that the size of any set is smaller than the size of the set of all its subsets.)

PROOF   As stated in the theorem, A is a reference set for the cardinal number $\mathscr{A}$ and S, the set of all subsets of A, is a reference set for the cardinal number $\mathscr{S}$. By the definition of "larger than," to prove that $\mathscr{S}$ is larger than $\mathscr{A}$ we must show that

1. A can be put in one-to-one correspondence with a proper subset of S, and
2. A cannot be put in one-to-one correspondence with all of S.

1. Among the subsets of A are all the single-element subsets of A; that is, for each element x in A there is the subset $\{x\}$ in S. The matching of each x with $\{x\}$ is obviously a one-to-one correspondence between all elements of A and some but not all of the elements of S (in particular, $\varnothing$ is an element of S not used in this correspondence), so condition (1) is satisfied.
2. To prove that there cannot exist a one-to-one correspondence between A and all of S, we use an indirect argument; that is, we assume such a correspondence exists and derive a contradiction from that assumption. Since the contradiction is obtained by logically valid steps, the only possible explanation for the contradiction must be that our assumption is false, and hence no such correspondence can exist. Specifically, assume that each subset of A corresponds to a unique element of A, and vice versa. Since each element of A is matched with a subset of A by this presumed correspondence, it makes sense to ask whether or not an element is contained in the set it is matched with. We select each element of A that is *not* contained in the set it is matched with, and denote the set of all such elements by W. Clearly, W is a subset of A, and hence it is an element of S. Since A and S are in one-to-one correspondence, there must be some element z in A which is matched with the set W. Now, either z is contained in W or it is not; let us examine both of these alternatives. If z is contained in W, then z is contained in the set it is matched with, and hence the definition of set W implies that z *cannot* be contained in it, which is an outright contradiction. On the other hand, if we suppose that z is not contained in W, then z is not contained in the set it is matched with, so the definition of W implies that z *must* be in W; again we have a contradiction. Since there are no other alternatives, we are forced to conclude that our argument contains a flaw somewhere. But each step follows logically from the one before it, except for our initial supposition that a one-to-one correspondence between A and S exists; hence, this supposition must be false, so condition (2) of the definition of "larger than" is

satisfied. Thus, we have proved that $\mathscr{S}$, the size of the set of all subsets of A, is larger than $\mathscr{A}$, the size of A.

EXERCISES 5.9

1. Write out the set of all subsets of $\{a, b, c\}$.

2. Consider the set $A = \{p, q, r, s, t\}$, and let S be the set of all subsets of A.
   (a) Give a three-element subset of A.
   (b) Give two elements of S.
   (c) Give a three-element subset of S.
   (d) What cardinal number represents the size of S? Why?

3. (a) Let $A = \{u, v, w, x, y, z\}$. Set up a one-to-one correspondence between A and six of its subsets. (You may choose any six subsets.)
   (b) In the proof of Cantor's Theorem we constructed a set W of all elements of A that are not in their corresponding subsets. Using the set A and your correspondence from part (a) of this exercise, list the elements of such a set W.

4. Consider a proposed correspondence between the set of natural numbers and its subsets which begins

$$
\begin{array}{rcl}
1 & \text{———} & \{3, 7, 11\} \\
2 & \text{———} & \{2, 4\} \\
3 & \text{———} & \{1, 2, 3, 4, 5\} \\
4 & \text{———} & \{5, 10, 15, \ldots, 5n, \ldots\} \\
5 & \text{———} & \{1\} \\
6 & \text{———} & \{2, 4, 6, 8, \ldots, 2n, \ldots\} \\
\vdots & & \vdots
\end{array}
$$

Describe how to construct a set that cannot appear anywhere in this correspondence. Which of the numbers 1, 2, 3, 4, 5, and 6 will be in this set?

5. Let $A = \{a, b\}$. Write out the set S of all subsets of A, then write out the set of all subsets of S. What size is each of these sets?

5.10 THE CONTINUUM HYPOTHESIS

Successive applications of Cantor's Theorem, first to a set A, then to the set S of all subsets of A, then to the set of all subsets of S, and so forth, provides a way of getting larger and larger sizes of infinite sets, thus guaranteeing an infinite string of transfinite numbers. However, the theorem does *not* guarantee that the numbers obtained in this way are successive; that is, if a set A represents a particular cardinal number, Cantor's Theorem does not tell us whether the set of all subsets of A represents the *next* cardinal number, or even if the idea of "next" makes sense in this context. Exercise 5.9.5 illustrates that we do not get successive numbers in the finite case—a two-element set has four subsets, a four-element set has sixteen subsets, and so on—so there is no reason to expect that the infinite situation must be different, although our previous experience with basing expectations for infinite sets on finite examples should warn us against assuming the situations are alike. Cantor established several useful facts about transfinite numbers which we state without proof:

1. $\aleph_0$ is the smallest transfinite cardinal number.
2. All the transfinite cardinal numbers can be arranged in a sequence of increasing sizes, starting with $\aleph_0$; that is, we may list the transfinite cardinals as $\aleph_0 < \aleph_1 < \aleph_2 < \cdots < \aleph_n < \cdots$.
3. c is the size of the set of all subsets of the natural numbers.

One of Cantor's most intriguing statements about transfinite numbers is an assertion that he could *not* prove. Having observed that a set of the smallest infinite size, $\aleph_0$, has a set of all subsets whose size is c, Cantor conjectured that c is the second smallest infinite size; that is, Cantor guessed that $c = \aleph_1$. However, he was unable to prove his conjecture, and by 1900 this question had become one of the most famous unsolved mathematical problems.* It is known as the "Continuum Hypothesis," and it asserts that c is the second smallest size of infinity. More formally:

CONTINUUM   If a set has size $\aleph_0$, then the set of all its subsets has size $\aleph_1$. In general, if a set has
HYPOTHESIS   size $\aleph_n$ for some n, then the set of all its subsets has size $\aleph_{n+1}$.

During the first half of this century there were many fruitless efforts to prove the Continuum Hypothesis. The elusiveness of this problem was rendered even more frustrating by the fact that Cantor's set theory had rapidly become respectable in most areas of mathematics and by the 1930s had been put on a solid axiomatic foundation. Many people felt that such a rigorously logical approach to set theory would quickly yield a solution for the Continuum Hypothesis, but the problem was more difficult than they anticipated and the eventual solution came in a surprising form. In 1940, Kurt Gödel, an Austrian logician now at Princeton, proved that the Continuum Hypothesis is consistent with the axioms for set theory; that is, he proved that the *assumption* that the Continuum Hypothesis is *true* will not lead to any contradictions within set theory. Of course, this did not *prove* the Continuum Hypothesis; rather, it showed that the Hypothesis could not be proven false by an argument based on the given axioms for set theory. The problem remained unresolved until 1963, when Paul Cohen of Stanford University proved that the *assumption* that the Continuum Hypothesis is *false* will not lead to any contradictions within set theory! This means that the Continuum Hypothesis is indepen-

---

* This was one of 23 outstanding problems for the twentieth century posed in 1900 by the great French mathematician David Hilbert at the Second International Congress of Mathematicians in Paris.

dent of the axioms for set theory; in other words, we may treat it as a separate axiom. If we assume it to be true, we get one kind of set theory; if we assume it to be false, we get another kind of set theory, different from the first but equally valid from a logical viewpoint. This discovery provides a striking illustration to support the modern view of mathematics as a study that is independent of the physical world. If there were a single "true" theory of sets waiting for discovery by mathematicians, we could not have conflicting but equally consistent theories of sets. It must be, then, that mathematics is invented by human beings and then applied to the world around them, imposing on their universe a convenient order which is useful for explaining observed phenomena but which is not determined by those phenomena. In the words of Cantor himself,* "mathematics is entirely free in its development and its concepts are restricted only by the necessity of being noncontradictory and coordinated to concepts previously introduced by precise definitions. . . . The essence of mathematics lies in its freedom."

EXERCISES 5.10

For the purpose of these exercises, assume that the general form of the Continuum Hypothesis is true.

1. (a) Give an example of a set A of size $\aleph_1$.
    (b) List three specific elements of A.
    (c) Find a subset of A that has size $\aleph_0$.

2. (a) Give an example of a set B of size $\aleph_2$.
    (b) List three specific elements of B.
    (c) Find a subset of B that has size $\aleph_0$.
    (d) Find a subset of B that has size $\aleph_1$.

3. (a) Give an example of a set C of size $\aleph_3$.
    (b) List three specific elements of C.

5.11 TRANSFINITE ARITHMETIC—ADDITION (optional)

Since the transfinite cardinal numbers are in fact numbers, it seems reasonable to expect that they can be added, multiplied, and so on, in much the same way as we manipulate finite cardinal numbers. As we have already seen, the behavior of cardinal numbers is governed by the behavior of their reference sets, whether the numbers be finite or infinite. Thus, to extend the operations of arithmetic from finite to transfinite numbers, we must carefully describe the behavior of the finite

---

* Georg Cantor, 1883, as quoted in Morris Kline, *Mathematical Thought from Ancient to Modern Times* (New York: Oxford University Press, 1972), p. 1031.

numbers in terms of their reference sets, *define* the arithmetic operations in those terms, then apply the definitions to the infinite case exactly and rigorously, without regard to any intuitive misgivings about the results obtained. We have already done this once, when we defined "larger than" as a concept applicable to all cardinal numbers. Now let us try to define a similarly general notion of addition.

The process of adding 2 and 3 can be described in terms of reference sets as follows: Choose $\{a, b\}$ as a reference set for 2, and $\{x, y, z\}$ as a reference set for 3. Form the union cf these sets, $\{a, b\} \cup \{x, y, z\} = \{a, b, x, y, z\}$. The resulting set belongs to the cardinal number 5, which is the sum $2 + 3$.

Using this example as a guide, we can now give a general prescription for finding the sum of two cardinal numbers.

DEFINITION
5.11.1

Let $\mathscr{A}$ and $\mathscr{B}$ be two cardinal numbers. Choose reference sets A for $\mathscr{A}$ and B for $\mathscr{B}$ such that A and B have no elements in common. The **sum** of $\mathscr{A}$ and $\mathscr{B}$ (written $\mathscr{A} + \mathscr{B}$) is the cardinal number that contains the set $A \cup B$.

This definition applies equally well to finite and transfinite cardinal numbers. However, some of the results involving transfinite numbers are surprisingly unlike the finite case. Let us look at a few examples.

Consider first the addition problem $3 + \aleph_0$. Applying the definition of sum, choose $\{a, b, c\}$ as a reference set for 3 and choose N as a reference set for $\aleph_0$. The union of these two sets is $\{a, b, c, 1, 2, 3, \ldots, n, \ldots\}$, which can be put in one-to-one correspondence with N itself, as follows:

$$\{a, b, c\} \cup N = \{a, b, c, 1, 2, \ldots, \quad n, \quad \ldots\}$$
$$N = \{1, 2, 3, 4, 5, \ldots, n + 3, \ldots\}$$

Hence, $\{a, b, c\} \cup N$ has size $\aleph_0$, so $3 + \aleph_0 = \aleph_0$.

Now let us look at $\aleph_0 + \aleph_0$. If we choose N^{even} and N^{odd} as the disjoint reference sets for the two summands $\aleph_0$, it is easy to see that $N^{\text{even}} \cup N^{\text{odd}} = N$ has size $\aleph_0$, so the definition of sum dictates that $\aleph_0 + \aleph_0 = \aleph_0$.

The problem $\aleph_0 + \mathbf{c}$ is a little harder, but a judicious choice of reference sets and a little ingenuity can dispose of it. Let N^{even} represent $\aleph_0$ and let $[0, 1]$ represent $\mathbf{c}$. Observe that $[0, 1]$ can be rewritten as $\{1, \frac{1}{2}, \frac{1}{3}, \ldots, 1/k, \ldots\} \cup S$, where S is the set of all real numbers in $[0, 1]$ that are not in $\{1, \frac{1}{2}, \frac{1}{3}, \ldots, 1/k, \ldots\}$. Now we can define a one-to-one correspondence between $N^{\text{even}} \cup [0, 1]$ and $[0, 1]$ itself:

$$N^{\text{even}} \cup [0, 1] = \{2, 4, 6, \ldots, 2n, \ldots\} \cup \{1, \frac{1}{2}, \frac{1}{3}, \ldots, \frac{1}{k}, \ldots\} \cup S$$
$$= \{1, 2, \frac{1}{2}, 4, \frac{1}{3}, 6, \frac{1}{4}, 8, \ldots\} \cup S$$

$$[0, 1] = \{1, \frac{1}{2}, \frac{1}{3}, \frac{1}{4}, \frac{1}{5}, \frac{1}{6}, \frac{1}{7}, \frac{1}{8}, \ldots\} \cup S$$

It is easily seen from this listing that the correspondence matches any even natural number $2n$ with the fraction $1/2n$ in $[0, 1]$, and it matches any fraction of the form $1/k$ in $N^{\text{even}} \cup [0, 1]$ with the fraction $1/(2k - 1)$ in $[0, 1]$; each element of S is matched with itself to complete the correspondence. Thus, $\aleph_0 + \mathbf{c} = \mathbf{c}$.

In general, it can be proved that the sum of two unequal cardinal numbers is just the larger one, provided that at least one of them is a transfinite number, and the sum of two equal transfinite numbers is that number itself. This peculiar behavior of transfinite addition suggests some other differences between it and finite addition. For instance, we know that given an equation of the form $n + 5 = m + 5$, where n and m are finite numbers, we can "cancel" the 5's and conclude that $n = m$. However, $3 + \aleph_0 = \aleph_0 + \aleph_0$, but clearly $3 \neq \aleph_0$! A few other properties of transfinite addition are left for exploration in Exercise 10 below.

EXERCISES 5.11

1. In the definition of sum of two cardinal numbers, why must we stipulate that A and B have no elements in common? Give an example to illustrate your answer.

2. Prove that $3 + 4 = 7$.

3. Prove that $\aleph_0 + 5 = \aleph_0$.

4. Prove: If k is any finite cardinal number, then $k + \aleph_0 = \aleph_0$.

5. Prove that $10 + \mathbf{c} = \mathbf{c}$.

6. Prove: If k is any finite cardinal number, then $k + \mathbf{c} = \mathbf{c}$.

7. Prove that $\mathbf{c} + \mathbf{c} = \mathbf{c}$.

8. Without giving any proofs, compute the following.
   (a) $7 + \aleph_0$ (b) $25 + \mathbf{c}$
   (c) $\aleph_0 + \aleph_2$ (d) $\aleph_0 + 19 + \mathbf{c}$
   (e) $\aleph_3 + \aleph_5$ (f) $\aleph_0 + \aleph_0 + 283$
   (g) $195 + \mathbf{c} + \aleph_0$ (h) $\mathbf{c} + \mathbf{c} + \mathbf{c} + \mathbf{c}$
   (i) $\aleph_0 + \aleph_1 + \aleph_2 + \aleph_3$ (j) $\aleph_4 + \aleph_2 + \aleph_4$

9. Find a number x such that $\mathbf{c} + x > \mathbf{c}$. Justify your answer. (Be careful.)

10. In dealing with finite numbers, we can define subtraction by saying:

$$a - b = c \quad \text{means that} \quad b + c = a$$

Is this definition reasonable for transfinite numbers as well? Why?

5.12   TRANSFINITE ARITHMETIC—MULTIPLICATION (optional)

As in the case of addition, an appropriate general definition for multiplication of cardinal numbers can be obtained from a careful examination of a finite example.

The process of multiplying 4 and 3 can be described in terms of reference sets as follows: Choose reference sets for 4 and 3, say $\{a, b, c, d\}$ and $\{x, y, z\}$, respectively. Form the Cartesian product of these two sets. (Recall that the Cartesian product $A \times B$ of two sets A and B is the set of all ordered pairs whose first elements are in A and whose second elements are in B.) $\{a, b, c, d\} \times \{x, y, z\} = \{(a, x), (a, y), (a, z), (b, x), (b, y), (b, z), (c, x), (c, y), (c, z), (d, x), (d, y), (d, z)\}$, a set belonging to the cardinal number 12, and 12 is the product $4 \cdot 3$. Hence, we are led to the following general definition.

DEFINITION
5.12.1

Let $\mathscr{A}$ and $\mathscr{B}$ be two cardinal numbers, with reference sets A and B, respectively. The **product** of $\mathscr{A}$ and $\mathscr{B}$ (written $\mathscr{A} \cdot \mathscr{B}$) is the cardinal number that contains the set $A \times B$.

As a first example involving a transfinite number, consider $2 \cdot \aleph_0$. By choosing the reference sets carefully, we can solve this problem with a small touch of elegance. Choose $\{0, 1\}$ as the reference set for 2, and N^{odd} as the reference set for $\aleph_0$. Then $\{0, 1\} \times N^{\text{odd}}$ is the set $\{(0, 1), (1, 1), (0, 3), (1, 3), (0, 5), (1, 5), \ldots, (0, 2n - 1), (1, 2n - 1), \ldots\}$, and the sum of the two terms in each ordered pair yields a natural one-to-one correspondence between this set and N:

$$\{0, 1\} \times N^{\text{odd}} = \{(0, 1), (1, 1), (0, 3), (1, 3), (0, 5), (1, 5), \ldots, (0, 2n - 1), (1, 2n - 1), \ldots\}$$

$$N = \{\ 1, \quad 2, \quad 3, \quad 4, \quad 5, \quad 6, \quad \ldots, \quad 2n - 1, \quad 2n, \ldots\}$$

Thus, $\{0, 1\} \times N^{\text{odd}}$ has size $\aleph_0$, so $2 \cdot \aleph_0 = \aleph_0$.

To compute $\aleph_0 \cdot \aleph_0$, we can use N as the reference set for both copies of $\aleph_0$ and then determine the size of $N \times N$ by using a variation of an earlier argument. $N \times N$, the set of all ordered pairs of natural numbers, can be arranged in an array

$$(1, 1), \ (1, 2), \ (1, 3), \ \ldots, \ (1, n), \ \ldots$$
$$(2, 1), \ (2, 2), \ (2, 3), \ \ldots, \ (2, n), \ \ldots$$
$$(3, 1), \ (3, 2), \ (3, 3), \ \ldots, \ (3, n), \ \ldots$$
$$\vdots \qquad \vdots \qquad \vdots \qquad \qquad \vdots$$
$$(m, 1), \ (m, 2), \ (m, 3), \ \ldots, \ (m, n), \ \ldots$$
$$\vdots \qquad \vdots \qquad \vdots \qquad \qquad \vdots$$

where the first number of a pair denotes its row position and the second number denotes its column. By again applying the zigzag matching pattern that we used before to match the integers with the rationals (see the illustration on page 169), this time without skipping any pairs, we get a one-to-one correspondence between N and $N \times N$. Thus, $\aleph_0 \cdot \aleph_0 = \aleph_0$.

The computation of $\aleph_0 \cdot \mathbf{c}$ provides another opportunity for elegance. It is easy to show that the interval of all real numbers greater than or equal to 0 and strictly less than 1, which we shall denote by [0, 1), is equivalent to [0, 1]; hence, [0, 1) can be used as the reference set for R. Choose I as the reference set for $\aleph_0$. The set $I \times [0, 1)$ is the set of all ordered pairs whose first elements are integers and whose second elements are nonnegative decimals strictly less than 1. A moment's reflection should reveal that by adding the two elements in each pair we obtain all the real numbers, and hence we have an obvious one-to-one correspondence between $I \times [0, 1)$ and R! (See Excercise 5.12.6 for a related example.) Thus, by definition of product, $\aleph_0 \cdot \mathbf{c} = \mathbf{c}$.

We have observed in an earlier section that $R \times R$ is equivalent to R, and hence $\mathbf{c} \cdot \mathbf{c} = \mathbf{c}$, by the definition of product. In general, it can be shown that the product of a transfinite cardinal number with itself is just that number, and the product of two unequal transfinite cardinals is the larger one. Thus, we have the unusual situation that addition and multiplication of cardinal numbers yield exactly the same results when transfinite numbers are used, and neither process provides a way of finding numbers larger than the ones used!

EXERCISES 5.12

1. In the definition of the product of two cardinal numbers, must the two reference sets chosen be disjoint? Why?

2. Prove that $3 \cdot 5 = 15$.

3. Prove that $3 \cdot \aleph_0 = \aleph_0$.

4. Prove: If k is any finite cardinal number, then $k \cdot \aleph_0 = \aleph_0$.

5. Prove that [0, 1) is equivalent to [0, 1]. (*Hint:* See Exercise 5.7.2.)

6. (a) Using the method of adding pairs discussed in this section, set up a natural one-to-one correspondence between the set $\{2, 3, 4, 5\} \times [0, 1)$ and some interval on the real line.
   (b) Prove that $4 \cdot \mathbf{c} = \mathbf{c}$.
   (c) Prove: If k is any finite cardinal number, then $k \cdot \mathbf{c} = \mathbf{c}$.

7. Does a cancellation hold for transfinite multiplication? That is, if $\mathscr{A}$, $\mathscr{B}$, and $\mathscr{C}$ are transfinite cardinal numbers such that $\mathscr{A} \cdot \mathscr{C} = \mathscr{B} \cdot \mathscr{C}$, must $\mathscr{A} = \mathscr{B}$? Give examples to support your answers.

8. Without giving any proofs, compute the following.
   (a) $27 \cdot \aleph_0$ (b) $\mathbf{c} \cdot 19$
   (c) $\aleph_1 \cdot \aleph_3$ (d) $\aleph_0 \cdot \aleph_0 \cdot \aleph_0$
   (e) $5 \cdot \aleph_0 \cdot \mathbf{c}$ (f) $\aleph_0 \cdot \aleph_1 \cdot \aleph_2$
   (g) $5 \cdot \mathbf{c} \cdot \mathbf{c} \cdot \mathbf{c}$ (h) $\aleph_0 \cdot \aleph_1 \cdot \mathbf{c}$

5.13   THE FOUNDATIONS OF MATHEMATICS

Although Cantor's theory of sets was well received in many parts of the mathematical community, acceptance was by no means universal. Cantor's set-theoretic treatment of infinity generated heated opposition from some of his foremost contemporaries, notably Leopold Kronecker. Kronecker's approach to mathematics was based on the premise that a mathematical entity does not exist unless it is actually constructible in a finite number of steps. From this point of view, infinite sets do not exist because it is clearly impossible to construct an infinite number of elements in a finite number of steps. The collection of natural numbers is "infinite" only in the sense that the finite collection of natural numbers that we have constructed to date may be extended as far as we please, but "the set of all natural numbers" is not a legitimate mathematical concept. To Kronecker and those who shared his views, Cantor's work was a dangerous mixture of heresy and alchemy that introduced lethal dosages of fantasy into the bloodstream of mathematics.

Kronecker's apprehensions for the safety of mathematical consistency were at least partially justified by the appearance of several paradoxes in set theory. Among the most renowned of these is the self-contradictory notion of the "set of all sets." Cantor's concept of set was extremely general: "By a set we are to understand any collection into a whole of definite and separate objects of our intuition or our thought."* Since any set is itself a "definite and separate object of our thought," it would seem sensible to consider the set $\mathscr{S}$ of all sets. $\mathscr{S}$ would have to contain more elements than any other set by its very nature; that is, the cardinality of $\mathscr{S}$ would have to be greater than the cardinality of any other set. But Cantor's Theorem states that there is no greatest cardinal number; hence there must be a set whose cardinality is greater than that of $\mathscr{S}$.

Perhaps the most famous set-theoretic paradox of all was formulated by Bertrand Russell in 1902. It depends solely on the notion of set, and therefore strikes at the very heart of set theory. The paradox begins by observing that all sets may be classified according to whether or not they are elements of themselves. For example, the set of abstract ideas is an abstract idea and hence is an element of itself, but the set of all elephants is hardly an elephant. Let us call any set that is not an element of itself *normal*, and consider the set $\mathscr{N}$ of all normal sets. In symbols, $\mathscr{N} = \{S \mid S \notin S\}$. Question: Is $\mathscr{N}$ normal? If we answer *yes*, then $\mathscr{N}$ is in the set of all normal sets; that is, $\mathscr{N}$ is an element of itself, implying that $\mathscr{N}$ is not normal. If we answer *no*, then $\mathscr{N}$ is an element of itself, and hence must be normal because $\mathscr{N}$ is the set of all normal sets. Thus, either choice leads to a contradiction.†

Dilemmas such as this resulting from an unrestricted use of the seemingly harmless concept of set forced mathematicians of the late nineteenth and early

---

* *Contributions to the Founding of the Theory of Transfinite Numbers*, trans. P. E. B. Jourdain (La Salle, Ill.: Open Court Publishing Company, 1915).

† There are many popularized versions of this paradox. Russell himself gave one in 1919: A (beardless) barber in a certain village claims that he shaves all those villagers who do not shave themselves, and that he shaves no one else. If his claim is true, who shaves the barber?

twentieth centuries to undertake a thorough reappraisal of the foundations of mathematics in an attempt to free it from the dangers of self-contradiction. This in turn led to the formulation of several different philosophies of mathematics. A philosophy of mathematics might be described as a viewpoint from which the various bits and pieces of mathematics can be organized and unified by some basic principles. There have been many philosophies of mathematics throughout history. As the body of mathematical knowledge grew and was changed by the results of new investigations, its philosophies underwent similar mutations. The advent of set theory in all its unifying simplicity and then the discovery of serious flaws in its fundamental structure, all within less than half a century, brought about a violent upheaval in mathematical philosophy. Its contemporary development may be separated into three branches, each attempting in its own way to safeguard mathematics from internal contradictions. In the brief space remaining we shall attempt to summarize a few of the basic tenets that characterize each of these schools of thought.

Logicism regards mathematics as a branch of logic, claiming that all mathematical principles are completely reducible to logical principles. Several attempts have been made to reduce all mathematics to a symbolic logical system, culminating in Bertrand Russell and Alfred North Whitehead's *Principia Mathematica* (1910). The *Principia* bases all of mathematics on a logical system derivable from five primitive logical propositions whose truth is founded on basic intuition. Refinements of Russell and Whitehead's work made by a number of people during the first half of this century have succeeded in ironing out many of the minor difficulties in the *Principia*. Nevertheless, there are some fundamental objections to the logistic standpoint as a whole. It is claimed that some primitive mathematical ideas must be used to develop the system in an orderly fashion, so the system is not completely self-contained. Some have also objected that logicism implies that all mathematical ideas are contained in five primitive statements, and the rest of mathematics is merely an exercise in redundancy, a formalized restatement of these five principles involving no additional information whatever.

Intuitionism proceeds from the premise that mathematics must be based solely on the intuitively given notion of a succession of things, exemplified by the sequence of natural numbers. The intuitionists claim that mathematics is dependent neither on language nor on classical logic. The symbols in mathematics are used for communication only, but they are incidental since mathematics is essentially an individual matter and need not be communicated to exist. Moreover, the rules of mathematical reasoning which are arrived at intuitively form a "logic" that differs from classical logic and is applicable only to mathematics. Like Kronecker, the intuitionists reject the idea that infinitely many elements can be treated as a single thing (set). They define a set as a law that generates a succession of elements. Thus, the law $1/n$ for any $n \in N$ is a set whose elements are $1, \frac{1}{2}, \frac{1}{3}, \ldots$, but the collection of its elements cannot be completed. The intuitionist logic does not accept the Law of the Excluded Middle, implying that proof by contradiction is an invalid procedure. The existence of a mathematical object can be proven only by establishing a procedure for its construction in a finite number of steps. These restrictions successfully eliminate the contradictions stemming from Cantor's set theory, but they also eliminate sizable portions of generally accepted classical mathematics. It remains to be seen whether the intuitionists can refine their approach sufficiently to obtain all of classical mathematics while retaining their advantage of working in a contradiction-free system.

Formalism claims that mathematics is concerned solely with the development of systems of symbols. A formal mathematical system is a collection of abstract statements expressing relationships among undefined terms which are subject to a variety of interpretations. Because these systems have no necessary relation to reality, the formalist must guard against potential contradictions by proving that his systems are internally consistent. Attempts were made to find one provably consistent axiom system for all of mathematics, but in 1931 Kurt Gödel proved this goal to be unattainable. Formalism is thus restricted to a piecemeal verification of the consistency of separate parts of mathematics and is often forced to be content with the tentative assurance that one mathematical system is as likely to be consistent as another.

It would be inaccurate and misleading to suggest that all or most mathematicians adhere rigidly to one of the three mathematical philosophies described above. The fact is that most mathematicians work in their respective fields "doing mathematics" and concern themselves very little with questions of philosophy. Each one has formulated an opinion about what constitutes mathematics that is sufficient to guide him in his research, and these opinions are often mixtures of the viewpoints expressed above. Even those who do concern themselves directly with mathematical philosophy seldom agree in every detail. The best we can say for the classifications described above is that they represent the general trends of current thought regarding the foundations of mathematics, especially the mathematical treatment of infinity.

For Further Reading

Bell, E. T. *Men of Mathematics*. New York: Simon and Schuster, Inc., 1937, Chap. 29.

Cantor, Georg. *Contributions to the Founding of the Theory of Transfinite Numbers*, trans./ed. P. E. B. Jourdain. La Salle, Ill.: Open Court Publishing Company, 1952.

Eves, Howard, and Carroll V. Newsom. *An Introduction to the Foundations and Fundamental Concepts of Mathematics*. New York: Holt, Rinehart and Winston, 1965, Chaps. 8 and 9.

Kline, Morris. *Mathematical Thought from Ancient to Modern Times*. New York: Oxford University Press, 1972, Chap. 41, 43, 51.

———. *Mathematics in Western Culture*. New York: Oxford University Press, 1964.

Newman, James R., ed. *The World of Mathematics*. New York: Simon and Schuster, Inc., 1956, Vol. 3, Part X.

Reid, Constance. *From Zero to Infinity*, 3rd ed. Thomas Y. Crowell Company, New York: 1964, Chap. . . . .

6

ELEMENTARY
PROBABILITY *

6.1 INTRODUCTION

The theory of probability is a mathematical discipline dealing with random experiments. Although its original purpose was to describe the exceedingly narrow domain of experience connected with the games of chance, it has found an increasing number of applications not only in business, biology, medicine, the insurance industry, and the like, but also in many scientific fields of human endeavor involving matters as remote as the study of various characteristics in a secret military strategy, observing the life span of radioactive atoms, transmission of signals in presence of noise, the density of telephone traffic in an election year, the quality of White House tapes recorded in the Watergate era, and the motion of particles immersed in a liquid or a gas. The simplest random experiences in everyday life are provided by tossing a coin, throwing a die, arranging a deck of cards, distributing balls in certain cells, or drawing a ball from an urn containing balls of several colors.

All these experiments with their unpredictable results are rather vague descriptions and we must look for a common characteristic possessed by the various "experiments" that we have described. Each experiment or **random phenomenon** is empirical in nature in the sense that its observation under a given set of conditions does not always yield the same result. Physical occurrences frequently considered to be random phenomena are the sex of an unborn baby, the number of telephone trunk lines in use at a certain hour, the number of cars registered in the state of Massachusetts in a certain year, the number of automobiles passing through a

---

* The reader may review, if necessary, basic set-theoretic concepts provided in Appendix B.

certain toll booth, the number of accidents on a certain highway, the number of lives lost in plane crashes in the western hemisphere, the number of students registered in a mathematics course in a college, the number of patients admitted to a certain hospital in a given year, the number of lottery tickets sold in the state of Connecticut, the number of passengers riding a commuter train between New Jersey and New York, the number of voters eligible for nomination in a mayoral election, and the like. Whether the Democrats will win the next presidential election or whether a given student will pass a certain course in economics are also random experiments, the outcomes of which can hardly be predicted with certainty.

To bring out in more detail what we mean by a random phenomenon, let us consider two random experiments. These are both the type of simple random experiment that most people enjoy trying on their own.

EXAMPLES 6.1

1. Suppose that we take a telephone directory, open it to any page, and record the last digit of 100 telephone numbers. Do you expect these digits to appear with equal frequency?

   Table 6.1 gives the result of one sample of 100. The digits are about equally frequent. Repeat the same experiment and record the last digit of 250 telephone numbers, then of 500 telephone numbers, then of 1000 telephone numbers. How are the digits distributed now?

TABLE 6.1

Digit	Frequency
0	11
1	12
2	7
3	8
4	10
5	9
6	10
7	11
8	10
9	12
	100

2. Consider a young man waiting for a young lady who is late. To amuse himself, he decides to take a walk under the following set of rules. He tosses a coin, which we may assume is balanced. He walks 10 yards north if the coin falls heads; he walks 10 yards south if the coin falls tails. He repeats this process every 10 yards and thus executes a **random walk**.

The gentleman (anxious to see his date) "expects" to be back at the starting point after completing his random walk of 100 yards. The following table shows the results of a simulated experiment:

Toss number	1	2	3	4	5	6	7	8	9	10
Direction	N	S	N	N	S	N	N	S	N	S

Thus, he is 20 yards north of his starting position. Surprised? A little reflection will convince you that this experiment is not deterministic in nature and that no one can predict whether the coin will fall heads or tails on a particular throw. Of course, we will make every effort to control the experimental conditions by always throwing the coin with the same force, in the same direction, always shaking the coin the same number of times, and so on. Despite all efforts, the results will continue to be variable and unpredictable.

6.2   THE EMPIRICAL BASIS

Closely related to the concept of a random experiment or a random phenomenon are the concepts of random event and the probability of a random event. A **random event** is one whose proportion of occurrence in a "very large number of repeated trials" approaches a stable limit. Let us consider further the random phenomenon of throwing a balanced coin. In the preceding example, a coin was thrown 10 times. Heads turned up in trials 1, 3, 4, 6, 7, and 9, and tails appeared in the remaining trials. Thus, the number of trials in which heads appeared was six. Dividing this number by the total number of trials, 10, we obtain the fraction 0.6, which is called the **relative frequency** of the number of heads. In general, if an experiment is repeated n times and a particular event occurs in r trials, the ratio r/n is the relative frequency of occurrence of that particular event.

We have been discussing a sequence of 10 throws of a coin. In a similar manner, nine additional sequences of 10 trials each were carried out. The proportion of the number of heads in each sequence is as follows:

Sequence number	1	2	3	4	5	6	7	8	9	10
Proportion of the number of heads	0.6	0.4	0.5	0.5	0.7	0.6	0.3	0.5	0.4	0.4

This experience suggests that the proportion of the number of heads ranges from 0.3 to 0.7. What will be the variability if each of the 10 sequences is based on 100 throws of a coin? Do you have a thoughtful guess? Do you expect the variation to be of the same magnitude as before if you actually perform the experiment?

In Table 6.2 the results of 1000 independent flips of a fair coin are given. It is evident that in each sequence of 100 trials, the proportion of the number of heads is approximately equal to $\frac{1}{2}$, with the variability in the proportion ranging from 0.46

TABLE 6.2

Trial	Number of Heads	Proportion
1–100	52	0.52
101–200	53	0.53
201–300	52	0.52
301–400	47	0.47
401–500	51	0.51
501–600	53	0.53
601–700	48	0.48
701–800	46	0.46
801–900	52	0.52
901–1000	54	0.54

to 0.54. Consequently, we might be tempted to assert that in reasonably long sequences based on larger and larger numbers of throws, the variation in proportion will be exceedingly small and the number of heads will be approximately half the number of trials. If we succumb to this temptation, we will have asserted that the outcome of this experiment is a random phenomenon and that the relative frequency of the number of heads approaches closer and closer to $\frac{1}{2}$ as the number of trials is increased.

6.3   THE SAMPLE SPACE

We talk about probabilities only in relation to a certain idealized experiment and its possible outcomes. Consider again the example of tossing a balanced coin. In actuality, the coin may not necessarily fall heads or tails; it can stand on an edge or it may roll away and drop through a crack in the floor. Nevertheless, we agree to regard "head" and "tail" as the only possible outcomes, because at least intuitively either outcome is "equally likely" and any other outcome is much less likely. If these outcomes are denoted by H and T, respectively, then each outcome of a throw of a balanced coin corresponds to exactly one element of the set {H, T}.

DEFINITION
6.3.1

A **sample space**, usually denoted by the letter S, is the set of all possibilities in an experiment. That is, to each element of the set S corresponds an outcome of the experiment; conversely, to each outcome of the experiment there corresponds one and only one element in the set S.

Notice especially the use of "a" rather than "the" in Definition 6.3.1. Usually, many models will describe an experiment; choosing an appropriate sample space is part of the skill needed to apply probability concepts to real-life problems. In general, it is safe to include as much detail as possible when deciding what to consider as an outcome of the random experiment. Because we are concerned with the collection of *all* possible outcomes of an experiment, our set is, in a sense, a universal set of outcomes. Everything that can occur will be represented in a sample space. Thus, the sample space simply provides a model of an experiment in the sense that every possible outcome of the experiment is completely described by one and only one element in the set S.

EXAMPLES 6.3

1. For the experiment of throwing a die, we frequently let each outcome correspond to one of the elements of the set $\{1, 2, 3, 4, 5, 6\}$. Six possible outcomes make up the sample space. If the die is not loaded, these outcomes are equally likely; that is, there is no reason to believe that one of the outcomes of the sample space is likely to occur more frequently than any of the others.

2. Consider an experiment of tossing a balanced coin three times. Suppose that our interest lies only in the number of heads obtained. The possible outcomes are 0, 1, 2, or 3 heads, and "a" sample space can be described as $S = \{0, 1, 2, 3\}$. We may observe that by merely recording the number of heads obtained, we lost some valuable information and our classification technique was rather coarse. We can achieve a finer classification by recording whether the heads appeared in all three tosses (HHH), heads appeared in the first two trials followed by a tail on the third toss (HHT), the coin fell tails on all three trials (TTT), and so on. Each outcome of the experiment corresponds to one and only one element of the set $S = \{$HHH, HHT, HTH, HTT, THH, THT, TTH, TTT$\}$ shown in the tree diagram of Figure 6.1. Clearly, the eight points in the set constitute a sample space different from the one in which the number of heads is recorded.

FIGURE 6.1

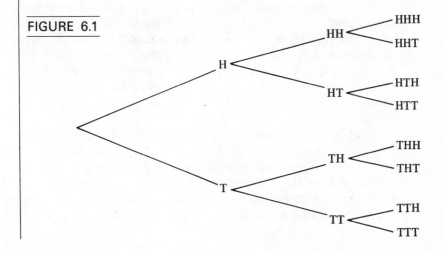

3. Consider rolling a pair of dice. The set S consists of three elements, 0, 1, and 2, if we are interested in recording the number of times "five" has appeared. This set can serve as a sample space, because it satisfies the definition. Another model can be described if the concern is simply the total number of spots on the uppermost faces when the dice come to rest. The sample space consists of 11 numbers 2, 3, 4, 5, 6, 7, 8, 9, 10, 11, and 12. But neither of the two sample spaces is sufficient to answer the question: Is the number on the first die less than the number on the second die? If the dice are distinguishable (for instance, if one is white and the other green), we are led to take as a sample space the set S consisting of 36 different outcomes for the pair. We will need a notation for recording these outcomes. Suppose that the white die bears the number 3 and the green die bears the number 4; we write this outcome as (3, 4). Further, (3, 4) and (4, 3) represent different possible outcomes in as much as the first number of each pair represents the number of spots on the white die and the second number in the pair indicates the number of spots on the green die. The order of the numbers is important and the symbol (x, y) is thus called an **ordered pair**. In terms of this notation, the sample space S consisting of 36 ordered pairs is as follows:

TABLE 6.3

$$S = \begin{cases} (1, 1), (1, 2), (1, 3), (1, 4), (1, 5), (1, 6), \\ (2, 1), (2, 2), (2, 3), (2, 4), (2, 5), (2, 6), \\ (3, 1), (3, 2), (3, 3), (3, 4), (3, 5), (3, 6), \\ (4, 1), (4, 2), (4, 3), (4, 4), (4, 5), (4, 6), \\ (5, 1), (5, 2), (5, 3), (5, 4), (5, 5), (5, 6), \\ (6, 1), (6, 2), (6, 3), (6, 4), (6, 5), (6, 6) \end{cases}$$

It should be emphasized that the sample space of an experiment is capable of being described in more than one way. Observers with different conceptions of what could possibly be observed may arrive at different sample spaces. In general, it is desirable to use sample spaces whose points represent equally likely outcomes that cannot be subdivided further; that is, in general, each point in the sample space should not represent two or more possibilities which are further distinguishable. Unless there are special reasons to the contrary, we shall find this rule extremely helpful.

All of the sample spaces we have considered so far are of finite size. However, there is no logical necessity for sample spaces to be finite. We shall briefly mention one example. Suppose that you call your father to wish him a happy birthday. The number of rings before he picks up the telephone is a random experiment. It is conceivable (at least in theory) that you can hear an unending sequence of telephone rings if he is not at home and you allow the phone to keep ringing. If he answers, you record the number of telephone rings before he picked up the receiver. The sample space $S = \{1, 2, 3, \ldots\}$ is clearly an infinite set. Although the theory of probability deals with both finite and infinite sample spaces, we shall restrict ourselves to *finite* sample spaces only.

The Events

For each experiment there is a sample space, and for each sample space there are events. Having explained what we mean by a sample space, the next logical step is to define the term "event." This is important because probabilities are always associated with the occurrence or nonoccurrence of events.

DEFINITION
6.3.2

An element of a sample space is called a **sample point** or a **simple event**.

DEFINITION
6.3.3

An **event** is a subset of the sample space S.

DEFINITION
6.3.4

A **compound event** is the union of simple events.

Unless stated otherwise, it is generally assumed that all simple events are *equally likely*.

To illustrate the notion of an event, consider the experiment of tossing a balanced coin twice. A sample space for this experiment is given by

$$S = \{HH, HT, TH, TT\}$$

The following events may be of interest:

1. There is exactly one head.
2. There is at least one head.

Each of these events is associated with a set whose elements are in the sample space. For instance, the event that there is exactly one head is associated with the set

$$\{HT, TH\}$$

Similarly, the set

$$\{HT, TH, TT\}$$

represents the event that there is at least one head.

EXAMPLE 6.3.4

Suppose that we plan to conduct a survey of families having three children and wish to record the sex of each child in the order of their births. Describe a suitable sample space and list each of the following events:

1. The first child is a boy.
2. All children are of the same sex.
3. There are more girls than boys in a family.
4. There are exactly two girls in a family.
5. There are at least two boys in a family.

SOLUTION   Let us denote a boy and a girl by the letters B and G, respectively. If we use three letters to represent the first, the second, and the third child, in that order, then an appropriate sample space for a single family is given by

$$\{BBB, BBG, BGB, BGG, GBB, GBG, GGB, GGG\}$$

We list below the correspondence between various events and subsets of S.

Description of Event	Corresponding Subset of S
(a) first child is a boy	$\{BBB, BBG, BGB, BGG\}$
(b) all children are of the same sex	$\{BBB, GGG\}$
(c) more girls than boys	$\{BGG, GBG, GGB, GGG\}$
(d) exactly two girls in a family	$\{BGG, GBG, GGB\}$
(e) at least two boys in a family	$\{BBB, BBG, BGB, GBB\}$

EXAMPLE 6.3.5

A committee of five members, A, B, C, D, and E, decides to appoint a subcommittee of two members to study the mathematics curriculum in a college. List the 10 elements of the appropriate sample space S and find subsets of S containing elements in which

1. A is selected.
2. A or B is selected.
3. C is not selected.

SOLUTION   We are interested in the number of possible subcommittees that can be formed from five members taken two at a time. Remember that AB and BA represent the same subcommittee. The sample space consisting of 10 elements is given by

$$S = \{AB, AC, AD, AE, BC, BD, BE, CD, CE, DE\}$$

We list below the correspondence between various events and subsets of the sample space S.

Description of Event	Corresponding Subset of S
(a) A is selected	{AB, AC, AD, AE}
(b) A or B is selected	{AB, AC, AD, AE, BC, BD, BE}
(c) C is not selected	{AB, AD, AE, BD, BE, DE}

EXAMPLE 6.3.6

Consider an urn containing four balls numbered 1, 2, 3, and 4. A blindfolded person draws two balls from the urn; the ball drawn on the first draw is not replaced before the second draw. Describe a suitable sample space and list each of the following events.

1. The sum of the numbers on the balls is five.
2. Each ball bears an even number.
3. One ball bears a number greater than or equal to four.

SOLUTION   Each outcome of the experiment can be represented by an ordered pair (x, y), where x is the number on the first ball and y is the number on the second ball drawn from the urn. An appropriate sample space S with 12 points is as follows:

$$S = \begin{cases} & (1, 2), (1, 3), (1, 4), \\ (2, 1), & (2, 3), (2, 4), \\ (3, 1), (3, 2), & (3, 4), \\ (4, 1), (4, 2), (4, 3) & \end{cases}$$

The subset consisting of the elements (4, 1), (3, 2), (2, 3), and (1, 4) represents the event that the sum of the numbers on the two balls is five. The event that each ball has an even number is associated with the subset {(2, 4), (4, 2)}. Similarly, the event that one ball shows a number greater than or equal to four can be represented by the subset {(1, 4), (2, 4), (3, 4), (4, 1), (4, 2), (4, 3)}.

EXERCISES 6.3

1. Give an example of a random experiment that would be studied by
   (a) an economist
   (b) a psychologist
   (c) an internal revenue staff
   (d) a college registrar
   (e) a television station
   (f) an airline
   (g) a hospital
   (h) an insurance company

2. Discuss whether or not each of the following occurrences would be classified as a random experiment.
   (a) the average daily temperature in Tempe, Arizona
   (b) the number of shares traded in one day on the New York Stock Exchange
   (c) the number of television sets sold by Zenith Corporation in 1979
   (d) the number of students applying for admission to a college
   (e) the number of cars passing through a certain toll station
   (f) the number of passengers in a bus
   (g) the number of hours a businessperson sleeps on a weekend
   (h) drawing a card from a well-shuffled deck
   (i) a man asking a woman for a date
   (j) a winner and a runner-up in a beauty contest in Connecticut

   In the Exercises 3–10, describe a suitable sample space.

3. Two coins are tossed.

4. A die is rolled and then a coin is tossed.

5. A coin is tossed and then a die is rolled.

6. A coin is tossed until a head appears or three tosses have been made.

7. A president and a secretary of a club are to be elected from six students.

8. A subcommittee of two members is to be formed from the six members of a college mathematics club.

9. Two boys and two girls are arranged in a row for a panel discussion.

10. Four persons, A, B, C, and D, are to be seated at a circular table.

11. Three couples are attending a dance in which three women choose their partners at random among the three men. Describe a suitable sample space showing six possibilities that a man is dancing with his or someone else's date.

12. Susan can study 1, 2, 3, or 4 hours for a certain course on any given evening. Describe a suitable sample space showing 12 possibilities in which she can plan 7 hours of study for the test on three consecutive evenings.

13. A coin is tossed three times. Describe a suitable sample space and list the elements in each of the following events:
    (a) The first toss is a tail.
    (b) There are exactly two heads.
    (c) There are at least two heads.
    (d) There are all tails.

14. A social worker plans to conduct a survey of families having four children and wishes to record the sex of each child in the order of their births. Describe a suitable sample space and list each of the following events:
    (a) There is exactly one boy in a family.
    (b) There are exactly three boys in the family.

    (c) There are more boys than girls in a family.

    (d) There are the same number of boys and girls in a family.

15. For the experiment of tossing a pair of dice (one white and the other green) and with the sample space described in Table 6.3, list each of the following events:

    (a) The number on at least one die is five.

    (b) The number on the green die is odd.

    (c) The sum of the numbers on the two dice is six.

    (d) Each of the two dice shows the same number.

16. An urn contains five balls, numbered 1 through 5. The ball drawn on the first draw is *not* returned before the second draw. Using the ordered pair (x, y) to indicate that x is the number on the first ball and y the number on the second ball drawn from the urn, describe a suitable sample space with 20 elements and list each of the following events:

    (a) The sum of the numbers on the balls drawn is six.

    (b) Each ball bears an odd number.

    (c) At least one of the balls drawn has the number five on it.

6.4 THE CONCEPT OF PROBABILITY

We are now ready to discuss the meaning of the word "probability." In conducting a survey of the families having two children, we generally record the sex of each child in the order of their births. An appropriate sample space in this case is given by the set **BB**, **BG**, **GB**, **GG**; since each of these simple events is equally likely, we say that the probability of each simple event is $\frac{1}{4}$. The compound event that the children are of the same sex consists of two simple events, **BB** and **GG**, and the probability of the occurrence of this compound event is $\frac{2}{4} = \frac{1}{2}$. Similarly, the probability of drawing a spade from a well-shuffled deck of cards is $\frac{13}{52} = \frac{1}{4}$, because the sample space consists of 52 cards, each of which is equally likely, and there are 13 spades in the deck. As another example, consider the experiment of drawing a ball from an urn containing 10 balls, numbered 1 to 10, of which 6 balls, numbered 1 to 6, are colored red and the remaining 4 balls are colored green. A suitable sample space is given by $S = \{1, 2, 3, 4, 5, 6, 7, 8, 9, 10\}$. The event that the ball drawn is red can be represented by a subset $E = \{1, 2, 3, 4, 5, 6\}$. If all the simple events of S are equally likely (the balls are of the same size and same weight), we may arrive at the result that the probability of drawing a red ball is $\frac{6}{10} = \frac{3}{5}$.

DEFINITION 6.4.1   If the simple events of a sample space are equally likely, then the **probability** of an event E is simply the ratio of the number of simple events in E to the number of all simple events in the sample space S.

Thus, the computation of the probability of an event E defined in a finite sample space of equally likely outcomes reduces to a simple computation of the number of simple events in E and the number of all simple events in S. We assign the probability to an event E by

$$P(E) = \frac{n(E)}{n(S)} = \frac{\text{number of elements in } E}{\text{number of elements in } S} \qquad (6.1)$$

This method of assigning the probability to event E has two immediate consequences: if E is an impossible event, then

$$P(E) = 0 \quad \text{because} \quad n(E) = 0 \qquad (6.2)$$

If E is an inevitable event, then

$$P(E) = 1 \quad \text{because} \quad n(E) = n(S) \qquad (6.3)$$

It follows that the probability of an event E varies between 0 and 1; that is,

$$\boxed{0 \le P(E) \le 1 \quad \text{for any event } E \text{ in } S} \qquad (6.4)$$

EXAMPLE 6.4.1

An ordinary die is rolled. What is the probability of getting a number greater than three?

SOLUTION   The sample space consists of six sample points $\{1, 2, 3, 4, 5, 6\}$. If E denotes the event that the number on the uppermost face is greater than three, then E consists of three sample points $\{4, 5, 6\}$. Hence,

$$P(E) = \frac{3}{6} = \frac{1}{2}$$

EXAMPLE 6.4.2

Two coins are tossed. Using $S = \{HH, HT, TH, TT\}$ as the sample space of this experiment, find the probability of getting

1. exactly one head
2. at least one head

SOLUTION   Let E denote the event that there is exactly one head and F denote the event that there is at least one head. Then,

$$E = \{HT, TH\} \quad \text{and} \quad F = \{HT, TH, HH\}$$

Clearly,

$$n(E) = 2 \quad \text{and} \quad n(F) = 3$$

Thus,

1. $P(E) = \dfrac{n(E)}{n(S)} = \dfrac{2}{4} = \dfrac{1}{2}$

2. $P(F) = \dfrac{n(F)}{n(S)} = \dfrac{3}{4}$

EXAMPLE 6.4.3

Using the sample space of Table 6.3 and assuming that all the sample points are equally likely, determine the following probabilities:

1. The event that the sum of the numbers on the dice is seven.
2. The number on the white die is greater than the number on the green die by more than two.
3. The event that the number on the white die is a multiple of three.
4. The event that the number on at least one die is six.

SOLUTION   Let E, F, G, and H be the events of S. Then,

$$E = \{(1, 6), (2, 5), (3, 4), (4, 3), (5, 2), (6, 1)\}$$

$$F = \{(4, 1), (5, 1), (5, 2), (6, 1), (6, 2), (6, 3)\}$$

$$G = \{(3, 1), (3, 2), (3, 3), (3, 4), (3, 5), (3, 6),$$
$$(6, 1), (6, 2), (6, 3), (6, 4), (6, 5), (6, 6)\}$$

and $H = \{(6, 1), (6, 2), (6, 3), (6, 4), (6, 5), (6, 6),$
$$(1, 6), (2, 6), (3, 6), (4, 6), (5, 6)\}$$

Since all sample points of S are equally likely, we have

1. $P(E) = \dfrac{n(E)}{n(S)} = \dfrac{6}{36} = \dfrac{1}{6}$

2. $P(F) = \dfrac{n(F)}{n(S)} = \dfrac{6}{36} = \dfrac{1}{6}$

3. $P(G) = \dfrac{n(G)}{n(S)} = \dfrac{12}{36} = \dfrac{1}{3}$

4. $P(H) = \dfrac{n(H)}{n(S)} = \dfrac{11}{36}$

EXAMPLE 6.4.4

Assuming that in a three-child family the eight simple points BBB, BBG, BGB, BGG, GBB, GBG, GGB, GGG are equally likely, determine the probabilities of the following events:

1. exactly one boy
2. exactly two boys
3. all boys
4. all girls

SOLUTION   The verbal description of the events, the algebraic conditions, the solution sets, and the required probabilities are given in Table 6.4, where x represents the number of boys in a three-child family.

TABLE 6.4

Verbal Description of the Event	Algebraic Condition	Event (subset of S)	Probability of the Event
(a) exactly one boy	$x = 1$	{BGG, GBG, GGB}	$\frac{3}{8}$
(b) exactly two boys	$x = 2$	{BBG, BGB, GBB}	$\frac{3}{8}$
(c) all boys	$x = 3$	{BBB}	$\frac{1}{8}$
(d) all girls	$x = 0$	{GGG}	$\frac{1}{8}$

EXERCISES 6.4

1. A single die is rolled. Find the probability of getting
   (a) an even number
   (b) an odd number
   (c) a number greater than four
   (d) a number less than five
   (e) a two, three, four, or five
   (f) a seven

2. A card is drawn at random from a well-shuffled deck of 52 playing cards. Find the probability that the card drawn is
   (a) a diamond
   (b) a red card
   (c) a queen
   (d) a jack or a king
   (e) a 2, 3, 4, 5, 6, or 7
   (f) neither a spade nor a club

3. A telephone number is chosen at random from a telephone directory and the final digit is recorded. Assuming that the digits appear equally frequently, find the probability that the last digit recorded is
   (a) an even number
   (b) greater than six
   (c) less than four
   (d) a multiple of three

4. Mrs. Talbot's cookie jar contains 10 chocolate chip cookies, 15 butter cookies, 8 fudge cookies, and 12 gingerbread cookies. While Mrs. Talbot is busy in her living room, her son reaches into the jar and picks one cookie at random. Find the probability that he picked
   (a) a butter cookie
   (b) a gingerbread cookie
   (c) a chocolate chip or a fudge cookie
   (d) a butter or a gingerbread cookie

5. An urn contains 15 balls, of which 5 are black, 6 are red, and 4 are white. A ball is drawn at random. Find the probability that it is
   (a) a white ball
   (b) a red or a white ball
   (c) a white or a black ball
   (d) not a white ball

6. An urn contains 30 balls, of which 10 are red, 8 are white, 7 are blue, and 5 are black. If one ball is drawn from the urn at random, find the probability that the ball is
   (a) white or blue
   (b) red, white, or blue
   (c) neither white nor black
   (d) neither red nor blue

7. A drawer contains four white, six brown, seven gray, and eight green socks. If a sock is selected at random, find the probability that the sock is
   (a) brown
   (b) white or gray
   (c) gray or green
   (d) white, brown, or green

8. The name of a state is selected at random from the 50 states. Assuming that all states have the same chance of being selected, find the probability that the name of the state selected
   (a) begins with the letter A
   (b) begins with the letter C
   (c) begins with the letter W
   (d) borders on Canada
   (e) contains four letters of the alphabet

9. A fair coin is tossed three times. Find the probability of getting
   (a) all heads
   (b) exactly one head
   (c) exactly two heads
   (d) at least one head
   (e) at least two heads

10. Determine all possible outcomes if a fair coin is tossed four times. Find the probability of getting
    (a) exactly one head
    (b) exactly two heads
    (c) exactly four heads
    (d) at least three heads

11. A pair of dice is rolled. Find the probability of getting
    (a) a total score of 8
    (b) a total score of 7 or 11
    (c) a total score of at least 10
    (d) the same score on both dice

12. In a single throw of three dice, find the probability of getting
    (a) a total score of 3
    (b) a total score of 5
    (c) a total score of at least 20
    (d) the same number on each die

13. Closing prices of the 15 most active stocks traded on the New York Stock Exchange one Friday were (to the nearest dollar) as follows:

    21, 51, 62, 31, 63, 60, 40, 27, 32, 42, 36, 25, 28, 75, 16

    Find the probability that a stock selected from this list was sold
    (a) for less than $30
    (b) anywhere from $16 to $32, inclusive
    (c) more than $50

14. The license plates of automobiles registered in a state are numbered serially begining with 1. Assuming that there are 2,000,000 automobiles registered in the state, what is the probability that the first digit on the license plate of an automobile selected at random will be the digit 3?

15. Two boys and two girls are arranged in a row for a panel discussion. What is the probability that the boys and girls alternate in a row?

16. Three letters and three corresponding envelopes are typed by a tipsy typist, and letters are put into the envelopes in such a way that although each envelope contains one letter, any letter is equally likely to be in any of the three envelopes. Find the probability that
    (a) all the letters were inserted correctly
    (b) none of the letters was inserted correctly
    (c) only one letter was inserted correctly
    (d) only two letters were inserted correctly

17. Three college students who share the same apartment arrive home one evening so tired that each student chooses at random a bed in which to sleep. What is the probability that no student is sleeping in their own bed?

6.5 SOME RULES OF PROBABILITY

<table>
<tr><td>DEFINITION
6.5.1</td><td>Two events are said to be **mutually exclusive** if they cannot occur simultaneously.</td></tr>
</table>

In the language of sets, two or more events are mutually exclusive if they have no points in common. In other words, the intersection of two or more mutually exclusive events is the empty set.

The discussion in Section 6.4 makes it natural to develop certain assumptions about how probabilities should always behave. We state the most basic of these assumptions and call them **axioms**.

Axiom 1. $P(E) \geqslant 0$ for any event E in the sample space S.　　　　(6.5)

Axiom 2. $P(S) = 1$.　　　　(6.6)

Axiom 3. If E and F are mutually exclusive events, then

$$P(E \cup F) = P(E) + P(F) \tag{6.7}$$

In other words, these axioms say that the probability of any event defined on the sample space S is a nonnegative number between 0 and 1, inclusive; that the probability of an entire sample space is 1.00; and that the probability of the union of two mutually exclusive events is the sum of their separate probabilities.

The result stated in equation (6.7) can be extended to any finite number of mutually exclusive events. Thus, if $E_1, E_2, E_3, \ldots, E_m$ are all mutually exclusive events in the sample space S, then

$$P(E_1 \cup E_2 \cup E_3 \cup \cdots \cup E_m) = P(E_1) + P(E_2) + P(E_3) + \cdots + P(E_m)$$

EXAMPLE 6.5.1

Three studies were recently made to analyze inflationary trends in the cost of living. The first study claims that the cost of living will go up, remain unchanged, or go down with probability 0.79, 0.16, and 0.07, respectively; the second study reveals that the respective probabilities are 0.83, 0.15, and 0.02; and the third study shows that the respective probabilities are 0.76, 0.18, and 0.04. Are these three sets of figures acceptable? Why or why not in each case?

SOLUTION   Let E, F, and G be three mutually exclusive events corresponding to the cost of living going up, remaining steady, or going down, respectively. Note that the probability of each of the events E, F, and G should be nonnegative and that $P(E) + P(F) + P(G)$ should equal 1.00. Such is the only case in the second study.

The other cost-of-living studies are not reliable because the sum of the probabilities *exceeds* 1.00 in the first case, whereas in the third case the sum of the probabilities *is less than* 1.00, contradicting Axiom 2.

EXAMPLE 6.5.2

The probabilities that a student in an economics course will receive an A, B, C, D, or F in a final examination are 0.40, 0.25, 0.20, 0.10, and 0.05, respectively. Find the probability that the student will receive

1. at least a B
2. at most a C
3. neither an A nor an F
4. a C or D

SOLUTION   Note that the events of receiving an A, B, C, D, or F in an examination are all mutually exclusive. Further, the probabilities of each of these events is a nonnegative number, and the probabilities add to 1.00. Thus, the axioms are satisfied and we have

1. Probability of at least a $B = P(A) + P(B) = 0.40 + 0.25 = 0.65$
2. Probability of at most a $C = P(C) + P(D) + P(F)$
   $$= 0.20 + 0.10 + 0.05 = 0.35$$
3. Probability of neither an A nor an $F = P(B) + P(C) + P(D)$
   $$= 0.25 + 0.20 + 0.10 = 0.55$$
4. Probability of a C or a $D = P(C) + P(D) = 0.20 + 0.10 = 0.30$

THEOREM 6.5.1

For any event E in the sample space S,

$$P(E^c) = 1 - P(E)$$

Thus, if the probability that the Democrats will win the next election is 0.65, then the probability that they will lose the election is 0.35. Similarly, if the probability is 0.78 that Miss Robinson will survive a heart attack, then the probability is 0.22 that she will not survive the attack; if the probability is 0.13 that Mr. Buckley will pass the Civil Service Examination, then the probability is 0.87 that he will not make it; and if the probability is 0.20 that there will be showers tonight, then the probability is 0.80 that there will be no showers.

EXAMPLE 6.5.3

A student is worried about his grades in a statistics course. However, he estimates that he will pass the course with a probability of 0.8, and the probability is 0.6 that he will get a grade of C or lower. What is the probability that he will get a grade of C or D?

SOLUTION   Since $P(A \text{ or } B \text{ or } C \text{ or } D) = 0.8$, it follows that $P(F) = 0.2$. Also, $P(C \text{ or } D \text{ or } F) = 0.6$. Hence,

$$P(C \text{ or } D) = P(C \text{ or } D \text{ or } F) - P(F) = 0.6 - 0.2 = 0.4$$

EXAMPLE 6.5.4

Two events, E and F, are mutually exclusive and $P(E) = 0.45$, $P(F) = 0.30$. Find the probabilities of

(a) $P(E^c)$

(b) $P(F^c)$

(c) $P(E \cap F)$

(d) $P(E \cup F)$

(e) $P(E^c \cup F^c)$

(f) $P(E^c \cap F^c)$

SOLUTION

(a) $P(E^c) = 1 - P(E) = 1.00 - 0.45 = 0.55$

(b) $P(F^c) = 1 - P(F) = 1.00 - 0.30 = 0.70$

(c) $P(E \cap F) = 0$, since E and F are mutually exclusive events

(d) From Axiom 3, we have

$$P(E \cup F) = P(E) + P(F)$$

$$= 0.45 + 0.30 = 0.75$$

From DeMorgan's laws, we see that

$$E^c \cup F^c = (E \cap F)^c \qquad \text{and} \qquad E^c \cap F^c = (E \cup F)^c$$

Hence,

(e) $P(E^c \cup F^c) = P[(E \cap F)^c] = 1 - P(E \cap F) = 1.00 - 0.00 = 1$

(f) $P(E^c \cap F^c) = P[(E \cup F)^c] = 1 - P(E \cup F) = 1.00 - 0.75 = 0.25$

THEOREM 6.5.2

For any two events E and F defined on the sample space S,

$$P(E \cap F^c) = P(E) - P(E \cap F)$$

PROOF   The Venn diagram of Figure 6.2 shows that $(E \cap F)$ and $(E \cap F^c)$ are two mutually exclusive events and that their union is the event E. Thus, by (6.7),

$$P(E \cap F^c) + P(E \cap F) = P(E)$$

from which the assertion follows.

FIGURE 6.2

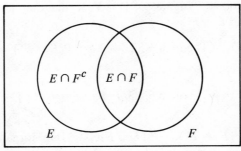

THEOREM 6.5.3

For any two events E and F defined on the sample space S,

$$P(E \cup F) = P(E) + P(F) - P(E \cap F)$$

PROOF   The union of two events E and F may be represented as the union of two disjoint events F and $E \cap F^c$ (see Figure 6.3); that is,

$$E \cup F = F \cup (E \cap F^c)$$

FIGURE 6.3

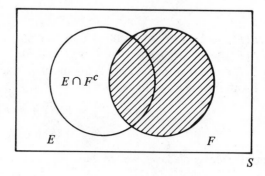

Then, by (6.7), we have

$$P(E \cup F) = P(F) + P(E \cap F^c)$$
$$= P(F) + P(E) - P(E \cap F) \qquad \text{(Theorem 6.5.2)}$$

The following examples illustrate Theorems 6.5.2 and 6.5.3.

EXAMPLE 6.5.5

Given that $P(E) = 0.37$, $P(F) = 0.28$, and $P(E \cap F) = 0.13$, compute

(a) $P(E \cap F^c)$
(b) $P(E \cap F)$
(c) $P(E^c \cap F^c)$

SOLUTION   Figure 6.4 is a Venn diagram showing the relationship among E, F, and $E \cap F$.

(a) From Theorem 6.5.2, we have $P(E \cap F^c) = P(E) - P(E \cap F)$

$$= 0.37 - 0.13 = 0.24$$

(b) From Theorem 6.5.3, $P(E \cup F) = P(E) + P(F) - P(E \cap F)$
$$= 0.37 + 0.28 - 0.13 = 0.52$$

(c) $P(E^c \cap F^c) = P[(E \cup F)^c]$ (DeMorgan's laws)

$$= 1 - P(E \cup F) = 1 - 0.52 = 0.48$$

FIGURE 6.4

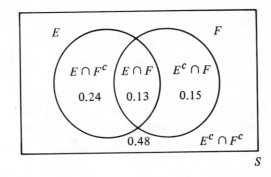

To demonstrate Theorem 6.5.3 in more detail, let us consider another example.

EXAMPLE 6.5.6

The probability that Gary, who is visiting a contractor, will sign a contract for remodeling his kitchen is 0.27, the probability that he will have his bathroom remodeled is 0.22, and the probability that he will have his kitchen as well as his bathroom remodeled is 0.03. Find the probability that he will have

(a) at least one room, the bathroom or the kitchen, remodeled
(b) neither the kitchen nor the bathroom remodeled
(c) the kitchen or the bathroom remodeled, but not both

SOLUTION   Let E be the event that Gary will have his kitchen remodeled and let F represent the event that he will have his bathroom remodeled. Then,

$$P(E) = 0.27 \qquad P(F) = 0.22 \qquad P(E \cap F) = 0.03$$

A Venn diagram in Figure 6.5 shows the relationship among E, F, and $E \cap F$.

FIGURE 6.5

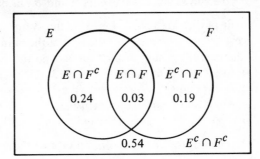

(a) Using Theorem 6.5.3, we have

$$P(E \cup F) = P(E) + P(F) - P(E \cap F)$$
$$= 0.27 + 0.22 - 0.03$$
$$= 0.46$$

(b) $P(E^c \cap F^c) = P[(E \cup F)^c]$

$$= 1 - P(E \cup F)$$

$$= 1 - 0.46$$

$$= 0.54$$

(c) The probability that he will have his kitchen or bathroom (but not both) remodeled corresponds to the two mutually exclusive events $(E \cap F^c)$ and $(E^c \cap F)$, respectively. Thus,

$$P(E \cap F^c) + P(E^c \cap F) = 0.24 + 0.19$$

$$= 0.43$$

EXERCISES 6.5

1. A coin is tossed once. A is the event "getting heads" and B is the event "getting tails." Are events A and B mutually exclusive?

2. Two coins are tossed. A is the event "getting two heads" and B is the event "getting two tails." Are events A and B mutually exclusive?

3. Three coins are tossed. A is the event "getting two heads," B is the event "getting two tails," and C is the event "getting one head."
   (a) Are events A and B mutually exclusive?
   (b) Are events B and C mutually exclusive?
   (c) Are events A and C mutually exclusive?

4. A die is rolled. A is the event "the die shows an odd number" and B is the event "the die shows six." Are events A and B mutually exclusive?

5. A pair of dice are rolled. Let A be the event "the sum of the numbers on the dice is seven," B the event "the sum of the numbers on the dice is eleven," and C the event "the sum of the numbers on the dice is even."
   (a) Are A and B mutually exclusive?
   (b) Are A and C mutually exclusive?
   (c) Are B and C mutually exclusive?
   (d) Evaluate $P(A \cup B)$, $P(A \cup C)$, and $P(B \cup C)$.

6. A fair coin is tossed three times. Let A be the event "getting all tails" and B be the event "getting at least one head." Are A and B mutually exclusive?

7. Explain why the following statements are false.
   (a) The probabilities that a student will register for zero, one, two, three, or four or more courses next semester are 0.08, 0.29, 0.35, 0.19, and 0.07, respectively.

(b) The probabilities that Mr. Black will receive a grade of A, B, C, D, or F in a statistics course are 0.21, 0.26, 0.32, 0.18, and 0.13, respectively.

(c) The probability that Mr. Carlson will register for exactly two courses next fall is 0.67 and the probability that he will register for at least two courses next fall is 0.33.

8. Explain in each case why the following assignments of probability to the five mutually exclusive outcomes A, B, C, D, and E are not permissible.

(a) $P(A) = 0.23$, $P(B) = 0.28$, $P(C) = 0.28$, $P(D) = 0.12$, $P(E) = 0.13$

(b) $P(A) = 0.31$, $P(B) = 0.17$, $P(C) = 0.18$, $P(D) = 0.24$, $P(E) = 0.14$

(c) $P(A) = 0.08$, $P(B) = 0.12$, $P(C) = 0.50$, $P(D) = 0.40$, $P(E) = -0.10$

(d) $P(A) = 0.23$, $P(B) = P(C) = 0.29$, $P(D) = 0.06$, $P(E) = 0.11$

(e) $P(A) = 0.26$, $P(B) = P(C) = P(D) = 0.25$, $P(E) = -0.01$

9. Mrs. Hawkes often uses her credit cards when shopping. Sometimes she uses credit card 1, sometimes she uses credit card 2, sometimes she uses credit card 3, sometimes she pays by personal check, and, rarely, she pays cash for merchandise purchased. The probabilities for these alternatives are 0.28, 0.24, 0.25, 0.17, and 0.06, respectively. Find the probability that she will

(a) not use her credit cards

(b) not pay cash for her merchandise

(c) use credit card 1 or pay by check or pay cash

(d) neither pay cash nor pay by check

10. A company vice-president often visits Kansas City on business. Sometimes she drives her own car, sometimes she drives a company car, sometimes she takes a bus, sometimes she takes a train, and sometimes she goes by plane. The probabilities for these alternatives are 0.28, 0.21, 0.17, 0.23, and 0.11, respectively. Find the probability that

(a) she will go by car

(b) she will catch a plane or a bus or drive her own car

(c) she will not catch a plane

(d) she will catch neither a plane nor a train

11. Of several mutually exclusive possibilities, the probability that Roger will entertain his friends on a Friday evening is 0.24, the probability that he will take his wife out for dinner is 0.36, the probability that he will go bowling with his daughter is 0.16, the probability that he will work around his house is 0.11, and the probability that he will go to bed early is 0.13. Find the probability that Roger will

(a) not go out for bowling

(b) neither fix his house nor go out

(c) not go out

(d) entertain his friends or work around his house

12. Let A and B be two *mutually exclusive* events such that $P(A) = 0.52$ and $P(B) = 0.43$. Find each of the following probabilities.

(a) $P(A^c)$ (b) $P(B^c)$
(c) $P(A \cap B^c)$ (d) $P(A^c \cap B)$
(e) $P(A \cup B)$ (f) $P(A^c \cup B)$
(g) $P(A^c \cap B^c)$ (h) $P(A^c \cup B^c)$

13. Let A and B be the two events defined on the same sample space such that $P(A) = 0.55$, $P(B) = 0.37$, and $P(A \cap B) = 0.19$. Find each of the following probabilities.
    (a) $P(A^c)$ (b) $P(B^c)$
    (c) $P(A \cap B^c)$ (d) $P(A^c \cap B)$
    (e) $P(A \cup B)$ (f) $P(A^c \cup B)$
    (g) $P(A^c \cap B^c)$ (h) $P(A^c \cup B^c)$

14. A card is drawn at random from a well-shuffled deck of 52 playing cards. Find the probability that the card drawn is
    (a) a king
    (b) a heart
    (c) a king of hearts
    (d) a king or a heart
    (e) neither a king nor a heart

15. A candidate runs for two political offices. She assesses her ability of winning as 0.45 and 0.35, respectively, and thinks she only has a chance of 0.08 of winning both. Find the probability that she wins
    (a) none of the offices
    (b) at least one of the two offices
    (c) one or the other office, but not both

16. The probability that a student at a certain college buys *Newsweek* is 0.63, the probability that he buys *Time* magazine is 0.59, and the probability that he buys both is 0.32. Find the probability that a student selected at random buys
    (a) neither magazine
    (b) at least one of the magazines
    (c) one or the other magazine, but not both

17. A group of 500 New Yorkers contains 210 persons who read the *New York Times* and 60 who read the *Daily News*. Among these, 32 read both the *New York Times* and the *Daily News*. Find the probability that a person chosen at random from this group reads
    (a) neither paper
    (b) at least one of the two papers
    (c) one or the other, but not both

18. The probability that a resident of a certain affluent community owns a swimming pool is 0.75, the probability that he owns a tennis court is 0.60, and the probability that he owns both is 0.45. Find the probability that a resident selected at random from this community owns
    (a) at least a swimming pool or a tennis court

   (b) a swimming pool or a tennis court, but not both

   (c) neither a swimming pool nor a tennis court

19. A job applicant estimates her probability of getting a job at Widget United to be 0.60; the probability of being offered a job at Universal Manifold, 0.35; and the probability of being offered a job at both companies, 0.23. Find the probability that she will be offered a job
    (a) in neither company
    (b) in one or the other company, but not both
    (c) in at least one of the two companies

20. To raise money for a good cause, an organization sells raffle tickets numbered serially from 1 to 600. Find the probability that the number selected at random is
    (a) divisible by 3
    (b) divisible by 4
    (c) divisible by 12
    (d) divisible by 3 or 4

21. For married couples living in a certain suburb of Detroit, the probability that the husband will vote in a mayoral election is 0.43, the probability that his wife will vote is 0.34, and the probability that both will vote is 0.17. Find the probability that
    (a) neither of them will vote
    (b) at least one of them will vote
    (c) exactly one of them will vote

22. The probability that a person taking his car to a garage will have the front-end alignment checked is 0.23, the probability that he will have his exhaust system checked is 0.10, and the probability that he will have both of these checked is 0.04. Find the probability that a person selected at random among the customers in the garage will have
    (a) front-end alignment or exhaust system checked, but not both
    (b) front-end alignment or exhaust system checked or possibly both
    (c) none of these checked

23. The probability that a truck entering a given state park has Connecticut license plates is 0.34, the probability that it is a pick-up truck is 0.29, and the probability that it is a pick-up truck with Connecticut license plates is 0.18. What is the probability that such a truck is neither a pick-up truck nor has Connecticut plates?

24. A Washington, D.C., lobbyist estimates that the probability of a certain bill passing the House of Representatives is 0.65, the probability that the bill will pass the Senate is 0.75, and the probability that it will be passed by the House or by the Senate is 0.95. What is the probability that the bill will be passed by the House and the Senate before it comes before the President of the United States for signing?

6.6   CONDITIONAL PROBABILITY

We shall now examine how the occurrence or nonoccurrence of one event affects the probability of subsequent events and how the probability of the intersection of two events is related to their separate probabilities. Frequently, we wish to find the probability of an event F but have additional information about an event E, the knowledge of which affects the probability we attach to F. Some illustrative examples will be helpful in forming an intuitive understanding of conditional probability.

EXAMPLE 6.6.1

Suppose that two balls are drawn without replacement from an urn containing five white and three red balls. What is the probability that the first ball drawn is white? Clearly, the answer is $\frac{5}{8}$. What is the probability that the second ball drawn is white if the first ball is not replaced in the urn? The answer to this question depends on what you know about the result of the first drawing. If the first ball is white, then the probability for a subsequent ball to be white is $\frac{4}{7}$, because there are only four white balls left in the remaining seven balls. If the first ball was red, we have still five white balls in the remaining seven balls in the urn, so the desired probability is $\frac{5}{7}$.

EXAMPLE 6.6.2

A pair of dice is rolled. What is the probability that the sum of the upturned faces is 10, given that at least one of the dice has a six?

SOLUTION   Let F be the event that the sum of the upturned faces is 10 and E be the event that at least one of the dice has a six on it. Using the sample space S in Table 6.3, we have

$$E = \{(6, 1), (6, 2), (6, 3), (6, 4), (6, 5), (6, 6),$$
$$(1, 6), (2, 6), (3, 6), (4, 6), (5, 6)\}$$

and $F = \{(6, 4), (4, 6)\}$

Relative to the restriction placed on event F by the occurrence of E, $P(F) = \frac{2}{11}$ but relative to the original sample space of 36 sample points, $P(F) = \frac{2}{36}$. This is because we are considering E as a *reduced* sample space and F as an event of E.

EXAMPLE 6.6.3

A lot of 8000 parts produced on two machines, I and II, was graded as shown in Table 6.5. Find the probability that a part selected at random

TABLE 6.5

Grade	Machine I	Machine II
Satisfactory	4000	2700
Unsatisfactory	700	600

1. is produced on machine I
2. is considered satisfactory
3. is produced on machine I and is satisfactory
4. is satisfactory, given that it is produced on machine I

SOLUTION   Let E represent the event that a part is produced on machine I and F denote the event that a part is satisfactory. Clearly, $n(S) = 8000$. Then we have

1. $P(A) = \dfrac{n(E)}{n(S)} = \dfrac{4000 + 700}{8000} = \dfrac{47}{80}$

2. $P(B) = \dfrac{n(F)}{n(S)} = \dfrac{4000 + 2700}{8000} = \dfrac{67}{80}$

3. $P(A \cap B) = \dfrac{n(E \cap F)}{n(S)} = \dfrac{4000}{8000} = \dfrac{1}{2}$

4. We are now interested in the probability that a part chosen at random is satisfactory assuming that it is produced on machine I. Since we know the number of parts produced by machine I, we need to restrict our attention to these parts. Thus, the reduced sample space that concerns us has 4700 possible outcomes (in the first column) and the required probability is given by the ratio

$$\frac{n(E \cap F)}{n(E)} = \frac{4000}{4700} = \frac{40}{47}$$

This is called the *conditional probability* of an event F given that event E has already occurred and is denoted by $P(F|E)$. In this example the conditional probability can also be expressed as

$$P(F|E) = \frac{4000/8000}{4700/8000} = \frac{47}{80} \quad \text{or} \quad P(F|E) = \frac{P(E \cap F)}{P(E)}$$

DEFINITION
6.6.1

Let E and F be two events defined on the sample space S. The **conditional probability** of the event F, given the event E, is defined by

$$P(F|E) = \frac{P(E \cap F)}{P(E)} \qquad P(E) > 0$$

and if $P(E) = 0$, then $P(F|E)$ has no meaning.

Venn diagrams can also be used to illustrate the conditional probability problems. From Figure 6.6, one can determine $P(F|E)$ by reducing the sample space to the set E and then recognizing the fact that $P(F|E)$ is simply the ratio of the probability of *that part of F which is also in E* to the total probability of the event E.

FIGURE 6.6

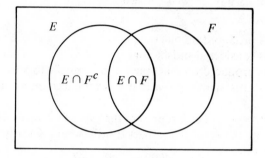

EXAMPLE 6.6.4

The probability that a customer visting a department store will buy a movie camera is 0.46, the probability that she will buy a movie projector is 0.35, and the probability that she will buy both is 0.18. Find the probability that the customer will buy

1. a movie camera, given that she has already purchased a movie projector
2. a movie projector, given that she has already purchased a movie camera
3. a movie camera and a projector, given that she will buy at least one of the two

SOLUTION   Let C denote the event that a customer will buy a movie camera and M represent the event that she will buy a movie projector (see Figure 6.7).

FIGURE 6.7

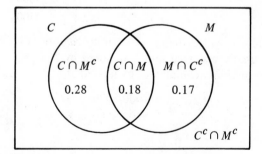

1. $P(C|M) = \dfrac{0.18}{0.35} = \dfrac{18}{35}$

2. $P(M|C) = \dfrac{0.18}{0.46} = \dfrac{9}{23}$

3. Note that since

$$P(M \cup C) = P(M) + P(C) - P(M \cap C)$$
$$= 0.35 + 0.46 - 0.18$$
$$= 0.63$$

we have

$$P(M \cap C \mid M \cup C) = \frac{0.18}{0.63} = \frac{2}{7}$$

As an immediate consequence of Definition 6.6.1, we have the following theorem.

THEOREM 6.6.1

For any events E and F,

$$P(E \cap F) = P(E) \cdot P(F \mid E)$$

EXAMPLE 6.6.5

A box contains 10 good and 4 defective items. If a sample of 2 items is selected at random, what is the probability that both items selected are nondefective?

SOLUTION   Let E be the event that the first item selected is nondefective. Obviously, $P(E) = \frac{10}{14}$. Next, let F denote the event that the second item selected is nondefective. Then $P(F \mid E) = \frac{9}{13}$, because there are only 9 nondefective items left in a total of 13. Now, by using Theorem 6.6.1, we have

$$P(E \cap F) = \frac{10}{14} \cdot \frac{9}{13} = \frac{45}{91} = 0.4945$$

This rule of multiplication can easily be extended to more than two events.

THEOREM 6.6.2

For any events E, F, and G,

$$P(E \cap F \cap G) = P(E) \cdot P(F \mid E) \cdot P(G \mid E \cap F)$$

EXAMPLE 6.6.6

In the meat section of a self-service market there are 100 chickens, 80 of which are fresh and 20 of which are one-week old. If three chickens are selected at random, what is the probability that they are all fresh?

SOLUTION   Let E, F, and G denote, respectively, the events that the first, second, and third chickens selected from those available are fresh. Then

$$P(E) = \tfrac{80}{100} \qquad P(F\,|\,E) = \tfrac{79}{99} \qquad P(G\,|\,E \cap F) = \tfrac{78}{98}$$

Using the extended rule of multiplication, we have

$$P(E \cap F \cap G) = \tfrac{80}{100} \cdot \tfrac{79}{99} \cdot \tfrac{78}{98} = \tfrac{4108}{8085} = 0.5081$$

EXERCISES 6.6

1. One card is drawn at random from a standard deck of cards. Find the probability that the card drawn is
   (a) a spade
   (b) a spade, given that it is a black card
   (c) a spade, given that it is not a diamond

2. A man tosses a fair coin twice. What is the probability of getting two heads given that there is at least one head?

3. A family has three children. Assume that each child is as likely to be a boy as it is to be a girl. Find the probability that all three children are girls given that
   (a) at least one of the children is a girl
   (b) at least two children are girls

4. Two dice, one green and the other red, are rolled. Find the probability that the sum of faces that turn up is nine, given that
   (a) at least one of the dice has yielded a five
   (b) the red die has turned up a four

5. The probability that a student will enroll in a mathematics course is 0.40, the probability that she will enroll in an economics course is 0.35, and the probability that she will enroll in both courses is 0.23. Find the probability that a student will enroll in
   (a) exactly one course
   (b) at least one course
   (c) an economics course, given that she is also enrolled in mathematics
   (d) a mathematics course, given that she is also enrolled in economics
   (e) both courses, given that she is enrolled in at least one of the courses

6. A recent survey claims that 30 percent of households in a certain city have electric dryers, 40 percent have electric stoves, and 25 percent of those who have electric stoves also have electric dryers. Find the probability that a randomly selected household will have
   (a) an electric stove, given that it has an electric dryer
   (b) an electric dryer, given that it has an electric stove
   (c) an electric dryer, given that it has an electric stove or an electric dryer or possibly both

7. The probability that a person is visiting her dentist to have her teeth cleaned is 0.43, the probability that she is going to have a cavity filled is 0.35, and the probability that she will have her teeth cleaned and a cavity filled is 0.26. Find the probability that the patient will have
   (a) her teeth cleaned, given that she is having a cavity filled
   (b) a cavity filled, given that she is having her teeth cleaned
   (c) a cavity filled, given that she may have her teeth cleaned or a cavity filled or possibly both

8. The probability that a person visiting a chain store will buy a color television is 0.35, the probability that he will buy a stereo complex is 0.26, and the probability that he will buy both is 0.17. Find the probability that the customer will buy
   (a) a color television, given that he has purchased a stereo complex
   (b) a stereo complex, given that he has purchased a color television
   (c) a stereo complex, given that he has purchased at least one of the two items
   (d) a stereo complex, given that he has purchased a color television or a stereo complex, but not both

9. A recent survey claims that 35 percent of Canadians vacationing in the United States visit New York, 24 percent visit Washington, D.C., and 11 percent visit both New York and Washington. Find the probability that a randomly selected Canadian vacationing in the United States will visit
   (a) New York, given that he has visited Washington, D.C.
   (b) Washington, D.C., given that he has visited New York

10. A person who has been having trouble starting her car in the morning takes it to a nearby garage for a checkup. The mechanic estimates that the probability that the voltage regulator needs replacement is 20 percent, the probability that the battery needs to be replaced is 45 percent, and the probability that both the battery and the voltage regulator need replacement is 6 percent. Find the probability that
   (a) the battery needs to be replaced, given that the voltage regulator needs replacement
   (b) the voltage regulator needs to be replaced, given that the battery needs replacement

11. Mrs. Smith, who is suffering from a severe congestion problem, visits her doctor. After a preliminary examination, her doctor estimates that the probability that she has a cold is 0.60, the probability that she has an allergy is 0.30, and the probability that she has both is 0.15. Find the probability that
   (a) she has a cold, given that she has an allergy
   (b) she has an allergy, given that she has a cold
   (c) she has an allergy, given that she has a cold or an allergy or possibly both

12. A recent survey of department stores produced the following information:

Geographic	Store Size ($ Volume)			
Location	Large	Medium	Small	Total
East Coast	30	60	110	200
West Coast	70	150	80	300
Total	100	210	190	500

One of the stores is selected at random for a spot check. Let A denote the event that the store is located on the East Coast and let L, M, and S represent that the size of the store is large, medium, or small, respectively. Find each of the following probabilities:

(a) $P(A)$

(b) $P(L)$

(c) $P(A \cap L)$

(d) $P(A^c \cap M)$

(e) $P(A \mid L)$

(f) $P(L \mid A)$

(g) $P(A^c \mid M)$

(h) $P(M \mid A^c)$

(i) $P(S \mid A)$

(j) $P(A \mid S)$

13. A recent survey of 500 employees in a large company produced the following information on the employees:

Employment Level	With M.B.A. Degree	Without M.B.A. Degree
Managerial	100	45
Nonmanagerial	125	230

Find the probability that an employee chosen at random

(a) has an M.B.A. degree

(b) is working in a managerial position

(c) has an M.B.A. degree and is employed in a managerial position

(d) is working in a managerial position, given that he has an M.B.A. degree

(e) has an M.B.A. degree, given that he is working in a managerial position

14. Anita has 15 cassette tapes, 6 with rock music and 9 with classical music. If 2 tapes are selected at random, find the probability that

(a) both tapes contain rock music

(b) both tapes contain classical music

15. A shipment consists of 15 color television sets and 10 black-and-white sets. If 2 sets are selected at random, find the probability that both are

(a) color sets

(b) black-and-white sets

16. A grievance committee of 2 members is to be selected at random from 12 labor and 8 management personnel. Find the probability that the committee so formed does not have

    (a) labor representation
    (b) management representation

17. A grievance committee of 3 members is to be chosen at random from a group of 6 men and 4 women. Find the probability that the committee so formed has no
    (a) male representative
    (b) female representative

18. A box contains 7 good and 5 defective items. If a sample of 3 items is to be selected at random, find the probability that
    (a) all items selected are defective
    (b) all items selected are nondefective

19. A package contains 8 seeds for red flowers and 4 seeds for white flowers. If 3 seeds are selected at random, find the probability that
    (a) all seeds germinate in red flowers
    (b) all seeds germinate in white flowers

20. A box of 16 ballpoint pens contains 6 that are defective. If 4 pens are selected at random, find the probability that
    (a) none are defective
    (b) at least one is defective
    (c) all are defective

6.7 INDEPENDENT EVENTS

When E and F are two events, each with a positive probability, we have observed that in general the conditional probability of the event F given the occurrence of the event E differs from the unconditional probability of the event F. However, the case when we have the equality

$$P(F|E) = P(F) \tag{6.8}$$

is of special significance. Equation (6.8) expresses the fact that event F is independent of E in the sense that the occurrence or nonoccurrence of event E in no way affects the probability of event F. This leads us to a formal definition.

DEFINITION 6.7.1

Let E and F be two events defined on the probability space S. The event F is said to be **independent** of the event E if and only if

$$P(F|E) = P(F)$$

Notice that

$$P(F) \cdot P(E|F) = P(E \cap F) = P(E) \cdot P(F|E) \tag{6.9}$$

If F is independent of E, then $P(F|E) = P(F)$ and equation (6.9) implies that $P(E|F) = P(E)$, which in turn means that E is independent of F. Thus, if F is independent of E, it follows that E is independent of F. It also follows from (6.9) that if E and F are independent events, then

$$P(E \cap F) = P(E) \cdot P(F) \qquad\qquad (6.10)$$

EXAMPLES 6.7

1. A fair coin is tossed twice. Let E be the event "head on the first toss" and F be the event "exactly one head." The usual sample space is given by {HH, HT, TH, TT}. Then

$$E = \{\text{HH, HT}\} \qquad F = \{\text{HT, TH}\} \qquad E \cap F = \{\text{HT}\}$$

Assigning a probability of $\frac{1}{4}$ to each simple event, we find that

$$P(E) = \tfrac{2}{4} \qquad P(F) = \tfrac{2}{4} \qquad P(E \cap F) = \tfrac{1}{4}$$

Hence,

$$P(E \cap F) = P(E) \cdot P(F)$$

In view of (6.10), events E and F are independent.

2. Toss a pair of dice, one white and the other green. The sample space S consists of 36 ordered pairs and is given in Table 6.3. Let E be the event "first die shows an even number" and F be the event "the second die shows an odd number." Then

$$E \cap F = \{(2, 1), (2, 3), (2, 5), (4, 1), (4, 3), (4, 5),$$
$$(6, 1), (6, 3), (6, 5)\}$$

and the probabilities are as follows

$$P(E \cap F) = \tfrac{9}{36} = \tfrac{1}{4} \qquad P(E) = \tfrac{18}{36} = \tfrac{1}{2} \qquad P(F) = \tfrac{18}{36} = \tfrac{1}{2}$$

Since $P(E \cap F) = P(E) \cdot P(F)$, it follows that events E and F are independent.

3. Consider a family of three children. The sample space is given by

$$S = \{\text{BBB, BBG, BGB, BGG, GBB, GBG, GGB, GGG}\}$$

Let E be the event "the first child is a boy" and F the event "there is exactly one boy." Then

$$E = \{\text{BBB, BBG, BGB, BGG}\} \qquad F = \{\text{BGG, GBG, GGB}\}$$

and

$$E \cap F = \{\text{BGG}\}$$

Assigning a probability of $\frac{1}{8}$ to each simple event in the sample space S, we find that

$$P(E) = \tfrac{4}{8} \qquad P(F) = \tfrac{3}{8} \qquad P(E \cap F) = \tfrac{1}{8}$$

Now $P(E \cap F) \neq P(E) \cdot P(F)$ and condition (6.10) does not hold. Consequently, events E and F as defined above are *not* independent.

EXERCISES 6.7

1. A fair coin is tossed. Show that the events "head on the first toss" and "tail on the second toss" are independent.

2. A green and a red die are rolled. Show that the events "three on green die" and "six on red die" are independent.

3. A card is drawn at random from a deck of 52 cards. Let A be the event "the card is a club," B be the event "the card is an ace," and C be the event "the card is an ace or a king." Are A and B independent? Are A and C independent?

4. In a group of 500 persons, it was revealed that the distribution of blood types is as follows.

Blood Type	Sex		Total
	Male	Female	
O	110	110	220
A	100	100	200
B	25	25	50
AB	15	15	30
Total	250	250	500

If a person is selected at random from this group, find the probability that
(a) the person selected will have blood type O
(b) the person selected will have blood type AB
(c) the person selected is male
Would you say that sex and blood type are independent? Demonstrate by computing appropriate probabilities.

5. A business executive has two secretaries. The probability that one will not show up in the office on a cold morning is 0.2 and the probability that the other will not show up is 0.3. On a cold morning, find the probability that
(a) both will show up
(b) neither of them will show up
(c) at least one will show up
(d) exactly one will show up

6. The probability that machine A will break down on a particular day is 0.2; the probability that machine B will break down is 0.1. Assuming independence, find the probability that on a particular day

(a) both machines will break down
(b) neither will break down
(c) at least one will break down
(d) exactly one will break down

7. The probabilities that two snowplowing trucks in a maintenance department will not operate on a cold morning are 0.15 and 0.25, respectively. Find the probability that
(a) both trucks will operate
(b) none will operate
(c) at least one will operate
(d) exactly one will operate

8. A couple owns an apartment in the city and a cabin at the lake. In one year the probability of the apartment being burglarized is 0.03, and the probability of the cabin being burglarized is 0.06. For any one year, find the probability that
(a) both will be burglarized
(b) neither will be burglarized
(c) at least one will be burglarized
(d) either the cabin or the apartment (but not both) will be burglarized

9. Two squadrons of parachutists are on a maneuver with each squadron to land at a different location. The probabilities that squadrons A and B will have successful landing are 0.95 and 0.90, respectively. Assuming independence, find the probability that
(a) both landings are successful
(b) none of the landings is successful
(c) at least one of the landings is successful

10. The guidance system of a new type of satellite is composed of two subsystems, A and B. The probabilities that subsystems A and B will operate successfully are 0.98 and 0.97, respectively. Assume that each subsystem works independently of the other but if any subsystem fails, the entire system would fail. What is the probability that the entire system will operate successfully? What is the probability that the system will fail to operate successfully?

6.8   RANDOM VARIABLES

A random variable is a rule that assigns to each simple event in the finite sample space one and only one numerical value. The number of cars in a parking lot, the number of overseas calls handled by an operator, the number of customers waiting to be served in a restaurant, the scores of students on an examination, the number of books sold in the college bookstore, and the velocity of a particle immersed in a liquid are all examples of a random variable. For instance, if a coin is thrown three times, the number of heads X is a random variable; the probability that $X = 2$ is $\frac{3}{8}$. The sum of the numbers X on two dice is a random variable; the probability that X assumes the value 4 is $\frac{3}{36}$ and the probability that X assumes the value 7 or 11, $P(X = 7 \text{ or } 11)$, is $\frac{8}{36}$. Since only events are associated with probabilities, it is reasonable that we set up a structure within which $\{X = 4\}$ or $\{X = 7 \text{ or } 11\}$ are events. This can be accomplished by defining a random variable that assigns a number to each simple event in the sample space. The sample space on which this rule operates is called the **domain** of the random variable, and the set of values that a random variable assumes at each simple event in S is called the **range**.

DEFINITION 6.8.1

A rule that assigns to each simple event in the finite sample space S a numerical value is called a **random variable**. If X is a random variable defined on the sample space $S = \{e_1, e_2, \ldots, e_n\}$, then $X(e_j)$ denotes the value of the random variable for the simple event $\{e_j\}$.

Let us now look at some illustrative examples.

EXAMPLES 6.8

1. Consider a family of three children. The sample space S is given by

$$\{BBB, BBG, BGB, BGG, GBB, GBG, GGB, GGG\}$$

If the random variable X on this sample space is defined to be the number of boys in the family, then

$$X(BBB) = 3 \quad X(BBG) = X(BGB) = X(GBB) = 2$$
$$X(BGG) = X(GBG) = X(GGB) = 1 \quad X(GGG) = 0$$

and the range of X is $\{0, 1, 2, 3\}$.

Let us define another random variable Y on this sample space S to be $+1$ if there are more boys than girls, -1 if there are more girls than boys, and 0 if there are same number of boys and girls. In this case,

$$Y(BBB) = Y(BBG) = Y(BGB) = Y(GBB) = 1$$
$$Y(BGG) = Y(GBG) = Y(GGB) = Y(GGG) = -1$$

The range of Y is evidently $\{-1, +1\}$.

2. A pair of dice is rolled. The sample space is

$$S = \{(1, 1), (1, 2), (1, 3), \ldots, (5, 6), (6, 6)\}$$

containing 36 elementary events, as shown in Table 6.3. Let the random variable X defined on the sample space be the sum of the numbers on two dice. Then

$$X(1, 1) = 2$$
$$X(1, 2) = X(2, 1) = 3$$
$$X(1, 3) = X(2, 2) = X(3, 1) = 4$$
$$X(1, 4) = X(2, 3) = X(3, 2) = X(4, 1) = 5$$

. . . .

$$X(5, 6) = X(6, 5) = 11$$
$$X(6, 6) = 12$$

The range of X is the set

$$\{2, 3, 4, 5, 6, 7, 8, 9, 10, 11, 12\}$$

EXERCISES 6.8

Determine the range of X in each of the following exercises.

1. A fair coin is tossed twice. The random variable X is the number of heads obtained.

2. A fair coin is tossed three times. The random variable X is the number of heads obtained on the three tosses.

3. In a family of four children, let X be the number of boys minus the number of girls.

4. A pair of dice, one white and the other green, are rolled. Let X be the number on the white die minus the number on the green die.

5. Three dice are rolled once. Let X be the sum of the numbers on the three dice.

6. You match coins with Kerry once, winning a dollar if you match and losing a dollar if you do not. Let X denote your winning.

6.9 MATHEMATICAL EXPECTATION OF A RANDOM VARIABLE

An important concept in probability theory that had its origin in games of chance is the mathematical expectation of a random variable. In its simplest form, the mathematical expectation or expected value of a random variable is the sum of the products of a player's possible gains with their associated probabilities. Consider the experiment of rolling a balanced die. Mr. Talbot receives a number of dollars equal to the number of spots that turn up. What should Mr. Talbot pay to make this a fair game? Mr. Talbot may receive $1, $2, $3, $4, $5, or $6, each with a probability of $\frac{1}{6}$. Thus, Mr. Talbot "expects" to receive

$$\$1(\tfrac{1}{6}) + \$2(\tfrac{1}{6}) + \$3(\tfrac{1}{6}) + \$4(\tfrac{1}{6}) + \$5(\tfrac{1}{6}) + \$6(\tfrac{1}{6}) = \$3.50$$

This determines the fair price of the game.

As another illustration, suppose that Susan holds one of the 1000 raffle tickets, for which the first prize is an automobile worth $6000, the second prize is a color television worth $700, the third prize is a movie camera worth $250, the fourth prize is $150 cash, and the remaining 996 tickets pay nothing. What should Susan pay for each raffle ticket she buys? Table 6.6 summarizes this information. The

TABLE 6.6

Prize	Value of the Prize x	Probability of Winning $P(x)$
Automobile	$6000	0.001
Color television	$700	0.001
Movie camera	$250	0.001
Cash	$150	0.001
	0	0.996

last row provides the information that 996 tickets do not pay anything at all. Thus, the price Susan must pay for each raffle ticket is

$$\$6000(0.001) + \$700(0.001) + \$250(0.001) + \$150(0.001) = \$7.10$$

Note that each value of x in the second column is multiplied by the corresponding probabilities in the third column and that mathematical expectation is simply the sum of the products thus obtained.

Generalizing this example, we have the following definition.

DEFINITION 6.9.1

Let X be a random variable that assumes the set of values $x_1, x_2, \ldots, x_n$ with corresponding probabilities $P(x_1), P(x_2), \ldots, P(x_n)$. Then the **mathematical expectation** or the **expected value** of X, denoted by $E(X)$, is the number

$$E(X) = x_1 P(x_1) + x_2 P(x_2) + \cdots + x_n P(x_n)$$

EXAMPLE 6.9.1

A balanced coin is tossed three times. The probabilities for 0, 1, 2, or 3 heads are 0.125, 0.375, 0.375, and 0.125, respectively. Determine the mathematical expectation of the number of heads.

SOLUTION   Table 6.7 summarizes the information.

TABLE 6.7

Number of Heads, x	Probability, $P(x)$
0	0.125
1	0.375
2	0.375
3	0.125

The mathematical expectation of the number of heads is

$$E(X) = 0(0.125) + 1(0.375) + 2(0.375) + 3(0.125) = 1.5$$

Thus, one can "expect" 1.5 heads in three flips of a balanced coin.

We must underscore the fact that an actual observation of 1.5 heads in three throws of a balanced coin is an impossible event. The number of heads can only be 0, 1, 2, or possibly 3, but in no circumstances can we have 1.5 heads. The figure "1.5 heads" is interpreted as simply an *average* that one may compute in a reasonably large number of sequences each based on three throws of a balanced coin.

EXAMPLE 6.9.2

Table 6.8 gives the distribution of weekly unit sales and the associated probabilities. Determine the expected weekly sales.

TABLE 6.8

Units Sold, x	Probability of Selling These Units, $P(x)$
200	0.05
205	0.10
210	0.25
215	0.30
220	0.20
225	0.10

SOLUTION   The mathematical expectation of the weekly sales is

$$E(X) = 200(0.05) + 205(0.10) + 210(0.25) + 215(0.30)$$
$$+ 220(0.20) + 225(0.10) = 214 \text{ units}$$

EXAMPLE 6.9.3

The probability that a 59-year-old coal miner will survive next year is 0.98, and the probability that he will die next year is 0.02. What premium must be charged for a $25,000 one-year term policy if the insurance company is to make a profit of $100?

SOLUTION   The insurance company is liable for payment of $25,000 only if the insured dies next year. Since the probability is 0.02 that the claim against the company will be filed, the net cost to the company for issuing a one-year term policy is

$$\$25,000(0.02) = \$500$$

Thus, the premium the insured must pay is

$$\$500 + \$100 = \$600$$

if the company is to make a profit of $100 on this policy.

EXERCISES 6.9

1. In a game played with a balanced die, a player wins $5 if he rolls a "six" and loses a dollar if he does not. Is this a fair game?

2. A player wins $2.00 if a coin comes up "heads" and loses $2.00 if the coin falls "tails." Is this a fair game?

3. You match coins with Henry, winning a dollar if you match and losing a dollar if you do not. Is this a fair game?

4. In a game played with a pair of balanced dice, a player wins $28 if she throws a sum of 7 or 11 and loses $8 if she throws any other sum. Is this a fair proposition? If not, for whom is it favorable?

5. A man has two half-dollars, four quarters, and four nickels in his pocket, and he plans to buy a newspaper worth 25 cents. A newspaper boy offers to sell him a paper in exchange for a coin to be drawn from his pocket at random. Is this a fair proposition? If not, for whom it is unfavorable?

6. A coin is tossed four times. The probabilities of 0, 1, 2, 3, or 4 heads are 0.0625, 0.25, 0.375, 0.25, and 0.0625, respectively. Determine the "expected" number of heads.

7. The probabilities that 0, 1, 2, 3, or 4 employees in a grocery store are absent on a given day are 0.45, 0.26, 0.15, 0.12, and 0.02, respectively. Determine the number of employees expected to be absent on any one day.

8. The probabilities that a person entering a bookstore will buy 0, 1, 2, or 3 books are 0.62, 0.27, 0.10, and 0.01, respectively. Determine the number of books the store "expects" to sell per customer entering the store on a given day.

9. The probabilities that Bob will sell a piece of real estate property at a profit of $15,000, $13,000, $10,000, $8000, and $6000 are 0.05, 0.10, 0.20, 0.30, and 0.35, respectively. What is his expected profit?

10. Elaine holds one of 1000 raffle tickets, for which the first prize is a boat worth $8000, the second prize is an automobile worth $6000, the third prize is a color television set worth $750, and the fourth prize is a movie camera worth $250. What is the expected profit to the organization if each raffle ticket is sold for $17.50?

11. A couple must sell their cabin to raise immediate cash to defray some unforeseen expenses. The probability that the couple will sell the cabin at a profit of $2000 is 0.30, the probability that the couple will break even is 0.50, and the probability is 0.20 that their cabin will be sold at a loss of $500. What is the couple's expected gain or loss?

12. The probability that a man aged 55 will survive next year is 0.996 and the probability that he will die next year is 0.004. An insurance company offers to sell such a person a $50,000 one-year term policy. What should the premium be if the company is to make a profit of $150?

13. The probability that a 40-year-old truck driver will survive next year is 0.998 and the probability that he will die next year is 0.002. An insurance company offers to sell the truck driver a $20,000 one-year term policy for $75. What is the company's expected profit?

14. The accompanying table gives the distribution of the number of articles manufactured and sold in one year. Determine the "expected" sales.

Units Sold, x	Probability, $P(x)$
100	0.40
200	0.25
250	0.15
300	0.10
350	0.06
375	0.04

15. Each new car rolling off an assembly line is equipped with five new tires. The accompanying table gives the distribution of the number of defective tires on each new car in a certain plant. Determine the "expected" number of defective tires on each new car rolling off the assembly line in this plant.

Defective Tires, x	Probability of Defective Tires, $P(x)$
0	0.96
1	0.02
2	0.015
3	0.0044
4	0.0005
5	0.0001

For Further Reading

More advanced treatment of probability may be found, for instance, in the following books.

Feller, W. *An Introduction to Probability Theory and Some of Its Applications.* New York: John Wiley & Sons, Inc., 1957.

Goldberg, S. *Probability: An Introduction.* Englewood Cliffs, N.J.: Prentice-Hall, Inc., 1960.

Hodges, J. L., and Lehmann, E. L. *Elements of Finite Probability.* San Francisco: Holden-Day, Inc., 1965.

Parzen, J. *Modern Probability Theory and Its Applications.* New York: John Wiley & Sons, Inc., 1960.

Uspensky, J. V. *Introduction to Mathematical Probability.* New York: McGraw-Hill Book Company, 1937.

7

INTRODUCTION TO STATISTICS

7.1 WHAT IS STATISTICS?

Historically, statistics had its origin in the collection of data concerned with human mortality—listing the births and deaths and their causes in England during the first part of the seventeenth century. In fact, one present-day meaning of the term "statistics" is a collection of data. A second meaning of "statistics," also in the plural, is the totality of scientific methods used in the collection and analysis of data in any field of inquiry; in this sense, statistics is a discipline or a subject and is a branch of applied mathematics. In the singular sense, a "statistic" refers to a number that in some sense describes a larger collection of data. Today, however, the science of statistics goes far beyond the mere collection of data and its presentation in tables, charts, and graphs; it is involved in almost all phases of human activity in which it is necessary to make decisions under uncertainty. Scientists, actuaries, engineers, psychologists, economists, and physicians recognize the need of scientific tools for collecting, analyzing, and processing data in their respective fields. Business cycles, drug reactions, learning rates, drilling samples, and several other operations involve decision making in the face of uncertainties and random fluctuations. The objective is to apply logical reasoning that minimizes the degree of uncertainty or at least informs the decision maker of his chances of being in error. Whereas statistics cannot provide a satisfactory solution in every possible uncertain situation, it can provide a basis for a rational decision.

As we said earlier, statistics is used in almost all fields of human activity. Suppose that a particular college wished to determine the average summer earnings of all its full-time students. How could this task be completed? One conceivable but most unlikely solution would be to search the appropriate records in the college, obtain the particular figure for each full-time student, and then compute the average. The result would be an exact answer—a number that would describe the average summer earnings of all full-time students in the college. Obviously, there is no need for statistics here! What would happen if the college records do not have the desired information? A reasonable approach in that case would be to select a number of full-time students among all those enrolled, ask each of them for their summer earnings report, and then base the *estimate* for the average summer earnings of *all* students upon the reports of those students selected for the study. The use of such a selection procedure is called **statistical sampling**.

Statistical sampling lies at the root of statistics itself. If we need to study some specific characteristic of a **population** and do not have the resources or the time to examine all elements of the entire population, we can examine some part of that population, called a **sample**. Having chosen the sample, we collect the relevant data, analyze them, and then draw the conclusions about the characteristic of the entire population based on what we discover about the sample. Reasoning of this type is called **statistical inference**. The methods of statistical inference are required, for instance, to estimate the price per gallon of unleaded gasoline next summer, to predict the impact of inflationary forces on the sale of residential property in New York, to compare the average nicotine content in two kinds of cigarettes, or to decide upon an appropriate dosage of a cold vaccine.

We recognize the fact that the primary function of statistics is the estimation of population characteristics from samples. Suppose that the purchasing department of a large corporation receives a shipment of 1000 special fuses. The company has never used such fuses before and suspects that a large number of them will fail to burn out under current overloads of the type anticipated. How can one test this suspicion? Certainly, it will make no sense to overload all 1000 fuses; if they perform as advertised and do burn out, such a plan would ruin the entire shipment! Alternatively, one may choose a sample of, say, 5 fuses or 10 fuses and test to determine whether or not the sample contains any defective fuses. Based on the performance of the sample, conclusions can be drawn about the entire shipment of 1000 fuses. Of course, the company can feel more confident about its conclusions if it exercises due care in selecting the sample; perhaps, the supplier put all the defective fuses in the bottom layers of these boxes in hopes that the company officials would choose a sample from only those fuses on top!

It must be clear to the reader by now that by **population** we mean the set of *all* possible observations of a certain phenomena; if a set of data contains only part of these observations, we refer to it as a **sample**. A measure based on a sample is called a **statistic**, and a similar measure based on the population data is called a **parameter**. The science of decision making, therefore, concerns itself with the estimation of population parameters based on sample statistics. We must underscore the fact that such inferences depend on the particular set of instances being a fair sample of the population about which we wish to make statements. The results obtained are not certain always to be correct; they are probable, relative to certain body of evidence.

7.2   FREQUENCY DISTRIBUTIONS AND HISTOGRAMS

The human mind is unable to assimilate a mass of complicated data, whether they be the sample or the population. It may, therefore, be necessary to organize the data and present it so as to bring out its essential characteristics. Then the data may not only become understandable; it may lend itself to further statistical treatment.

Consider a newspaper dealer who has kept a record of his sales during a 100-day period that he considers to be representative. The data are given in Table 7.1. We

TABLE 7.1   Sales of Newspapers

49	51	50	53	45	49	50	56	46	54
48	51	60	46	52	46	52	50	57	51
50	52	49	51	40	54	50	55	47	53
49	50	51	54	53	50	52	51	48	44
52	51	47	49	61	52	39	52	57	45
47	48	46	46	50	47	50	51	47	53
50	51	50	48	52	61	49	45	54	60
56	44	45	50	46	47	55	47	53	49
50	41	54	53	50	52	51	45	51	50
49	51	53	46	49	48	46	51	50	49

will use these data to provide a concrete illustration of a frequency distribution. The smallest of these values is 39 and the largest is 61. How are the remaining 98 observations distributed? Do all these observations cluster around a particular value, or are they evenly distributed over the interval from 39 to 61? To answer this question, we perform the actual tally and count the number of values that correspond to the daily sales of newspapers. The results are shown in Table 7.2. The first column in Table 7.2 refers to the 18 distinct values that correspond to the daily sales of the newspapers, the second column provides the actual tallies, and the third column lists the number of days each number of sales occurred.

To further condense the data, if necessary, we may divide the interval from 39 to 61 into an arbitrary number of equal subintervals—the actual number of subintervals depending on the amount of data available. As a general rule, the number of subintervals ranges from 5 to 15; the more data that are available, the greater the number of subintervals needed to group or classify the information. The subintervals are chosen to ensure that they accommodate all the data, and the points subdividing the subintervals are such that it is impossible for any observation to fall exactly on the point of division. This would essentially eliminate any ambiguity regarding the classification of a particular observation or measurement. For the data in Table 7.1, it seems reasonable to choose six classes: 37.5 to 41.5, 41.5 to 45.5, 45.5 to 49.5, . . . , 57.5 to 61.5. These six subintervals would allow the data to be presented on a continuous scale. The distribution of 100 observations in these six subintervals is given in Table 7.3. The numbers 37.5 to 41.5, 41.5 to 45.5 . . . ,

TABLE 7.2   Distribution of Sales of Newspapers

Daily Sales of Newspapers	Tally	Frequency
39	/	1
40	/	1
44	//	2
45	++++	5
46	++++ ///	8
47	++++ //	7
48	++++	5
49	++++ ++++	10
50	++++ ++++ ++++ /	16
51	++++ ++++ ////	14
52	++++ ////	9
53	++++ //	7
54	++++	5
55	//	2
56	//	2
57	//	2
60	//	2
61	//	2

57.5 to 61.5 are called **class boundaries**. The numbers shown in the second column of this table are called **class frequencies**; they give the number of observations falling into each class.

Once we have constructed a frequency distribution, we present it graphically in a form that makes the statistical data readily comprehensible. There are several types of graphical representations used as aids in the study of a frequency distribution, but the most common among all graphical presentation of data is the **histogram**. A histogram is a vertical bar graph, with no space between bars, that

TABLE 7.3   Frequency Distribution of Sales of Newspapers

Daily Sales of Newspapers	Frequency
37.5–41.5	2
41.5–45.5	7
45.5–49.5	30
49.5–53.5	46
53.5–57.5	11
57.5–61.5	4

has class boundaries as points on the horizontal axis and frequencies as units along the vertical axis. The frequency corresponding to a class interval is represented by the height of a rectangle whose base is the class interval. The histogram for the data of Table 7.3 is shown in Figure 7.1.

FIGURE 7.1

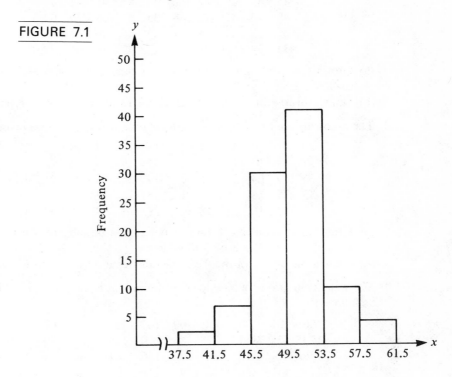

EXERCISES 7.2

1. A salesperson for Richmond and Naylor, Inc., has gathered the following daily sales for its principal product:

$$
\begin{array}{cccccccccc}
28 & 25 & 18 & 23 & 28 & 34 & 31 & 16 & 24 & 26 \\
26 & 21 & 23 & 27 & 21 & 22 & 23 & 22 & 25 & 18 \\
29 & 27 & 27 & 28 & 23 & 27 & 26 & 27 & 28 & 16 \\
18 & 20 & 24 & 25 & 27 & 20 & 29 & 23 & 25 & 28 \\
20 & 21 & 16 & 18 & 26 & 29 & 21 & 27 & 18 & 26
\end{array}
$$

(a) Group these data into a table having class boundaries 15.5 to 18.5, 18.5 to 21.5, 21.5 to 24.5, ..., 33.5 to 36.5.

(b) Draw a histogram of the frequency distribution obtained in part (a).

2. The management of a restaurant recorded the following number of steak dinners served during the last 50 consecutive Saturdays:

```
48  35  38  41  46  45  34  39  36  32
45  42  48  42  40  48  45  42  41  40
48  40  41  38  48  35  41  38  39  34
39  43  44  34  44  38  40  40  34  44
34  39  45  41  42  43  47  38  35  39
```

(a) Group these data into a table having class boundaries 31.5 to 34.5, 34.5 to 37.5, ..., 46.5 to 49.5.

(b) Draw a histogram of the frequency distribution obtained in part (a).

3. The following are the scores 60 students obtained in a college placement test:

```
22  47   9  42  31  17  13  15  18  13   2  21
27  38  15  12  38  10  34  29  26  16  25  33
36  10  24  22  26  19  14  36  18  25  21  35
33  25  18  28  25  17  38  10  13  31  23  24
23  12  16  33  18  26  29  27  29  45  40  46
```

(a) Group these data into a table having class boundaries 1.5 to 7.5, 7.5 to 13.5, 13.5 to 19.5, ..., 43.5 to 49.5.

(b) Draw a histogram of the frequency distribution obtained in part (a).

4. The weights in grams of 50 apples picked out at random from a shipment are as follows:

```
106  107   76   82  109  107  115   93  187   95
123  125  111   92   86   70  126   68  130  129
139  119  115  128  100  186   84   99  113  204
111  141  136  123   90  115   98  110   78   90
107   81  131   75   84  104  110   80  118   82
```

(a) Group these data into a table having class boundaries 64.5 to 84.5, 84.5 to 104.5, ..., 184.5 to 204.5.

(b) Draw a histogram of the frequency distribution obtained in part (a).

5. The weights in grams of the livers of 80 hens are as follows:

```
26.9  26.0  39.6  32.3  30.4  35.6  29.0  29.4  26.3  34.9
29.5  41.2  29.8  32.4  25.9  24.9  24.1  22.8  23.4  28.3
34.1  41.5  25.8  32.5  29.5  30.3  27.3  35.0  27.3  31.7
23.8  29.8  24.7  28.3  26.6  23.4  25.5  29.7  28.2  32.8
26.3  23.1  29.3  31.6  31.2  27.0  34.6  27.4  30.0  26.1
33.3  25.9  26.0  25.2  27.3  32.2  37.7  27.5  36.2  26.7
30.6  27.7  39.1  25.6  22.6  28.3  29.3  28.9  30.6  40.3
44.7  26.4  28.2  22.9  31.0  30.9  21.5  32.3  32.2  31.7
```

(a) Group these data into a table having class boundaries 21.45 to 24.45, 24.45 to 27.45, 27.45 to 30.45, ..., 42.45 to 45.45.

(b) Draw a histogram of the frequency distribution obtained in part (a).

7.3 SUMMATION NOTATION

Symbols are often used to express mathematical formulas or operations. One such symbol, used frequently in mathematics, is the symbol for summation. It is customary to use the Greek letter Σ (capital sigma) for this purpose. The Σ is followed by an expression representing the typical term of the summation and containing a variable, called the *index of summation*, that takes on successive integral values to form the different terms of the sum. The equation below the Σ indicates the index of summation and the first integral value of the index, and the number above the Σ indicates the last value of the index.

EXAMPLE 7.3.1

(a) $\displaystyle\sum_{k=1}^{6} k = 1 + 2 + 3 + 4 + 5 + 6 = 21$

(b) $\displaystyle\sum_{k=1}^{5} k^2 = 1^2 + 2^2 + 3^2 + 4^2 + 5^2$

$\qquad = 1 + 4 + 9 + 16 + 25$

$\qquad = 55$

(c) $\displaystyle\sum_{k=1}^{4} k^3 = 1^3 + 2^3 + 3^3 + 4^3$

$\qquad = 1 + 8 + 27 + 64$

$\qquad = 100$

(d) $\displaystyle\sum_{k=1}^{4} k(k + 3) = 1(1 + 3) + 2(2 + 3) + 3(3 + 3) + 4(4 + 3)$

$\qquad\qquad = 1(4) + 2(5) + 3(6) + 4(7)$

$\qquad\qquad = 4 + 10 + 18 + 28$

$\qquad\qquad = 60$

Summation notation is also used to represent the sum of the terms that involve subscripts. Thus, the height of n individuals, for example, may be referred to as $x_1, x_2, x_3, \ldots, x_n$, where x_i represents the height of ith individual. Obviously, the subscript i can assume any of the values from 1 to n.

EXAMPLE 7.3.2

(a) $\displaystyle\sum_{i=1}^{6} x_i = x_1 + x_2 + x_3 + x_4 + x_5 + x_6$

(b) $\displaystyle\sum_{i=1}^{5} x_i^2 = x_1^2 + x_2^2 + x_3^2 + x_4^2 + x_5^2$

(c) $\displaystyle\sum_{i=1}^{4} i \cdot x_i = x_1 + 2x_2 + 3x_3 + 4x_4$

THEOREM 7.3.1

If c is any constant, then

$$\sum_{i=1}^{n} cx_i = c \sum_{i=1}^{n} x_i$$

PROOF

$$\sum_{i=1}^{n} cx_i = cx_1 + cx_2 + cx_3 + \cdots + cx_n$$

$$= c(x_1 + x_2 + x_3 + \cdots + x_n)$$

$$= c \sum_{i=1}^{n} x_i$$

EXAMPLE 7.3.3

$$\sum_{i=1}^{3} 4x_i = 4 \sum_{i=1}^{3} x_i$$

Let $x_1 = 3$, $x_2 = 5$, and $x_3 = 7$. Then

$$\sum_{i=1}^{3} 4x_i = 4x_1 + 4x_2 + 4x_3$$

$$= 4(3) + 4(5) + 4(7)$$
$$= 12 + 20 + 28$$
$$= 60$$

and

$$4 \sum_{i=1}^{3} x_i = 4(x_1 + x_2 + x_3)$$

$$= 4(3 + 5 + 7)$$
$$= 4(15)$$
$$= 60$$

Thus,

$$\sum_{i=1}^{3} 4x_i = 4 \sum_{i=1}^{3} x_i$$

THEOREM 7.3.2

$$\sum_{i=1}^{n} (x_i + y_i + z_i) = \sum_{i=1}^{n} x_i + \sum_{i=1}^{n} y_i + \sum_{i=1}^{n} z_i$$

PROOF $\displaystyle\sum_{i=1}^{n} (x_i + y_i + z_i) = (x_1 + x_2 + x_3 + \cdots + x_n)$

$$+ (y_1 + y_2 + y_3 + \cdots + y_n)$$
$$+ (z_1 + z_2 + z_3 + \cdots + z_n)$$

$$= \sum_{i=1}^{n} x_i + \sum_{i=1}^{n} y_i + \sum_{i=1}^{n} z_i$$

COROLLARY   If c is any constant, then

$$\sum_{i=1}^{n} c = nc$$

EXAMPLE 7.3.4

$$\sum_{i=1}^{15} (x_i^2 + 8x_i + 12) = \sum_{i=1}^{15} x_i^2 + \sum_{i=1}^{15} 8x_i + \sum_{i=1}^{15} 12$$

$$= \sum_{i=1}^{15} x_i^2 + 8 \sum_{i=1}^{15} x_i + 12(15)$$

$$= \sum_{i=1}^{15} x_i^2 + 8 \sum_{i=1}^{15} x_i + 180$$

Now we will show that

$$\sum_{i=1}^{n} x_i^2 \neq \left(\sum_{i=1}^{n} x_i \right)^2 \tag{7.1}$$

and

$$\sum_{i=1}^{n} x_i y_i \neq \left(\sum_{i=1}^{n} x_i \right)\left(\sum_{i=1}^{n} y_i \right) \tag{7.2}$$

EXAMPLE 7.3.5

Let $x_1 = 3$, $x_2 = 4$, $x_3 = 6$, $y_1 = 1$, $y_2 = 5$, and $y_3 = 8$. Then

$$\sum_{i=1}^{3} x_i = x_1 + x_2 + x_3 = 3 + 4 + 6 = 13$$

$$\sum_{i=1}^{3} x_i^2 = x_1^2 + x_2^2 + x_3^2 = 3^2 + 4^2 + 6^2 = 9 + 16 + 36 = 61$$

Notice that

$$\left(\sum_{i=1}^{3} x_i \right)^2 = (13)^2 = 169$$

Thus,

$$\sum_{i=1}^{3} x_i^2 \neq \left(\sum_{i=1}^{3} x_i \right)^2$$

It may also be observed that

$$\sum_{i=1}^{3} y_i = y_1 + y_2 + y_3 = 1 + 5 + 8 = 14$$

$$\sum_{i=1}^{3} x_i y_i = x_1 y_1 + x_2 y_2 + x_3 y_3$$

$$= 3(1) + 4(5) + 6(8)$$

$$= 3 + 20 + 48$$

$$= 71$$

but

$$\left(\sum_{i=1}^{3} x_i \right)\left(\sum_{i=1}^{3} y_i \right) = 13(14) = 182$$

It follows that

$$\sum_{i=1}^{3} x_i y_i \neq \left(\sum_{i=1}^{3} x_i \right)\left(\sum_{i=1}^{3} y_i \right)$$

EXERCISES 7.3

1. Express each of the following without the summation sign.

   (a) $\displaystyle\sum_{i=1}^{3} x_i$

   (b) $\displaystyle\sum_{i=1}^{4} x_i y_i$

   (c) $\displaystyle\sum_{i=1}^{5} (x_i + 4)$

   (d) $\displaystyle\sum_{i=1}^{4} (x_i + y_i + 4)$

2. Express each of the following in summation notation.

   (a) $x_1 + x_2 + x_3 + x_4 + x_5$
   (b) $x_1^2 + x_2^2 + x_3^2 + x_4^2$
   (c) $(x_1 + 2) + (x_2 + 2) + (x_3 + 2) + (x_4 + 2) + (x_5 + 2)$
   (d) $x_2 y_2 + x_3 y_3 + x_4 y_4 + x_5 y_5 + x_6 y_6$
   (e) $x_1 y_1 z_1 + x_2 y_2 z_2 + \cdots + x_n y_n z_n$

3. Let $x_1 = 2$, $x_2 = 4$, $x_3 = 5$, $x_4 = 1$, and $x_5 = -2$. Determine each of the following sums.

   (a) $\displaystyle\sum_{i=1}^{4} x_i$

   (b) $\displaystyle\sum_{i=3}^{5} (x_i - 3)$

   (c) $\displaystyle\sum_{i=1}^{5} x_i^2$

   (d) $\displaystyle\sum_{i=1}^{5} (x_i + 2)^2$

(e) $\displaystyle\sum_{i=2}^{4} (3x_i + 4)$

(f) $\displaystyle\sum_{i=1}^{5} i \cdot x_i$

4. Let $x_1 = 2$, $x_2 = 5$, $x_3 = 5$, $y_1 = 4$, $y_2 = 6$, and $y_3 = 5$. Determine each of the following.

(a) $\displaystyle\left(\sum_{i=1}^{3} x_i\right)^2$

(b) $\displaystyle\sum_{i=1}^{3} x_i^2$

(c) $\displaystyle\sum_{i=1}^{3} x_i y_i$

(d) $\displaystyle\left(\sum_{i=1}^{3} x_i\right)\left(\sum_{i=1}^{3} y_i\right)$

(e) $\displaystyle\sum_{i=1}^{3} y_i^2$

(f) $\displaystyle\left(\sum_{i=1}^{3} y_i\right)^2$

(g) Is $\displaystyle\sum_{i=1}^{3} x_i^2 = \left(\sum_{i=1}^{3} x_i\right)^2$?

(h) Is $\displaystyle\sum_{i=1}^{3} x_i y_i = \left(\sum_{i=1}^{3} x_i\right)\left(\sum_{i=1}^{3} y_i\right)$?

7.4   MEASURES OF CENTRAL TENDENCY

After a brief mathematical excursion in Section 7.3, we return to one of the main objectives of our study—describing a large set of data. Frequency distributions and other graphical devices can only point to the salient features and general trends in the data; yet these are inadequate and do not provide a summary description of the essential characteristics of a set of data. Something more precise, yet readily comprehensible, is needed to understand the significance of the data. In this section we shall study one measure of description, the method of statistical averages.

Statistical averages are commonly referred to as "measures of location," "measures of position," "measures of central value," or "measures of central tendency." These measures are also referred to as "averages" in the sense that they provide numbers that represent the "center" or "middle value" of a large set of data.

The word "average" has a loose connotation and different meanings: for example, when we speak of the average scores of students in an examination, the average shoe size sold in a store, the average rent of a two-bedroom apartment, a person of average intelligence, a problem of average difficulty, the average number of telephone calls received at a switchboard during the rush hour, the average salary of a mathematics teacher in a college, and so on. Nevertheless, an average is a single numerical expression in which the net result of a mass of complicated data is concentrated. Its main purpose is to provide an overall picture of the data and in doing so, it affords a basis of comparison with other sets of data.

There are several types of averages used to describe the data, and it is generally the purpose for which an average is to be employed that will determine its choice.

However, we shall restrict ourselves to three types of averages: mean, median, and mode.

The most common measure of location of data is the arithmetic average, arithmetic mean, or simply the mean.

DEFINITION 7.4.1

Let $x_1, x_2, x_3, \ldots, x_n$ be n observations in a sample. Then the sample **mean**, denoted by $\bar{x}$, is given by

$$\bar{x} = \frac{\sum\limits_{i=1}^{n} x_i}{n} \tag{7.3}$$

In other words, the mean of a finite set of n observations is simply their sum divided by the total number of observations. For example, if we are given the ages of 10 top-ranked tennis players in a city tournament:

$$18, \quad 31, \quad 26, \quad 18, \quad 34, \quad 23, \quad 18, \quad 32, \quad 25, \quad 27$$

then the mean or the average age of these tennis players is

$$\bar{x} = \frac{18 + 31 + 26 + 18 + 34 + 23 + 18 + 32 + 25 + 27}{10}$$

$$= \frac{252}{10} = 25.2 \text{ years}$$

Recall that we are always concerned with both the sample and the population, each of which possesses a mean. If we refer to x's as a sample, we represent the sample mean by $\bar{x}$, but if we look up x's as a population, we shall use the symbol μ (the Greek lowercase letter mu) to describe the population mean. The number of values in the sample, the sample size, is denoted by n, but if we substitute the population size N for the sample size n, we have

$$\mu = \frac{\sum\limits_{i=1}^{N} x_i}{N} \tag{7.4}$$

When a balanced die is rolled, the upturned faces will be 1, 2, 3, 4, 5, or 6. These six numbers, being all the possible values, constitute the population. Thus, the population mean is

$$\mu = \frac{1 + 2 + 3 + 4 + 5 + 6}{6} = 3.5 \text{ dots}$$

We have described the arithmetic mean as the most commonly used measure of location. Now we mention some of the properties that make the mean a desirable measure. The mean is widely used and is well understood by most people. It can always be computed for numerical data, makes use of all observations in the data, and is unique. It is amenable to further statistical manipulation.

One obvious disadvantage of the arithmetic mean is that a single extreme value can have a profound effect on the arithmetic mean to such an extent that it becomes questionable whether the arithmetic mean is a true representative of the data. Consider that four houses sold by a real estate agency last month were priced at $62,000, $70,000, $68,000, and $65,000. The average selling price is, therefore, given by

$$\frac{62,000 + 70,000 + 68,000 + 65,000}{4} = \$66,250$$

Had the real estate agency sold one more house last month at a price of, say, $250,000, the average selling price would have jumped to $103,000. It is, therefore, reasonable to conclude that some other measure of location or an additional descriptive measure is required to convey this information.

A second measure of location is the *median*.

DEFINITION 7.4.2

The **median** of a set of n observations $x_1, x_2, x_3, \ldots, x_n$ is the value of x that falls in the middle when the observations are arranged in an increasing or decreasing order of magnitude.

If the number of observations is odd, then there is always a middle item whose value is the median. Suppose, for example, that the students of a class, 31 in number, are asked to stand in order of their height. Then the 16th student from either side will be the one whose height will represent the median height of the class. This method of picking up the median can be symbolically expressed as

$$\text{median} = \text{size of } \left(\frac{n + 1}{2}\right)\text{th item}$$

where n is the number of observations in the data. Thus, the median of 13 observations is given by the $(13 + 1)/2 = 7$th largest, the median of 47 observations is given by the $(47 + 1)/2 = 24$th largest, and the median of 123 observations is given by the $(123 + 1)/2 = 62$nd largest observation in the data.

If we have an even number of observations, then the median is defined as the arithmetic mean of two middle values. Consider the set of observations

$$18, \quad 5, \quad 12, \quad 10, \quad 40, \quad 28$$

which, when arranged in order of magnitude, are

$$5, \quad 10, \quad 12, \quad 18, \quad 28, \quad 40$$

The median of these six observations is the mean of 12 and 18, that is, 15. Remember that the formula $(n + 1)/2$ gives the position of the median regardless of whether n is even or odd. If $n = 6$, then the $(6 + 1)/2 = 3.5$th observation, which lies halfway between the third and fourth largest values, represents the value of the median. For $n = 126$, the median is the $(126 + 1)/2 = 63.5$th largest item, and essentially this falls halfway between the 63rd and 64th largest observations in the data.

By Definition 7.4.2, there is always one median. Once the data are arranged in order of magnitude, the median is easy to locate. Like the mean, the median can always be located for numerical data, and it is unique. Unlike the mean, it is not affected significantly by the extreme values. Further, both measures of location describe the central position of the data. The median is central in the sense that it divides the data into two equal parts, one part consisting of all values less than or equal to the median and the other part consisting of all values greater than or equal to the median. The mean is central in a different sense: it describes the center of gravity or the point around which the data are concentrated. However, the median has a distinct advantage over the mean, as it can be used to study phenomena that do not permit a quantitative description.

Besides the mean and the median, there are several other measures of location used to describe the "center" of the data. We shall briefly consider one of these, the concept of *mode*.

DEFINITION
7.4.3

Mode is the value of that item in the data that occurs most frequently.

This means that when we speak of the average student, the average wage, or the average rent, we generally imply the modal student, the modal wage, and the modal rent. If we say that modal marks obtained by the students in a statistics examination is 68, we mean that 68 is the predominant score in the sense that more students scored 68 points than any other score. As high a score as 95 points or as low a score as 10 points in that examination are exceptions; they are less frequented.

A distinct disadvantage of the mode is that it does not always exist; if it does, it may not be unique. For example, the data

$$2, \quad 4, \quad 6, \quad 6, \quad 6, \quad 7, \quad 9$$

have one mode, 6. The data

$$3, \quad 6, \quad 7, \quad 8, \quad 10, \quad 7, \quad 12, \quad 10, \quad 13$$

have two modes, 7 and 10, because both occur with the same frequency. However, the data

$$12, \quad 13, \quad 13, \quad 13, \quad 14, \quad 14, \quad 14, \quad 15, \quad 16, \quad 16, \quad 16, \quad 17, \quad 18, \quad 18, \quad 18$$

have no mode.

EXERCISES 7.4

In Exercises 1–10, calculate the mean, median, and the mode for each of the following sets of data.

1. 8, 11, 11, 12, 11

2. 17, 9, 10, 10, 12

3. 27, 29, 33, 37, 30, 36

4. 54, 45, 46, 39, 44, 53, 41

5. 40, 52, 65, 70, 74, 76, 88, 76, 76, 90

6. 19, 25, 24, 25, 23, 22, 25, 27, 22

7. 14, 24, 22, 20, 25, 20, 17, 20, 21

8. 75, 55, 53, 65, 75, 72, 85, 90, 77, 88, 75

9. 78, 80, 79, 85, 78, 87, 86, 89, 91, 85, 78, 74, 60, 54

10. 55, 63, 60, 58, 53, 54, 45, 68, 42, 70, 41

11. The incomes for the seven employees of a small store are

    $8000, $10,000, $12,000, $15,000, $20,000, $22,000, $25,000

    Calculate the mean and median salaries.

12. The five boys in the Talbot family have the following heights (in inches):

    58, 72, 70, 68, 60

    Calculate the mean and median heights of the boys.

13. The following are the scores of 25 students on a statistics examination:

    81, 85, 73, 56, 74, 62, 84, 86, 78, 93, 77, 64
    68, 77, 75, 70, 78, 86, 73, 69, 72, 86, 83, 75, 84

    Find the mean and median scores.

14. The downtimes (length of periods of inactivity due to failure) of a certain computer (in minutes) are as follows:

    73, 82, 49, 10, 52, 50, 62, 52, 40

    Compute the mean downtime.

15. The following are the scores of 29 students on an examination:

    31, 96, 82, 69, 45, 82, 70, 33, 78, 54, 63
    30, 56, 52, 57, 38, 66, 55, 51, 62, 44, 46
    62, 51, 94, 66, 53, 40, 52

    Find the mean and the median scores.

16. In a class of 25 students, test grades have just been returned. What statistic is apt to give a particular student the best idea of his or her standing in the class—mode, the median, or the mean? Justify your answer.

17. A chicken farmer found that his hens "averaged" 350 eggs per day in a certain week.

    (a) To which "average" is the farmer probably referring; mode, median, or mean?

(b) The farmer's records for that week show

$$347, \quad 351, \quad 358, \quad 345, \quad 350, \quad 353$$

Records for one day have been lost. How many eggs were laid that day?

18. Anita has scores of 75, 50, 80, and 60 on the first four hour examinations in a mathematics course. Determine the score on her next hour examination if she needs an average of 72 in the course.

7.5   MEASURES OF VARIATION

Measures of location consider only the central position in a set of data; they provide no information as to how the individual observations in the data are spread out. It is reasonable to assume that all distributions are not similar; essentially, they differ in two respects. First, two sets of data may have the same mean, yet the observations may be scattered differently. Consider, for example, the following sets of data:

Data set 1:   8,   10,   11,   12,   13,   14,   16
Data set 2:   4,   6,   8,   10,   12,   14,   16,   18,   20

Both sets of data have the same mean, 12. In fact, one can visualize several sets of data that may have the same mean; but the mean or any other measure of location does not relay any information about the spread, scatter, dispersion, or variation among the individual observations in the data. Observe that data sets 1 and 2 have the same mean, yet they differ in some respects. The mean of data set 1 is based on seven values, whereas the mean of data set 2 takes nine values into account. Thus, the extent to which individual values are spread out is larger for data set 2 than it is for data set 1. As another illustration, consider

Data set 3:   11,   13,   14,   15,   16,   17,   19

The reader can verify that the mean of this data set is 15; that each observation in data set 3 is 3 units more than the corresponding observations in data set 1, but nevertheless any measure of variability that is proposed remains the same for data sets 1 and 3.

Thus, we recognize the need for more information in addition to what the mean can provide. The range (difference between the largest and smallest values in the data) is not a satisfactory measure because it is based on the extreme values and does not take into account *all* the observations in the data. Suppose that in a class of 30 students, the height of the tallest student is 73 inches, and that of the shortest student is 58 inches. Then the range would evidently be $73 - 58 = 15$ inches. Despite the obvious advantage that the range is easy to compute and easy to understand, it remains too indefinite to be used as a practical measure of variation. For instance, if a dwarf whose height is 42 inches were admitted to the class, the range would suddenly rise to 31 inches, although the average height of the class might not be materially affected. Consequently, the range is a quick, though not necessarily accurate, measure of variability of a set of data.

In searching for the alternative to the range for measuring variation in the data, we are essentially looking for a measure of variation that uses all the observations in the data and can be treated algebraically. It would, therefore, seem reasonable to define variation in terms of the distances by which observations in the data deviate from a central value such as the mean. This means that if we have a set of n observations

$$x_1, x_2, x_3, \ldots, x_n$$

whose mean is $\bar{x}$, we need to determine the amounts by which the individual observations differ from their mean, such as

$$x_1 - \bar{x}, x_2 - \bar{x}, x_3 - \bar{x}, \ldots, x_n - \bar{x}$$

These quantities are called the **deviations from the mean**. If we consider the sum of deviations, we find that since some of the deviations are positive and some are negative, the sum of the deviations

$$\sum_{i=1}^{n} (x_i - \bar{x}) = 0$$

This difficulty with deviations from the mean can be resolved if we square each of the deviations and then form the sum

$$\sum_{i=1}^{n} (x_i - \bar{x})^2$$

This leads us to the following definition of the sample variance.

DEFINITION 7.5.1

Let $x_1, x_2, \ldots, x_n$ be n observations with mean $\bar{x}$. The **sample variance**, denoted by s^2, is

$$s^2 = \frac{\sum_{i=1}^{n} (x_i - \bar{x})^2}{n - 1} \tag{7.5}$$

DEFINITION 7.5.2

The square root of the sample variance, denoted by s, is called the *sample standard deviation*.

In other words,

$$s = \sqrt{\frac{\sum_{i=1}^{n} (x_i - \bar{x})^2}{n - 1}} \tag{7.6}$$

Let us now look at an illustrative example.

EXAMPLE 7.5.1

Calculate the sample standard deviation for the following set of data:

$$8, \quad 10, \quad 11, \quad 12, \quad 13, \quad 14, \quad 16$$

SOLUTION   As a first step, we compute the sample mean $\bar{x}$ and then obtain the necessary information, as shown in Table 7.4. The sample mean is

$$\bar{x} = \frac{84}{7} = 12$$

The sample standard deviation is given by

$$s = \sqrt{\frac{42}{6}} = \sqrt{7} = 2.6457$$

TABLE 7.4

x_i	$x_i - \bar{x}$	$(x_i - \bar{x})^2$
8	−4	16
10	−2	4
11	−1	1
12	0	0
13	1	1
14	2	4
16	4	16
84	0	42

That the calculations of sample standard deviation were straightforward in Example 7.5.1 is due largely to the fact that all observations in the data, their mean $\bar{x}$, and the deviations from the mean were all whole numbers. If the values in the data or its mean are not in whole numbers, the following theorem would save us considerable time and effort otherwise spent in unnecessary calculations.

THEOREM
7.5.1

If $x_1, x_2, x_3, \ldots, x_n$ be n observations in a sample, then the sample variance is given by

$$s^2 = \frac{\sum\limits_{i=1}^{n} x_i^2 - \dfrac{\left(\sum\limits_{i=1}^{n} x_i\right)^2}{n}}{n - 1}$$

PROOF   By Definition 7.5.1 for the sample variance, we have

$$s^2 = \frac{\sum\limits_{i=1}^{n} (x_i - \bar{x})^2}{n - 1}$$

Observe that

$$\sum_{i=1}^{n} (x_i - \bar{x})^2 = \sum_{i=1}^{n} (x_i^2 - 2\bar{x}x_i + \bar{x}^2)$$

$$= \sum_{i=1}^{n} x_i^2 - \sum_{i=1}^{n} 2\bar{x}x_i + \sum_{i=1}^{n} \bar{x}^2$$

$$= \sum_{i=1}^{n} x_i^2 - 2\bar{x} \sum_{i=1}^{n} x_i + n\bar{x}^2$$

$$= \sum_{i=1}^{n} x_i^2 - 2 \frac{\left(\sum\limits_{i=1}^{n} x_i\right)}{n} \left(\sum_{i=1}^{n} x_i\right) + n \frac{\left(\sum\limits_{i=1}^{n} x_i\right)^2}{n^2}$$

$$= \sum_{i=1}^{n} x_i^2 - \frac{2\left(\sum\limits_{i=1}^{n} x_i\right)^2}{n} + \frac{\left(\sum\limits_{i=1}^{n} x_i\right)^2}{n}$$

$$= \sum_{i=1}^{n} x_i^2 - \frac{\left(\sum\limits_{i=1}^{n} x_i\right)^2}{n}$$

Thus, the sample variance is

$$s^2 = \frac{1}{n-1}\left[\sum_{i=1}^{n} x_i^2 - \frac{\left(\sum\limits_{i=1}^{n} x_i\right)^2}{n}\right]$$

A distinct advantage in using Theorem 7.5.1 is that it does not require that we actually find the deviations of the individual observations from the mean; instead, we simply compute

$$\sum_{i=1}^{n} x_i \quad \text{and} \quad \sum_{i=1}^{n} x_i^2$$

and then use Theorem 7.5.1.

As an illustration, let us consider the following example.

EXAMPLE 7.5.2

Calculate the sample variance for the following data:

27,   30,   31,   32,   34,   32,   30,   36

SOLUTION   Without using Theorem 7.5.1, we obtain the values shown in Table 7.5. The sample mean is

$$\bar{x} = \frac{252}{8} = 31.5$$

and the sample variance is

$$s^2 = \frac{52}{7} = 7.4285$$

Using Theorem 7.5.1, we get the values given in Table 7.6.

$$\sum_{i=1}^{8} x_i = 252 \qquad \sum_{i=1}^{8} x_i^2 = 7990$$

and the sample variance is given by

$$s^2 = \frac{7990 - \dfrac{(252)^2}{8}}{7} = \frac{7990 - 7938}{7} = \frac{52}{7} = 7.4285$$

as before.

TABLE 7.5		
x_i	$x_i - \bar{x}$	$(x_i - \bar{x})^2$
27	−4.5	20.25
30	−1.5	2.25
31	−0.5	0.25
32	0.5	0.25
34	2.5	6.25
32	0.5	0.25
30	−1.5	2.25
36	4.5	20.25
252	0	52.00

TABLE 7.6	
x_i	x_i^2
27	729
30	900
31	961
32	1024
34	1156
32	1024
30	900
36	1296
252	7990

We shall now show that these calculations can be further simplified by subtracting an arbitrary constant from each value x_i in the data. That this has no effect on the final result is illustrated by the fact that if we subtract a constant, say $c = 27$, from each of the observations x_i in Example 7.5.2, we would obtain $y_i = x_i - 27$, as follows:

$$y_i: \quad 0, \quad 3, \quad 4, \quad 5, \quad 7, \quad 5, \quad 3, \quad 9$$

The reader can verify that

$$\sum_{i=1}^{8} y_i = 36 \qquad \sum_{i=1}^{8} y_i^2 = 214$$

Using Theorem 7.5.1, we get

$$s^2 = \frac{214 - \dfrac{(36)^2}{8}}{7} = \frac{214 - 162}{7} = \frac{52}{7} = 7.4285$$

which is precisely what we had before.

Formulas (7.5) and (7.6) apply essentially to samples, but if we substitute population mean μ for the sample mean $\bar{x}$ and use N as a divisor instead of $(n - 1)$, we obtain corresponding formulas for the population variance and the population standard deviation.

DEFINITION 7.5.3

Let $x_1, x_2, \ldots, x_N$ constitute a finite population with mean μ. The **population variance**, denoted by σ^2, is

$$\sigma^2 = \frac{\displaystyle\sum_{i=1}^{N} (x_i - \mu)^2}{N} \tag{7.7}$$

The square root of the population variance, denoted by σ, is called the **population standard deviation**.

The reader may wonder about the apparent inconsistency in definition of the population and sample variances. In using $(n - 1)$ instead of n for the denominator in definition of the sample variance, we are not being arbitrary; there is a good reason for it. In all statistical studies, the purpose of calculating a sample statistic, such as the sample mean or the sample variance, is to estimate the corresponding population parameters. We use the sample mean as an estimate of the population mean. Although it was not stated specifically, we wished to convey the impression that the sample mean is a good estimate of the population mean, μ. In the same vein, the sample variance, s^2, which uses $(n - 1)$ in the denominator, provides an unbiased estimate of the population variance σ^2. Note, however, that this modification of using $(n - 1)$ instead of n is significant only when n is small and its effect becomes negligible for large value of n.

EXAMPLE 7.5.3

Calculate the population standard deviation for the data

$$6, \quad 8, \quad 10, \quad 12, \quad 14, \quad 16, \quad 18, \quad 20, \quad 22$$

SOLUTION   The population mean $\mu = 14$. To compute the sum of squares of the deviations of the observations from the population mean, we first obtain values as shown in Table 7.7. The population variance is

$$\sigma^2 = \frac{240}{9} = 26.67$$

TABLE 7.7

x_i	$x_i - \mu$	$(x_i - \mu)^2$
6	−8	64
8	−6	36
10	−4	16
12	−2	4
14	0	0
16	2	4
18	4	16
20	6	36
22	8	64
126	0	240

and the population standard deviation is

$$\sigma = \sqrt{26.67} = 5.164$$

If the population mean μ is *not* a whole number, we would prefer to use the formula

$$\sigma^2 = \frac{1}{N}\left[\sum_{i=1}^{N} x_i^2 - \frac{\left(\sum_{i=1}^{N} x_i\right)^2}{N}\right]$$

for calculating the population variance.

EXERCISES 7.5

In Exercises 1–8, calculate the sample mean and the sample standard deviation for each of the following sets of data.

1. 7, 7, 8, 9, 10, 11, 11

2. 15, 17, 12, 11, 13, 10, 13

3. 13, 10, 11, 15, 13, 8, 14

4. 7, 11, 1, 4, 10, 3, 6

5. 22, 26, 16, 19, 25, 18, 21

6. 75, 60, 50, 80, 71

7. 2.3, 2.1, 3.1, 2.7, 3.3

8. 72, 67, 69, 65, 64, 68, 77, 62, 71, 74, 73, 69

9. The incomes for the five employees of a store are as follows:

$8000, $10,000, $12,000, $14,000, $16,000

Calculate the mean and population standard deviation.

10. The following are the scores of 12 students on a mathematics examination:

81, 85, 73, 56, 74, 62, 84, 86, 78, 93, 77, 64

Calculate the sample mean and the sample standard deviation.

11. Calculate the sample standard deviation for the data in Exercise 7.4.14.

12. Calculate the population standard deviation for the data in Exercise 7.4.12.

7.6 CONTINUOUS RANDOM VARIABLES

Random variables are generally classified into two categories: discrete and continuous. A random variable is **discrete** if it assumes a finite set of values or as many values as there are integers. The number of customers in a department store, the number of defective televisions received by a dealer in a new shipment, the number of speeding tickets issued on Interstate 84 in the state of New York last month, and the number of passengers in a plane are examples of discrete random variables. We shall continue to use capital letters, say X, to denote the random variable and the corresponding lowercase letter, x, to denote the set of values assumed by the random variable X. Thus, if a coin is thrown three times, the number of heads X is a random variable that may assume values $x = 0$, $x = 1$, $x = 2$, and $x = 3$.

A **continuous** random variable, on the other hand, assumes values on a continuous scale and is associated with the sample space representing the infinitely large number of sample points in an interval. The height or weight of an individual, the blood pressure of a patient, the consumption of heating oil in a new home, the amount of alcohol in a wine, the amount of nicotine in a cigarette, and the time needed to reach a college campus are all typical examples of continuous random variables. Although we may round our answers to the nearest whole number or to a few decimal places, there remain a continuum of possibilities in each of these examples.

A discrete distribution is described by listing probabilities associated with the possible integer values of the discrete random variable. In the continuous case, the probabilities associated with the individual points on the real line are zero. Thus, if X is a continuous random variable, then $P(X = 2.0) = 0$, $P(X = 3.14) = 0$, and so on. At first this may seem startling, since the random variable must assume some values on the real line. We recognize that each point on a line segment has length zero, and the collection of all points in the interval taken together gives a segment of positive length. It is, therefore, reasonable to assume that if X is a continuous random variable, then

$$P(X = a) = 0$$

and $$P(a < X < b) > 0, \qquad b \neq a$$

Thus, we might ask for the probability that the length of life of an automatic washer will be less than three years, that a commuter train will arrive at the railroad station in the next 10 minutes, that the College Board verbal score of a randomly selected student will be below 600, and so on. To associate probabilities with continuous random variables, we will introduce the concept of a continuous distribution.

Curves like the one in Figure 7.2 represent graphs of functions that are referred to as continuous distributions. We assume that the total area under the curve is 1.00 and that the probability that a continuous random variable will have a value between a and b is given by the area bounded by the curve, the horizontal axis, and the vertical lines $x = a$ and $x = b$. In other words,

$$P(a < X < b) = \text{striped area in Figure 7.2.}$$

FIGURE 7.2

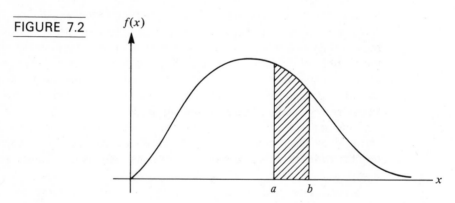

We wish to emphasize that there is no difference between

$$P(a < X < b), \quad P(a < X \leqslant b), \quad P(a \leqslant X < b), \quad \text{and} \quad P(a \leqslant X \leqslant b)$$

since the probability is zero that a continuous random variable X assumes a particular value a or b.

The mean and the standard deviation of a continuous distribution that approximates in some sense a distribution of a discrete random variable are approximately equal to those of the discrete distribution, with the essential difference that the constants in the continuous case are obtained through the use of advanced mathematics. Nevertheless, μ and σ still measure, respectively, the approximate center and the "spread" or dispersion of the distribution.

7.7 THE NORMAL DISTRIBUTION

One of the most important and useful continuous distributions in probability and statistics is the **normal distribution**. Historically, its study goes back to the seventeenth and eighteenth centuries and is associated with Abraham de Moivre (1667–1754), Carl Gauss (1777–1825), and Pierre Laplace (1749–1827). At that time, it received the attention of mathematicians and natural and social scientists. Its application to biological data was established by Sir Francis Galton (1822–1911).

The graph of the normal distribution, also called the *normal curve*, is a bell-shaped curve that is symmetrical about its mean μ. The curve extends infinitely in both directions, coming closer and closer to the horizontal axis but never quite touching it (see Figure 7.3). Notice that as we move along the normal curve away from the point P, we pass through two points A and B located on either side of the vertical line $x = \mu$. These points, where the curve changes its course from the **concave-down** to the **concave-up** position, are called **inflection points** for the curve. The horizontal distance from A (or B) to the line $x = \mu$ is 1 unit of standard deviation.

FIGURE 7.3

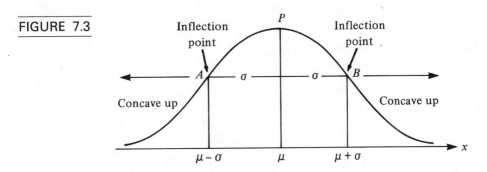

In actual practice, it is not necessary to extend the tails of the curve very far, because the area under the curve becomes negligible as we move more than 3 standard deviations away from the mean μ of the distribution. An important feature of a normal distribution is that it is completely determined by its mean μ and standard deviation σ. The curve has the following additional characteristics.

1. The curve has a single maximum at $x = \mu$.
2. The curve is symmetrical about the vertical line through $x = \mu$. Thus, the height of the curve at some point, say $x = \mu + \sigma$, is the same as the height of the curve at $x = \mu - \sigma$.
3. The curve is **concave downward** between $x = \mu - \sigma$ and $x = \mu + \sigma$, and **concave upward** for values of x outside that interval.
4. Since the curve is symmetrical about $x = \mu$ and the total area under the curve is 1.0, it follows that the area on either side of the vertical line $x = \mu$ is 0.5.
5. As x moves away on either side of the mean μ, the height of the curve decreases but remains nonnegative for all real values of x.

The distribution in Figure 7.3 having the foregoing characteristics is the normal distribution. Note that if X is a random variable with mean μ and standard deviation σ, then there is one and only one normal distribution that has the given mean μ and the given standard deviation σ.

Figure 7.4 shows two normal distributions that have the same mean but different standard deviations σ_1 and σ_2, and Figure 7.5 illustrates two normal curves that have the same standard deviation but different means μ_1 and μ_2. The two curves are identical in shape but are centered at different positions along the horizontal axis. Many random variables have distributions that can be described by the normal curve once μ and σ are specified. We will assume here that these two parameters, μ and σ, are known.

FIGURE 7.4

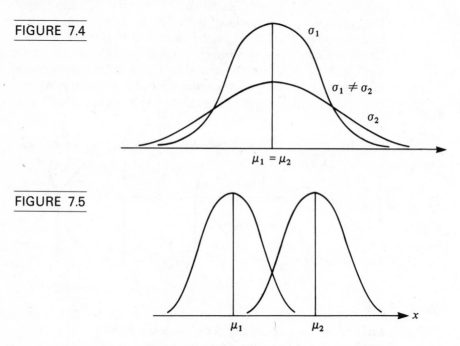

FIGURE 7.5

Since both parameters μ and σ can assume any finite value, we can generate an infinite number of normal curves, one for each pair of μ and σ. Because a separate table of areas for each of these bell-shaped curves is obviously impractical, we convert the unit of measurement of an arbitrary normal distribution into standard units by using the formula

$$Z = \frac{X - \mu}{\sigma} \qquad \text{or} \qquad X = \mu + \sigma Z$$

Figure 7.6 shows the original variable X as well as the transformed variable Z.

The distribution of a normal random variable with mean zero and standard deviation 1 is called the **standard normal distribution**. The probabilities for the

FIGURE 7.6

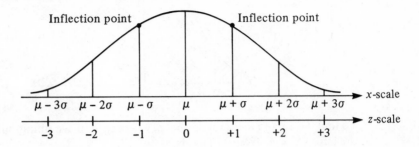

standard normal curve are given in the table in Appendix D, where z, correct to the nearest tenth, is recorded in the left-hand column while the second decimal place for z, corresponding to hundredths, is given across the top row in the table. The entries in the table give the area to the left of the number. We introduce the notation

$$\Phi(z) = P(Z < z)$$

$$= \text{area under the normal curve to the left of } z$$

where Z denotes the standard normal random variable.

Figure 7.7 shows the area to the left of $z = 1.32$. To determine this area, we find in Appendix D the entry in the row labeled 1.3 and the column labeled 0.02. Thus,

$$\Phi(1.32) = P(Z < 1.32) = 0.9066$$

FIGURE 7.7

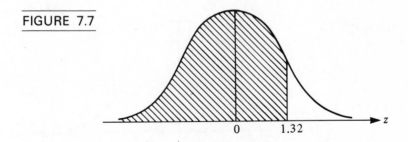

To find the area to the left of $z = -z_0$ where $-z_0 < 0$, we observe that the area to the right of z_0 is precisely the same as the area to the left of $-z_0$, as shown in Figure 7.8. Thus,

$$\Phi(-z_0) = P(Z < -z_0) = P(Z > z_0) = 1 - P(Z < z_0) = 1 - \Phi(z_0) \quad (7.9)$$

For example,

$$\Phi(-1.96) = 1 - \Phi(1.96)$$

$$= 1 - 0.975 = 0.025$$

FIGURE 7.8

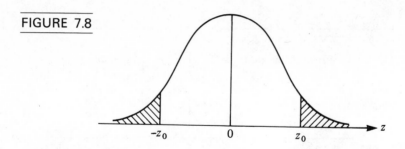

To find the area to the right of $z = z_0$ where $z_0 > 0$, we simply subtract the area to the left of $z = z_0$ from 1. That is,

$$P(Z > z_0) = 1 - P(Z < z_0) = 1 - \Phi(z_0) \qquad (7.10)$$

Thus, the area to the right of $z = 1.08$, for instance, is given by

$$P(Z > 1.08) = 1 - \Phi(1.08)$$
$$= 1 - 0.8599$$
$$= 0.1401$$

Again, the area between $z = -2.14$ and $z = -0.32$ is the same as the area between $z = 0.32$ and $z = 2.14$ (see Figure 7.9). Thus,

$$P(-2.14 < Z < -0.32) = P(0.32 < Z < 2.14)$$
$$= \Phi(2.14) - \Phi(0.32)$$
$$= 0.9838 - 0.6255$$
$$= 0.3583$$

FIGURE 7.9

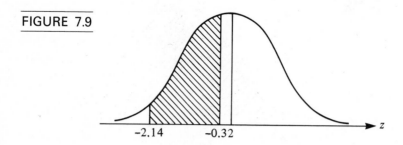

In general, for any pair $z_1 < z_2$, we have

$$P(z_1 < Z < z_2) = \Phi(z_2) - \Phi(z_1)$$

EXAMPLE 7.7.1

Find the area under the standard normal distribution that lies

(a) between $z = 0.25$ and $z = 1.68$

(b) between $z = -2.50$ and $z = 2.50$
(c) between $z = -1.64$ and $z = -0.28$
(d) to the right of $z = 1.46$
(e) to the left of $z = 0.88$

SOLUTION

(a)
$$P(0.25 < Z < 1.68) = \Phi(1.68) - \Phi(0.25)$$
$$= 0.9535 - 0.5987$$
$$= 0.3548$$

FIGURE 7.10

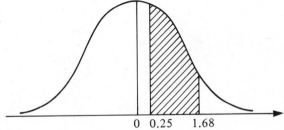

$$0 \quad 0.25 \quad 1.68$$

(b)
$$P(-2.5 < Z < 2.5) = \Phi(2.5) - \Phi(-2.5)$$
$$= \Phi(2.5) - [1 - \Phi(2.5)]$$
$$= 2\Phi(2.5) - 1$$
$$= 2(0.9938) - 1$$
$$= 0.9876$$

FIGURE 7.11

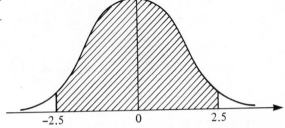

$$-2.5 \quad\quad 0 \quad\quad 2.5$$

(c)
$$P(-1.64 < Z < -0.28) = \Phi(-0.28) - \Phi(-1.64)$$
$$= [1 - \Phi(0.28)] - [1 - \Phi(1.64)]$$
$$= \Phi(1.64) - \Phi(0.28)$$
$$= 0.9495 - 0.6103$$
$$= 0.3392$$

FIGURE 7.12

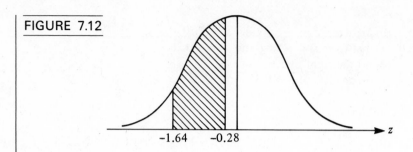

(d) $P(Z > 1.46) = 1 - P(Z < 1.46)$
$= 1 - \Phi(1.46)$
$= 1 - 0.9279$
$= 0.0721$

FIGURE 7.13

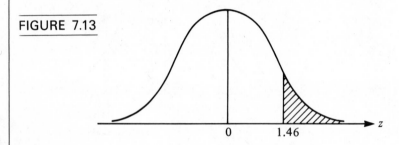

(e) $P(Z < 0.88) = \Phi(0.88)$
$= 0.8106$

FIGURE 7.14

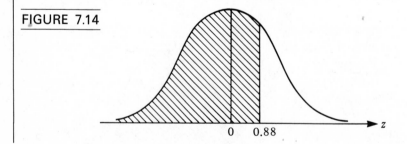

EXAMPLE 7.7.2

Find the value of z_0 if the area

(a) between 0 and z_0 is 0.4357
(b) to the right of z_0 is 0.8643
(c) to the left of z_0 is 0.1711
(d) between $-z_0$ and z_0 is 0.6826

SOLUTION

(a)

FIGURE 7.15

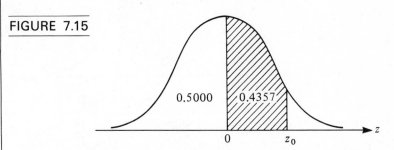

The area to the left of z_0 in Figure 7.15 is $0.4357 + 0.5000 = 0.9357$. The problem now reduces to: What is z_0 so that $\Phi(z_0) = 0.9357$? Reference to Appendix D shows that $z_0 = 1.52$.

(b)

FIGURE 7.16

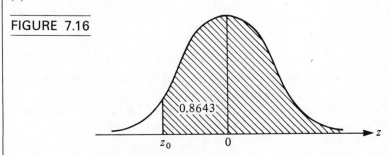

Using Appendix D, we see that $\Phi(1.1) = 0.8643$. Since z_0 happens to be to the left of 0, we conclude that $z_0 = -1.10$.

(c)

FIGURE 7.17

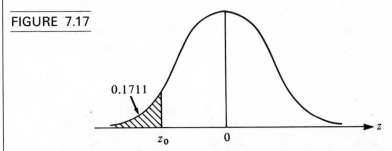

The area to the right of z_0 in Figure 7.17 is $1 - 0.1711 = 0.8289$. Using Appendix D, we note that $\Phi(0.95) = 0.8289$. Again, z_0 happens to be to the left of $z = 0$, so we conclude that $z_0 = -0.95$ is the required value.

(d)

FIGURE 7.18

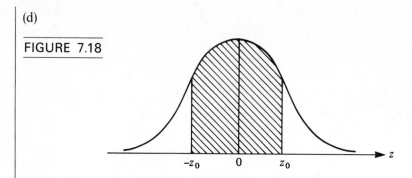

Here, we need to determine z_0 such that
$$P(-z_0 < Z < z_0) = 0.6826$$
Note that
$$P(-z_0 < Z < z_0) = \Phi(z_0) - \Phi(-z_0)$$
$$= \Phi(z_0) - [1 - \Phi(z_0)]$$
$$= 2\Phi(z_0) - 1$$
If $2\Phi(z_0) - 1 = 0.6826$, then $\Phi(z_0) = 0.8413$. Using Appendix D, we note that $z = 1.00$.

Let us now turn to the area under a normal curve whose mean and standard deviation are *not* zero and 1, respectively. Recall that in this case we simply convert the number on the x-scale in Figure 7.6 to the number on the z-scale and then use Appendix D to determine the area. For instance,
$$P(X < b) = P(\mu + \sigma Z < b)$$
$$= P\left(Z < \frac{b - \mu}{a}\right)$$
$$= \Phi\left(\frac{b - \mu}{a}\right)$$

Similarly,
$$P(a < X < b) = P(a < \sigma Z + \mu < b)$$
$$= P(a - \mu < \sigma Z < b - \mu)$$
$$= P\left(\frac{a - \mu}{\sigma} < Z < \frac{b - \mu}{\sigma}\right)$$
$$= \Phi\left(\frac{b - \mu}{\sigma}\right) - \Phi\left(\frac{a - \mu}{\sigma}\right)$$

Also,
$$P(X > b) = P(\sigma Z + \mu > b)$$
$$= P\left(Z > \frac{b - \mu}{\sigma}\right)$$

$$P(X > b) = 1 - P\left(Z < \frac{b - \mu}{\sigma}\right)$$

$$= 1 - \Phi\left(\frac{b - \mu}{\sigma}\right)$$

where $\Phi(z)$ represents the area to the left of z under the standard normal curve.

To give an illustration in which we must first convert the number on the x-scale to the number on the z-scale, let us consider the following example.

EXAMPLE 7.7.3

Given that X is a normal random variable with $\mu = 52.4$ and $\sigma = 2.6$, find

(a) $P(X < 55)$
(b) $P(50.32 < X < 55.52)$
(c) $P(X > 47.2)$
(d) $P(49.54 < X < 51.88)$

SOLUTION

(a) $P(X < 55) = P\left(\dfrac{X - 52.4}{2.6} < \dfrac{55.0 - 52.4}{2.6}\right)$

$\qquad = P(Z < 1)$

$\qquad = \Phi(1)$

$\qquad = 0.8413$

(b) $P(50.32 < X < 55.52) = P\left(\dfrac{50.32 - 52.4}{2.6} < \dfrac{X - 52.4}{2.6} < \dfrac{55.52 - 52.4}{2.6}\right)$

$\qquad = P(-0.80 < Z < 1.20)$

$\qquad = \Phi(1.20) - \Phi(-0.80)$

$\qquad = \Phi(1.20) - [1 - \Phi(0.80)]$

$\qquad = 0.8849 - 0.2119$

$\qquad = 0.6730$

(c) $P(X > 47.2) = P\left(\dfrac{X - 52.4}{2.6} > \dfrac{47.2 - 52.4}{2.6}\right)$

$\qquad = P(Z > -2)$

$\qquad = 1 - P(Z < -2)$

$\qquad = 1 - \Phi(-2)$

$\qquad = 1 - [1 - \Phi(2)]$

$\qquad = \Phi(2)$

$\qquad = 0.9772$

(d) $P(49.54 < X < 51.88) = P\left(\dfrac{49.54 - 52.4}{2.6} < \dfrac{X - 52.4}{2.6} < \dfrac{51.88 - 52.4}{2.6}\right)$

$= P(-1.10 < Z < -0.20)$

$= \Phi(-0.20) - \Phi(-1.10)$

$= [1 - \Phi(0.20)] - [1 - \Phi(1.10)]$

$= \Phi(1.10) - \Phi(0.20)$

$= 0.8643 - 0.5793$

$= 0.2850$

EXERCISES 7.7

1. Calculate each of the following probabilities for the standard normal distribution using the table in Appendix D.

   (a) $P(Z < 1.41)$ (b) $P(0 < Z < 1.23)$
   (c) $P(Z > 0.98)$ (d) $P(0.31 < Z < 0.73)$
   (e) $P(-1.96 < Z < 1.96)$ (f) $P(-1.64 < Z < 1.43)$

2. If Z is the standard normal random variable, use Appendix D to calculate each of the following probabilities.

   (a) $P(Z < 1.33)$ (b) $P(1.34 < Z < 1.92)$
   (c) $P(-1.03 < Z < 1.78)$ (d) $P(Z < 3.08)$
   (e) $P(-1.15 < Z < 1.15)$ (f) $P(-0.61 < Z < 1.61)$
   (g) $P(Z < 1.17)$ (h) $P(Z > -0.37)$
   (i) $P(-0.23 < Z < 0.75)$ (j) $P(Z > -0.98)$

3. Given that Z is the standard normal random variable, use the table in Appendix D to find z_0 in each of the following.

   (a) $P(-1.0 < Z < 1.0)$ (b) $P(-1.39 < Z < 1.39)$
   (c) $P(-2.0 < Z < 2.0)$ (d) $P(-2.38 < Z < 2.38)$
   (e) $P(-3.0 < Z < 3.0)$ (f) $P(-3.09 < Z < 3.09)$

4. Given that Z is the standard normal random variable, use the table in Appendix D to find z_0 in each of the following.

   (a) $P(Z > z_0) = 0.975$ (b) $P(-z_0 < Z < z_0) = 0.6826$
   (c) $P(Z < z_0) = 0.9251$ (d) $P(0 < Z < z_0) = 0.1255$
   (e) $P(Z > z_0) = 0.0017$ (f) $P(-1.5 < Z < z_0) = 0.7036$
   (g) $P(0 < Z < z_0) = 0.4394$ (h) $P(-0.62 < Z < z_0) = 0.6765$

5. Find z_0 when the normal curve

   (a) between 0 and z_0 is 0.4649   (b) to the right of z_0 is 0.0392
   (c) to the left of z_0 is 0.0495   (d) to the right of z_0 is 0.9525

6. Find z_0 when the normal curve area

   (a) between $-z_0$ and z_0 is 0.8788   (b) between $-z_0$ and z_0 is 0.8664
   (c) between $-z_0$ and z_0 is 0.9974   (d) between $-z_0$ and z_0 is 0.9634
   (e) between $-z_0$ and z_0 is 0.8882   (f) between $-z_0$ and z_0 is 0.7016

7. Suppose that X is a normal random variable with mean 49.5 and standard deviation 10. Compute each of the following.

   (a) $P(X > 64.5)$ (b) $P(X < 29.5)$
   (c) $P(X > 44.5)$ (d) $P(39.5 < X < 59.5)$
   (e) $P(44.5 < X < 54.5)$ (f) $P(29.5 < X < 69.5)$

8. Suppose that X is a normal random variable with mean 384.8 and standard deviation 50.4. Find each of the following.

   (a) $P(X < 334.4)$ (b) $P(X > 454.35)$
   (c) $P(334.40 < X < 435.20)$ (d) $P(284 < X < 485.6)$
   (e) $P(359.6 < X < 410)$ (f) $P(X < 460.4)$

9. A random variable X has a normal distribution with a mean of 120.5 and a standard deviation of 36. Find the probability that this random variable will assume a value

   (a) between 84.5 and 138.5   (b) more than 156.5
   (c) less than 84.5

10. A random variable X has a normal distribution with a mean of 45 and a standard deviation of 5. Find the probability that this random variable will assume a value

    (a) more than 53.2   (b) between 42.5 and 47.5
    (c) less than 37.5

11. A random variable X has a normal distribution with mean μ and standard deviation $\sigma = 2.5$. Determine the mean μ if $P(X > 74) = 0.0228$.

12. A random variable X has a normal distribution with mean μ and standard deviation $\sigma = 126.40$. Find the mean μ if the probability that X assumes a value less than 451.22 is 0.7019.

7.8 APPLICATIONS OF NORMAL DISTRIBUTIONS

In this section we shall consider some practical examples involving random variables that are approximately normally distributed with mean μ and standard deviation σ.

EXAMPLE 7.8.1

The annual snowfall in a region is approximately normally distributed, with a mean of 54.3 inches with a standard deviation of 4.2 inches. What is the probability that the snowfall in a particular year is more than 60.81 inches?

SOLUTION   Let X be a normal random variable with $\mu = 54.3$ inches and $\sigma = 4.2$ inches. To find $P(X > 60.81)$ we need to calculate the area under the normal curve to the right of $x = 60.81$ (see Figure 7.19). This is accomplished by finding the area to the right of the corresponding z-value. Thus,

$$P(X > 60.81) = P\left(\frac{X - 54.3}{4.2} > \frac{60.81 - 54.3}{4.2}\right)$$
$$= P(Z > 1.55)$$
$$= 1 - P(Z < 1.55)$$
$$= 1 - \Phi(1.55)$$
$$= 1 - 0.9394$$
$$= 0.0606$$

FIGURE 7.19

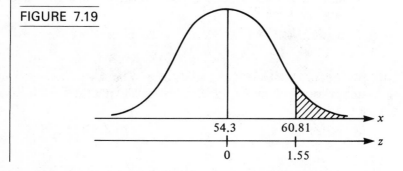

EXAMPLE 7.8.2

The lengths of steel bars produced by a steel company are approximately normally distributed, with a mean of 31.6 feet and a standard deviation of 0.45 foot. What is the probability that a steel bar selected at random is between 32.05 and 32.50 feet long?

SOLUTION   Let X be a random variable normally distributed with $\mu = 31.6$ feet and $\sigma = 0.45$ foot. We need to find the area between $x_1 = 32.05$ and $x_2 = 32.50$, as shown in Figure 7.20. The corresponding z-values are

$$z_1 = \frac{32.05 - 31.60}{0.45} = 1.0$$

$$z_2 = \frac{32.50 - 31.60}{0.45} = 2.0$$

FIGURE 7.20

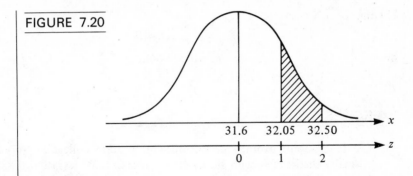

Hence,

$$P(32.05 < X < 32.50) = P(1.0 < Z < 2.0)$$
$$= \Phi(2.1) - \Phi(1.0)$$
$$= 0.9772 - 0.8413$$
$$= 0.1359$$

EXAMPLE 7.8.3

A soft drink machine on the college campus has been regulated so as to discharge an average of μ ounces per cup. Assuming that the amount of soft drink is approximately normally distributed with a standard deviation of 0.2 ounce, what is the upper limit for μ so that 8-ounce cup will overflow only 1 percent of the time?

SOLUTION   Since we begin with a known area or probability, we need to determine the z-value from the table in Appendix D and then compute the corresponding x-value using the formula $x = \mu + \sigma z$. An area of 1 percent corresponding to the overflow of the soft drink is shaded in Figure 7.21. The area to the left of z_0 is 0.99. A reference to the table in Appendix D shows that $z_0 = 2.33$. Hence,

$$\frac{8 - \mu}{0.2} = 2.33$$

FIGURE 7.21

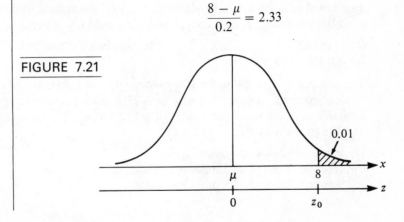

Solving for μ, we have

$$8 - \mu = 0.466$$

or

$$\mu = 7.534 \text{ ounces}$$

EXERCISES 7.8

1. The grade-point average of freshmen at a state college is approximately normally distributed with mean 2.34 and standard deviation 0.55. What is the probability that a freshman selected at random will have a grade-point average between 2.89 and 3.44?

2. Refer to problem 1. If a freshman possessing a grade-point average less than or equal to 1.64 is dropped from the college, what percentage of the freshman class will be dropped?

3. The weights of all quart cartons of milk from a certain dairy have a mean of 30 ounces and a standard deviation of 0.20 ounce. Assuming that these weights are approximately normally distributed, what is the probability that a quart of milk will contain less than 29.90 ounces of milk?

4. A soft drink machine has been regulated so as to discharge an average of 8 ounces per cup. If the amount of soft drink is normally distributed with a standard deviation of 0.4 ounce, what is the probability that a cup selected at random contains between 7.8 and 8.2 ounces of drink?

5. The heights of a large population of college students are approximately normally distributed, with $\mu = 66.5$ inches and $\sigma = 2.5$ inches. Find the probability that a college student selected at random will have a height

   (a) less than 64 inches   (b) between 69 and 71.5 inches
   (c) more than 72.5 inches

6. The charge account at a certain department store is approximately normally distributed with an average balance of $80 and a standard deviation of $30. Find the probability that a charge account randomly selected has a balance

   (a) over $125   (b) between $65 and $95
   (c) less than $50

7. The gasoline consumption for new automobiles of a certain make is approximately normally distributed, with $\mu = 23.3$ miles per gallon and standard deviation $\sigma = 2.5$ miles per gallon. If a new car of this make is purchased, find the probability that it will get

   (a) at least 20 miles per gallon
   (b) between 25.8 and 28.3 miles per gallon
   (c) between 20.8 and 25.8 miles per gallon

8. The average rainfall recorded in New York City for the month of August is 4.2 inches, with a standard deviation of 0.8 inches. Assuming that the rainfall is approximately normally distributed, find the probability that next August New York City will receive

   (a) less than 3 inches of rain   (b) more than 5 inches of rain

9. A department store uses fluorescent lights that have a mean of 3500 hours and a standard deviation of 600 hours. If the length of life of these lights is approximately normally distributed, what proportion of the lights would need replacement after 4400 hours?

10. The length of life of one type of automatic washer is approximately normally distributed, with a mean of 3.1 years and a standard deviation of 1.2 years. If this type of washer is guaranteed for one year, what fraction of original sales will require replacement?

11. The average length of time required for an examination is found to be approximately normally distributed, with $\mu = 70$ minutes and $\sigma = 12$ minutes. When should the examination be terminated if 90 percent of those taking the examination must complete it?

12. A new filling station estimates that the weekly demand for gasoline is approximately normally distributed, with $\mu = 1000$ and $\sigma = 50$ gallons. Assuming that the station is supplied with gasoline once a week, what must be the capacity of its tank if the probability that its supply will be exhausted in a given week is to be no more than 1 percent?

13. A coffee machine has been regulated so as to discharge an average of μ ounces per cup. If the amount of coffee is normally distributed with $\sigma = 0.3$ ounce, what is the upper limit for μ so that an 8-ounce cup will overflow only 1 percent of the time?

14. The grade on an examination taken by a large number of students is approximately normally distributed, with mean $\mu = 74$ and standard deviation $\sigma = 7$. If the instructor assigns A's to 10 percent of the class, above what numerical grade would A's be assigned?

15. The grades on a statistics examination are approximately normally distributed, with mean $\mu = 80$ and standard deviation $\sigma = 6$. Find the lowest passing grade if the lowest 10 percent of the students are given F's.

16. On a final examination in economics, the average grade was 70.0 and the standard deviation was 10.1. What is the probability that a student selected at random will have a score from 61 to 79, inclusive?

17. Assume that the IQ's of elementary schoolchildren as measured by a certain test have a mean of 100 and a standard deviation of 12. What is the probability that a student selected at random will have an IQ of 130 or higher?

18. The number of customers entering a certain restaurant in any given day is approximately normally distributed, with mean $\mu = 40$ and standard deviation $\sigma = 10$. Find the probability that during a given day

   (a) at least 45 customers arrive
   (b) at least 50 customers arrive
   (c) between 30 and 50 customers arrive

19. The time spent for breakfast by patrons in a restaurant is approximately normally distributed, with $\mu = 25$ minutes and $\sigma = 10$ minutes. Find the probability that a customer selected at random will stay at the table for more than 40 minutes.

20. The waiting time before a customer in a restaurant is served lunch is approximately normally distributed, with mean $\mu = 15$ minutes and standard deviation $\sigma = 5$ minutes. What is the probability that a customer selected at random will wait for at most 5 minutes?

7.9   PREDICTION AND FORECASTING

In this section we shall introduce the concept of **regression**, a relationship that exists between paired observations of two variables. Regression is used widely in many statistical investigations for predicting the value of the dependent variable y on the basis of an observable value of an independent variable x. In business and economic situations, we may like to predict, for example, the consumption of gasoline in terms of price per gallon, the potential sales of new cars in terms of auto industry rebates, the import of foreign crude oil in terms of new tax levies, the spending budget of a state in terms of its revenue, the salaries of employees in terms of their experience, the family savings in a bank in terms of its annual income, and so on. In education, we may want to predict the grade-point average of the students at the end of freshman year in relation to their verbal scores on the Scholastic Aptitude Test. In agriculture, a farmer may want to predict the yield of wheat in a particular state in terms of the seasonal rainfall. Of the many mathematical equations used for prediction, the simplest and most widely used is the linear equation

$$\hat{y} = mx + b \qquad (7.11)$$

where m is the slope of the line and the constant b is the value of $\hat{y}$ at the y-axis when $x = 0$.

An example will serve to introduce the concepts and techniques of regression. A survey was conducted to determine the relationship of the employees' test scores to their productivity after three months of practical training in a factory. The six pairs of scores shown in Table 7.8 are obtained by testing six randomly selected applicants and later measuring their productivity. We shall use these data to predict the productivity of an employee on the basis of his aptitude score. An initial approach

TABLE 7.8

Aptitude Score, x	Productivity, y
9	23
17	35
20	29
19	33
20	43
23	32

to the analysis of this information is to plot the data as points on a graph, as shown in Figure 7.22. Aptitude scores of the employees are an independent variable and are represented along the horizontal axis; their productivity is the dependent variable, shown along the vertical axis. Plotting these pairs of observations, we note that the points (x, y) on the graph are spread in an irregular pattern called a **scatter diagram**. However, the points are clustered along a straight line

FIGURE 7.22

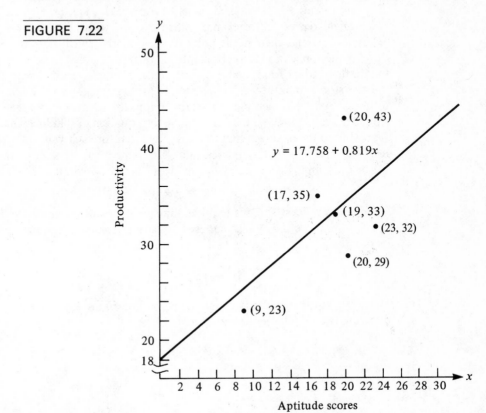

$y = 17.758 + 0.819x$

which may be used to estimate the realtionship of the employee's test scores to his productivity in the factory.

We want to determine the equation of the line $y = mx + b$, which in some sense provides the *best* fit to the set of points (x, y) in Figure 7.22. To do so, we need to determine the constants m and b in such a way that the points in the scatter diagram lie as close to the line as possible. Note that after the line of best fit has been determined, it will yield a certain y-value corresponding to each x-value. This y-value is an estimate of the actual value and is denoted by $\hat{y}$. To determine the line (7.11) of best fit, we use the **method of least squares**. According to this method, the constants m and b are chosen so as to minimize the sum of squares of the vertical distances from the plotted points to the line. Thus, the problem reduces to finding, for a given set of observations, the constants m and b in equation (7.11) in such a way that

$$\sum_{i=1}^{6} (y_i - \hat{y}_i)^2$$

is minimized.

To explain how the method of least squares is used, in general, to determine the line of *best* fit, suppose that we consider n measurements of a quantity x, obtaining values $x_1, x_2, \ldots, x_n$ and for each value of x corresponds a value y_i, $i = 1, 2, 3, \ldots, n$ of quantity y. This yields a set of n pairs $(x_1, y_1), (x_2, y_2), \ldots, (x_n, y_n)$, which might represent, for example, the students' verbal score on the Scholastic Aptitude Test and their grade-point average at the end of freshman year, test scores of the applicants for a certain job and their production rating, the population and the demand for food products, the number of employees in a department store and the store's monthly sales volume, or average height and weight of men in Southeast Asia. If, when plotted on the graph, the points (x_i, y_i) seem to cluster around a straight line, we must determine the equation of that line which provides the best fit to the data (see Figure 7.23). Note that for each value of x_i we have an **observed value** y_i and an **estimated value** $\hat{y}_i$ obtained from the line of best fit.

For the simple linear regression model (7.11), the deviation is

$$y_i - \hat{y}_i = y_i - (mx_i + b)$$

The method of least squares requires that the constants m and b be determined so as to minimize the sum of squared deviations for the n sample observations, denoted by z:

$$z = \sum_{i=1}^{n} (y_i - \hat{y}_i)^2 = \sum_{i=1}^{n} [y_i - (mx_i + b)]^2 \tag{7.12}$$

It can be shown that the two unknowns, m and b, are obtained by solving the following two simultaneous equations:

$$\sum_{i=1}^{n} y_i = nb + m \sum_{i=1}^{n} x_i \tag{7.13}$$

$$\sum_{i=1}^{n} x_i y_i = b \sum_{i=1}^{n} x_i + m \sum_{i=1}^{n} x_i^2 \tag{7.14}$$

FIGURE 7.23

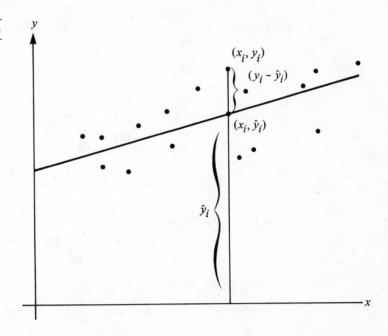

These equations are called **normal equations**. To solve these equations for m and b, we multiply Eq. (7.13) by $\sum_{i=1}^{n} x_i$, the second equation (7.14) by n, and then subtract the first equation from the second, to obtain

$$m = \frac{n \sum_{i=1}^{n} x_i y_i - \left(\sum_{i=1}^{n} x_i \right)\left(\sum_{i=1}^{n} y_i \right)}{n \sum_{i=1}^{n} x_i^2 - \left(\sum_{i=1}^{n} x_i \right)^2} \qquad (7.15)$$

Substituting this value of m in (7.13), we compute b as follows:

$$b = \frac{\sum_{i=1}^{n} y_i - m \sum_{i=1}^{n} x_i}{n} \qquad (7.16)$$

To compute m and b for our illustration of the aptitude scores x and the productivity y, we calculate

$$\sum_{i=1}^{6} x_i, \quad \sum_{i=1}^{6} y_i, \quad \sum_{i=1}^{6} x_i^2, \quad \sum_{i=1}^{6} x_i y_i$$

These calculations are shown in Table 7.9. Thus,

$$\sum_{i=1}^{6} x_i = 108 \qquad \sum_{i=1}^{6} y_i = 195$$

$$\sum_{i=1}^{6} x_i^2 = 2060 \qquad \sum_{i=1}^{6} x_i y_i = 3605$$

TABLE 7.9

x_i	y_i	x_iy_i	x_i^2
9	23	207	81
17	35	595	289
20	29	580	400
19	33	627	361
20	43	860	400
23	32	736	529
108	195	3605	2060

Using (7.15), we obtain

$$m = \frac{6(3605) - (108)(195)}{6(2060) - (108)^2} = \frac{570}{696} = 0.819$$

and, from (7.16), we get

$$b = \frac{195 - (0.819)(108)}{6} = 17.758$$

Thus, the equation of the line of best fit is

$$\hat{y} = 17.758 + 0.819x$$

We must caution the reader that this linear equation gives an *estimated* value of y corresponding to each value of x. These are the best predictions of the actual values in the sense that the sum of the squares of the deviations of the predicted value $\hat{y}$ from the actual value y is minimum.

To graph the line of best fit, we need to determine two points on the line. For $x = 17$, we have $\hat{y} = 31.681$ and for $x = 23$, we obtain $\hat{y} = 36.595$. Connecting these points, we obtain the graph of the line of the best fit as shown in Figure 7.22.

EXERCISES 7.9

1. A manufacturing company keeps a monthly record of sales per month and the corresponding costs. For the first five months of 1979, the data are as follows:

Advertising expenditures	1	2	3	4	5
Sales	2	4	7	9	11

Find the equation of the line of best fit for these data. What will be the estimated sales if advertising expenditures are set at 6 and 10 units, respectively?

2. A survey of used car dealers was conducted to determine the relationship between the amount of classified advertising of used cars and subsequent car sales. The data in the accompanying table show the amount of classified

advertising (in hundreds of lines) and number of cars sold for each of the six dealers randomly selected.

Advertising (hundreds of lines)	Number of Used Cars Sold
74	139
45	108
48	98
36	76
27	62
16	57

(a) Plot the data as a scatter diagram.
(b) Find the equation of the line of best fit that can be used to estimate car sales from the amount of advertising.
(c) Determine the estimated sales of a dealer if he advertises 50 hundred lines.

3. The following data show the advertising costs (x) and the weekly sales (y) in eight randomly selected weeks in 1978.

Advertising Costs, x (dollars)	40	20	25	30	50	40	50	25
Weekly Sales, y (dollars)	490	365	480	475	560	525	510	400

(a) Find the line of best fit for the data.
(b) What are the expected sales per month if advertising costs are set at 12 and 15 units, respectively?

4. The following data show the number of automobile registrations and the total consumption of gasoline in six towns.

Auto Registrations (thousands)	1.1	3.2	4.1	6.2	7.5	7.9	8.5
Gasoline Consumption (millions of gallons)	1.0	2.0	2.2	2.5	3.0	3.2	3.5

(a) Plot the data.
(b) Determine the line of best fit.
(c) What is the predicted consumption of gasoline if the auto registrations are set at 5700?

5. The following experimental data show the amount of water applied (x) and the corresponding yield of rice (y).

Water, x (inches)	12	18	24	30	36	42	48
Yield, y (tons/acre)	5.27	0.68	6.25	7.21	8.02	8.71	8.42

Obtain the line of best fit for the yield of rice as a function of the amount of water applied. Estimate the most probable yield of rice for 40 inches of water.

6. The following data show the intelligence test scores (x) and mathematics achievement scores (y) for 10 students randomly selected from a high school freshman class.

Intelligence Test Scores, x	45	55	56	58	60	65	68	70	75	80
Mathematics Achievement Scores, y	76	70	68	80	82	84	85	90	94	92

(a) Plot the data.

(b) Obtain the line that best fits the data.

(c) Estimate the mathematics achievement score for a student who has an intelligence test score of 50.

7. Ten students randomly selected from a college sophomore class were asked how many hours they had studied for their final examination in statistics (x). The responses, together with the students' scores on the final examination (y), are as follows:

Number of Hours Studied, x	2	3	4	5	6	7	8	9	10	12
Scores on the Final Examination, y	40	45	60	55	50	65	80	95	90	80

(a) Plot the data.

(b) Find the line of best fit which may be used to predict the scores on the final examination corresponding to the number of hours studied.

(c) Estimate the score for a student who studies for 7.5 hours.

8. A college administration collected the following data for the last six consecutive years with respect to the number of applications received by January 1 and the number of new students enrolled in the fall.

Year	Number of Applications, x	Number of New Students Registered, y
1974	2350	1500
1975	2190	1420
1976	2380	1520
1977	2570	1640
1978	2810	1890
1979	2590	1680

(a) Determine the line of best fit using the method of least squares.

(b) Estimate the number of new students registered in the fall of 1980 if 2700 students applied for admission by January 1, 1980.

9. Let x and y represent the stock market index and the sales of a department store, respectively. The following data show the calculations from a sample of 12 observations.

$$\sum_{i=1}^{12} x_i = 725 \qquad \sum_{i=1}^{12} x_i^2 = 44{,}475$$

$$\sum_{i=1}^{12} y_i = 1011 \qquad \sum_{i=1}^{12} x_i y_i = 61{,}865$$

Determine the line of best fit and estimate the sales if the stock market index is 9.2.

10. An automobile dealer collected data for the past three years on the number of inquiries about cars received during a week and the number of cars sold during the following week. The data collected can be summarized as follows:

x = number of inquiries received during given week
y = numbers of cars sold during the next week
n = 156 weeks

$$\sum_{i=1}^{156} x_i = 54{,}649 \qquad \sum_{i=1}^{156} y_i = 11{,}492$$

$$\sum_{i=1}^{156} x_i y_i = 4{,}047{,}975 \qquad \sum_{i=1}^{156} x_i^2 = 855{,}766$$

(a) Determine the line of best fit using the method of least squares.
(b) Estimate the number of sales during a week if the dealer received 400 inquiries during the preceding week.

For Further Reading

Freund, J. *Modern Elementary Statistics*, 5th ed. Englewood Cliffs, N.J.: Prentice-Hall, Inc., 1979.

Hamburg, M. *Basic Statistics*, 2nd ed. New York: Harcourt Brace Jovanovich, Inc., 1979.

Johnson, R. *Elementary Statistics*, 2nd ed. North Scituate, Mass.: Duxbury Press, 1979.

Mendenhall, W., and Reinmuth, J. *Statistics for Management and Economics*, 2nd ed. North Scituate, Mass.: Duxbury Press, 1977.

8

NUMBER THEORY

8.1 INTRODUCTION

The purpose of this chapter is to give the reader a taste of number theory and with it some idea of the flavor of "pure" mathematics. The main topic is perfect numbers, together with some indispensable preliminaries and several related topics to fill in the immediate neighborhood of perfect numbers.

As you wander among the trees, keep in mind that they are part of a forest. Take special note of the generalities pointed out along the way and of the specifics at hand which typify them.

Number theory deals with the set of natural numbers (or positive integers): 1, 2, 3, 4, 5, ... It uses familiar operations of arithmetic (addition, subtraction, multiplication, division), but more as the starting point of intriguing investigations than as topics of primary interest. Number theory is more involved in finding patterns, relations, and structures of natural numbers.

As an example of the distinction between arithmetic and number theory, consider the following:

$$1 + 3 = ? \qquad 1 + 3 + 5 = ? \qquad 1 + 3 + 5 + 7 = ?$$

If there were such persons as "arithmeticians," they would presumably be interested only in the answers to these simple addition problems: 4, 9, and 16, respectively. However, number theorists would note the similarity in the three problems (sums of odd numbers) and in their answers ($4 = 2 \times 2$; $9 = 3 \times 3$; $16 = 4 \times 4$). Struck by this curiosity, they would go on to ask: Are these isolated instances? Is there some general pattern? When odd numbers are added, is the result always a square? The mark of the number theorists, and indeed of mathematicians in general, is not primarily whether they can answer these or similar questions, but the fact that they pose the questions.

Returning to the example cited, we note that the last question surely has a negative answer (for example, $3 + 5 = 8$, not a square). But a negative answer to one question does not mean that we should stop questioning or seeking generalizations. In some instances, a negative answer may be what we expect or hope for. In others, such as this, the answer merely eliminates a simplistic generalization and suggests a more careful look at the examples that fostered the questions.

In this example, you may already have noticed a more specific pattern, namely, that the sum of consecutive odd numbers, starting with 1, is a square. If you test this conjecture for several cases, you will find it verified. You might also observe a more precise formulation of the pattern.

> *The sum of consecutive odd numbers, starting with 1, is the square of the number of odd numbers added.* (8.1.1)

Here we begin to run afoul of language; the statement above is wordy and thus may tend to confuse, rather than to clarify, the pattern we have observed. Fortunately, we can use letters to represent numbers. You must have encountered this process in algebra, where letters are frequently used to label unknown numbers ("Let Mary's age equal x.") The letters are then manipulated in equations using the rules of arithmetic, and the number (solution) is determined.

Letter representation can also convey information about a large, sometimes infinite, set of numbers. For example, we can abridge the verbal statement, "The sum of any two numbers is unchanged when we reverse the order in which the two are added," to "$x + y = y + x$." To be precise, the latter formulation should include a phrase, verbal or symbolic, indicating the set of numbers whose elements can replace x and y in the statement. In practice, such qualifiers are not always expressed, especially when the numbers represented by the letters have no restrictions. Alternatively, the restrictions (all real numbers, all integers, all positive fractions, and so forth) have frequently been agreed upon at the outset of the discussion.

In this chapter we shall make such an agreement. *Unless otherwise stated, any letter used to represent a number may be replaced by any natural number.* This means that any general statement or formula without explicit restriction is understood to be true for all natural numbers but is not necessarily true, or even meaningful, for other numbers.

EXAMPLES 8.1

1. Upon noting that $1 + 3 = 4$, $9 + 11 = 20$, $13 + 5 = 18$, plus a few more examples, we might write "$x + y$ is an even number." Of course, this would be incorrect; it is the type of incorrect generalization, or incorrect formulation of a correct idea, that we must watch for. A correct statement typifying the examples above would be, "If x and y are odd, $x + y$ is even."

2. Consider the statement, "xy is positive." This statement is not true when we make the replacement $x = 0$, $y = 3$, or $x = -2$, $y = 5$. Nonetheless, the statement is true for all natural numbers, and these are, by agreement, the only numbers that we are considering. Thus, for our purposes the statement "xy is positive" is true. The fact that it fails to be true for some numbers outside the set of interest to us does not matter. It also does not matter that this statement *is* true for many numbers other than natural numbers; we do not demand that our statements be true *only* for natural numbers.

Returning, then, to our earlier example, we can replace the verbal statement (8.1.1) by:

If the first n odd numbers are added, the answer is n · n   (8.1.2)

We can symbolize even further by determining the appropriate label for the nth odd number. Noting that the first odd number, 1, is one less than the first even number, 2, and that the second odd number, 3, is one less than the second even number, 4, and so on, it should be clear that the nth odd number is one less than the nth even number. It is also easy to see that the nth even number is just twice n. So the label for the nth odd number is $(2n - 1)$.

Finally, then, we can write the generalization as:

$$1 + 3 + 5 + \cdots + (2n - 1) = n \cdot n$$ (8.1.3)

Some would argue that the ellipsis (indicated by three dots) in this statement makes it ambiguous. Other notations avoid this ambiguity, but we must be careful not to lose sight of the meaning of the notation. The authors are tempted to reply that, in this case, any ambiguity is in the eye of the beholder.

There are also those who would maintain that this last form is less communicative of the observed pattern than the previous one. Statement (8.1.2) is simple and clear, but (8.1.3) can be manipulated using the formal rules of arithmetic or algebra. That makes it more useful.

This example expresses a generalization that is plausible because it is verified in all the instances we have checked. But to assert it as a factual property of all natural numbers, we must prove it. And our proof cannot consist of checking every number; we would be faced with an unending task. We must have a general means of proof to match the generality of what we claim to be true.

There are a variety of arguments available to demonstrate the universal truth of our statement on adding odd numbers. They vary from pictorial and intuitive to arguments of extreme formality. For example, a proof known by the Greeks in the second century B.C., basically inductive, uses a pictorial approach (see Figure 8.1). This figure shows, for example, that from a 3×3 square array of dots, one can form a 4×4 array by adding 3 dots on the right, 3 below, and 1 in the corner. But $3 + 3 + 1 = 7$, the fourth odd number. Clearly, the next "bent row" of dots contains 9 dots, the next 11, and so on; this process will necessarily continue for as many rows as we care to add. So a square array of dots, containing

FIGURE 8.1

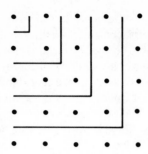

$n \cdot n$ dots, is the sum of 1 dot plus 3 dots plus 5 dots, and so on, up to and including $(2n - 1)$ dots. In other words, the sum of the first n odd numbers is $n \cdot n$.

An alternate proof relying on arithmetic and algebraic notation is available. Suppose we pretend we do not already know the sum of the first n odd numbers and represent this sum by some letter, say x. Then

$$x = 1 + 3 + 5 + 7 + \cdots + (2n - 1)$$

Then certainly

$$x = (2n - 1) + (2n - 3) + (2n - 5) + \cdots + 5 + 3 + 1$$

Don't let this step throw you. We are merely writing the same numbers in reverse order. Of course, we are writing in a few of the numbers implied but not written in the first step [found by decreasing $(2n - 1)$ by 2, by 4, and so on, following the pattern of odd numbers], but nothing is changed. If we now add the two equations, combining corresponding terms of each on the right, we obtain

$$2x = 2n + 2n + 2n + \cdots + 2n + 2n$$

On the right, there are n numbers (why?), and thus we can write

$$2x = n \cdot 2n$$

or
$$x = n \cdot n$$

and we are finished.

Before going any further, let us review the notation for exponents. Since we will be dealing only with natural numbers, we will restrict our review to that alone.

DEFINITION
8.1.1

For any number k, $k^1 = k$; $k^2 = k \cdot k$; $k^3 = k \cdot k \cdot k$; and so on.

In general, then, k^m is merely shorthand for the product of m factors of k.

This definition of exponents allows us to manipulate certain combinations of exponents formally. These manipulations are frequently termed "Laws of Exponents," although they are in no way arbitrary, as some laws tend to be. They are merely simple consequences of Definition 8.1.1.

For example, if 5^3 is multiplied by 5^4, we are merely multiplying the product of three 5's by the product of four 5's, the result being the product of seven 5's:

$$5^3 \cdot 5^4 = (5 \cdot 5 \cdot 5)(5 \cdot 5 \cdot 5 \cdot 5) = 5^7$$

This is an example of the first law of exponents. The others are explained just as simply, and this task is left to the reader.

THEOREM
8.1.1

Laws of Exponents

(1) $x^a \cdot x^b = x^{a+b}$

(2) $x^a \div x^b = x^{a-b}$ $(a \geqslant b)$

(3) $(x^a)^b = x^{ab}$

(4) $(xy)^a = x^a \cdot y^a$

A brief word on the second law is in order. The statement is true as long as a is greater than b. Note that there is not anything wrong with dividing x^a by x^b when b is greater than a. $2^3 = 8$ can be divided by $2^5 = 32$; but the result is a fraction, not a natural number. Since we have restricted our discussion to natural numbers, we exclude this case. But we can divide 2^3 by 2^3 or, in general, x^a by x^a; the result is 1. Since the second law suggests an exponent of $a - a = 0$, we extend Definition 8.1.1 to include $x^0 = 1$. We also extend our notational agreement to allow a letter to represent 0, if the letter is used as an exponent.

EXERCISES 8.1

1. What is the 100th odd number?

2. What is the 100th even number?

3. What is the sum of the first 100 odd numbers?

4. What is the sum of all odd numbers from 1 to 345 inclusive?

5. What is the sum of all odd numbers from 201 to 345 inclusive?

6. Consider the sequence of numbers that starts: 1, 4, 7, 10, 13, ... What is the 10th number in this sequence? What is the 200th number in this sequence? Determine an expression for the nth number in this sequence.

7. What is the sum of the first n natural numbers (that is, $1 + 2 + 3 + \cdots + n$)? [*Hint:* Follow the process of the second proof of (8.1.3) in the text.] Sums of consecutive natural numbers, starting with 1, were called *triangle numbers* by the Greeks. Thus, the first triangle number is 1, the second is 3, the third is 6, and so on. Why do you think they were so named?

8. What is the 100th triangle number? (See Exercise 7.)

9. What is the result when a pair of consecutive triangle numbers is added? (See Exercise 7.) Look at a few examples. Formulate a general statement. Prove your generalization, if possible.

10. Which of the following are true?
   (a) $3^4 \cdot 3^5 = 3^{20}$ (b) $2^5 \cdot 2^7 = 2^{12}$
   (c) $4^6 \cdot 3^3 = 12^9$ (d) $4^6 = 8^3$
   (e) $4^6 = 16^3$ (f) $4^6 \cdot 3^3 = 48^3$
   (g) $(7^3)^2 = 7^9$ (h) $(7^3)^2 = (7^2)^3$
   (i) $5^7 + 3^7 = 8^7$ (j) $5^7 + 5^3 = 5^{10}$

8.2   DIVISIBILITY

In this section we shall investigate a relationship among numbers which is the foundation for much of number theory. The concept is quite simple.

DEFINITION 8.2.1

b divides c if and only if there is some number, x, such that $bx = c$. In this case, we will also say **b is a factor of c**, or **b is a divisor of c**, or **c is a multiple of b**. We write $b \mid c$. If there is no number x such that $bx = c$, we write $b \nmid c$.

For example, $2 \mid 6$, $3 \mid 9$, $7 \mid 7$, $2 \nmid 5$, $4 \nmid 2$, and so on.

There is no problem in understanding the arithmetic connection between b and c in the definition. Confusion often arises, however, over the nature of the phrase "b divides c." A distinction must be made between the mathematical concepts of *relation* and *operation*. We shall not formally define these concepts here* but shall distinguish them by example, to clarify the notion of "divides" given in Definition 8.2.1.

Addition, subtraction, multiplication, and division are the familiar *operations* of arithmetic. When we see "$2 + 5$," the "answer," 7, quickly comes to mind. This is characteristic of operations, although the answer is not always so trivial.

Probably the only familiar *relation* of arithmetic† is that of "less than." When one sees "$2 < 5$," no "answer" is called for. Indeed, "$2 < 5$" is a complete statement, identifying a certain relation between 2 and 5. Divisibility is a relation. When we say, "2 divides 6," or symbolically, "$2 \mid 6$," we are not posing a problem of combining 2 and 6 in some way to produce an answer; we are stating a relationship between 2 and 6.

---

* A more formal treatment is available in Chapter 4.

† Familiar relations are more common in geometry: parallel, coincident, congruent, similar.

Of course, faced with two less familiar numbers, we might be forced to perform an operation to determine whether one divides the other.* For instance, it is not immediately obvious whether 37 divides 703. This is settled by dividing 703 by 37, obtaining a quotient 19. Thus, computation has determined the existence of a number, 19, which when multiplied by 37 produces 703. Note that the statement "37 | 703" does not contain the information that 19 is the quotient of $703 \div 37$. It only states that there *is* a (natural number) quotient.

In general, we should note:

THEOREM 8.2.1

$b \mid c$ if and only if the fraction c/b is a natural number.

PROOF   If $b \mid c$, then by definition, there is a natural number, x, such that $bx = c$. Then $c/b = bx/b = x$. Conversely, if c/b is a natural number x, then $bx = b(c/b) = c$, and thus $b \mid c$.

There are a number of additional formal properties of divisibility, some of which are given in the following theorems.

THEOREM 8.2.2

$1 \mid c$.

THEOREM 8.2.3

$c \mid c$.

THEOREM 8.2.4

If $b \mid c$, then $(c/b) \mid c$.

THEOREM 8.2.5

If $a \mid b$ and $b \mid c$, then $a \mid c$.

THEOREM 8.2.6

If $a \mid b$ and $a \mid c$, then $a \mid (b + c)$.

THEOREM 8.2.7

If $a \mid b$ and $a \mid c$ and $c < b$, then $a \mid (b - c)$.

THEOREM 8.2.8

If $a \mid b$, then $a \leqslant b$.

THEOREM 8.2.9

If $a \mid b$ and $b \mid a$, then $a = b$.

---

* The same is true of the relation "less than." For example, it is not immediately obvious whether 5^7 is less than 7^5, or, going outside the set of natural numbers, whether $\frac{12}{19}$ is less than $\frac{103}{163}$.

The proofs of most of these are simple matters of utilizing the definition with basic properties of arithmetic. For instance:

PROOF OF THEOREM 8.2.2 $1 \cdot c = c$. Therefore, there is a number which when multiplied by 1 yields c. In other words, $1 \mid c$.

PROOF OF THEOREM 8.2.5   If $a \mid b$, that means there is some number x such that $ax = b$. Similarly, since $b \mid c$, there is some number y such that $by = c$. Combining these, we obtain $c = by = (ax)y = a(xy)$. In other words, there is some number k ($= xy$) such that $ak = c$. Thus, $a \mid c$.

EXERCISES 8.2

1. Which of the following are true?
   (a) $9 \mid 99$
   (b) $7 \mid 59$
   (c) $12 \nmid 6$
   (d) $3^8 \mid 3^{11}$
   (e) $5^7 \mid 7^5$
   (f) $37 \nmid 444$
   (g) $2 \mid 123{,}456{,}789$
   (h) $5 \mid 24{,}680$

2. b and c are two numbers with the property that $b \mid c$ and $b \mid (c + 1)$. What is b?

3. If $5 \mid k$, what is the next number larger than k which is a multiple of 5?

4. Prove Theorem 8.2.3.

5. Prove Theorem 8.2.4.

6. Prove Theorem 8.2.6.

7. Prove Theorem 8.2.7.

8. Prove Theorem 8.2.8. (*Hint:* Recall that for any three natural numbers x, y, z, if $x \leqslant y$, then $xz \leqslant yz$; and use the fact that 1 is the smallest natural number.)

9. If "divides" is replaced by "less than" in Definition 8.2.1, what operation should replace multiplication?

10. Create a table with three columns. In the first column list the natural numbers from 1 to 30 (or more, if you like). In the second, next to each number in the first, enter all the divisors of that number. In the third column, tally the number of divisors entered in the second column. For example, the entries in the second and third columns corresponding to "18" would be "1, 2, 3, 6, 9, 18" and "6."

8.3 COUNTING DIVISORS

In Exercise 8.2.10, you were asked to tabulate the number of divisors of each integer from 1 to 30. In this section we shall pursue this notion. For brevity, we shall adopt the following notational definition:

DEFINITION 8.3.1

The number of divisors of n will be denoted by $D(n)$.* (This is read "D of n.")

For example, $D(10) = 4$, $D(36) = 9$, and so on.

Our goal is to develop the ability to determine $D(n)$ for any number n. There is, of course, the "brute force" method, whereby we test every number less than or equal to n, recording, or at least counting, those which divide n. (Theorem 8.2.8 points out that it is fruitless to test numbers larger than n.) This is precisely the method suggested in Exercise 8.2.10.

Unfortunately, this method is not very practical for large numbers. Imagine testing 10,000 numbers! Clearly, we must abandon this method for large values of n.

A great deal of time and effort can be saved by observing the import of Theorem 8.2.4. This theorem says, for example, that since $40 \mid 10{,}000$, then 250, which is $\frac{10000}{40}$, is also a divisor of 10,000. More simply put, when we test 40 as a divisor of 10,000 and obtain 250 as a quotient, not only do we establish 40 as a divisor, but also 250. In other words, divisors occur in pairs.

Thus, in testing for divisors of 10,000, we find pairs: 1, 10,000; 2, 5000; 4, 2500; and so on. Observe that there are no divisors of 10,000 between 2500 and 5000. For if there were, there would be a "companion" divisor between 2 and 4—which is clearly false.

To find all the divisors of 10,000, then, we need only test "halfway" locating pairs of divisors. Note that halfway does not mean up to 5000; that would be halfway if we were *adding* pairs of numbers to obtain 10,000. For pairs of numbers whose product is 10,000, halfway is 100—the square root of 10,000. Each divisor less than 100 must pair up with a divisor greater than 100, and vice versa; so when we have found all divisors from 1 to 100, with their paired divisors from 100 to 10,000, we have found all the divisors of 10,000.

In general, then, to find all the divisors of a number, say n, we can test all the numbers up to $\sqrt{n}$, identifying pairs of divisors. (Normally, $\sqrt{n}$ is not a natural number, of course, but the reasoning is just the same whether or not $\sqrt{n}$ is a natural number. Practically, if we test for divisibility by successive integers starting with 1, we merely keep going until we obtain a quotient less than or equal to the divisor being tested. It is not necessary to know or to compute $\sqrt{n}$.)

---

* This is not the standard notation; we opt to avoid the Greek alphabet as not central to the comprehension of mathematics.

EXAMPLE 8.3.1

Find all divisors of 200. Determine $D(200)$.

SOLUTION   We divide 200 by 1, 2, and so on, recording the results in a table:

Test Divisor	Quotient	Remainder
1	200	0
2	100	0
3	66	2
4	50	0
5	40	0
6	33	2
7	28	4
8	25	0
9	22	2
10	20	0
11	18	2
12	16	8
13	15	5
14	14	4

As noted above, there is no need to continue, so the divisors of 200 are 1, 200, 2, 100, 4, 50, 5, 40, 8, 25, 10, and 20; and $D(200) = 12$. (It is worth noting that we have also found that $\sqrt{200}$ is between 14 and 15.)

This method is a great improvement over our first idea of testing all numbers less than or equal to n. But it is not the best we can do.

Turning again to the table of Exercise 8.2.10, let us reorganize the information available by asking which numbers have exactly one divisor, which have exactly two, and so on.

Clearly, 1 is the only number with exactly one divisor. Several numbers have exactly two divisors: 2, 3, 5, 7, 11, 13, 17, 19, and others. These numbers play a very important role in number theory.

DEFINITION 8.3.2

A **prime number** is one that has exactly two divisors.

A prime number is often defined as a number divisible only by itself and 1. This is inaccurate, as it erroneously includes 1 as a prime. If 1 is specifically excluded, this definition becomes acceptable. Another equivalent definition is the following:

THEOREM 8.3.1

A number is prime if and only if it is greater than 1 and cannot be written as the product of two smaller numbers.

The proof is left as an exercise.

Note that we have not explained *why* certain numbers have exactly two divisors. We have only established a label for such numbers. There are many fascinating theorems and conjectures involving prime numbers. Some are very old, such as Euclid's proof that there exists an infinite number of primes. Others are very recent, as we shall see as we proceed. We shall not digress at this point, but continue our pursuit of a formula or process for $D(n)$.

Looking now at numbers with exactly three divisors, we find 4, 9, 25 (and others, if your list goes beyond 30). A bit of reflection shows that these numbers are just squares of the prime numbers just defined.

In general, it should be clear that if a number is the square of a prime ($n = p^2$), it will have three divisors: $1, p, p^2$. The question that should occur to you is: Could there be more than these three, for some primes? For example, 193 is prime, so its square, 37,249, has divisors 1, 193, and 37,249; but are there any others? To answer this question, we need a bit more terminology and one basic fact about numbers.

DEFINITION 8.3.3

n, or any representation of n as a product of two or more numbers, is called a **factorization** of n. If all the factors in a factorization of n are prime, it is called a **prime factorization.**

EXAMPLE 8.3.2

The following are all factorizations of 36: 36, $1 \cdot 36$, $4 \cdot 9$, $3 \cdot 4 \cdot 3$, $1 \cdot 2 \cdot 3 \cdot 6$, $2 \cdot 3 \cdot 2 \cdot 3$, 6^2, $2^2 \cdot 3^2$, $1 \cdot 2 \cdot 2 \cdot 1 \cdot 3 \cdot 3$. Of these, $2 \cdot 3 \cdot 2 \cdot 3$ and $2^2 \cdot 3^2$ are prime factorizations of 36. Note that $1 \cdot 2 \cdot 2 \cdot 1 \cdot 3 \cdot 3$ is not a prime factorization because 1 is not a prime number.

Observe that every number has factorizations, and every number except 1 has prime factorizations. The latter is true since we can omit 1 from a factorization of any number other than 1, and can write nonprime numbers as products of smaller numbers, which in turn are either prime or products of smaller numbers; eventually, the process of breaking up numbers into products must terminate in a product of primes. This is the easy part of the proof of a basic fact about numbers:

THEOREM 8.3.2

The Fundamental Theorem of Arithmetic: Every number greater than 1 has a prime factorization which is unique aside from the order in which the primes are written.

The proof of the second part of this theorem is somewhat involved and will not be given here. It is available in any standard text on number theory (see the references at the end of the chapter).

The theorem states that every number greater than 1 has a unique set of primes associated with it. The product of those primes is the number, and no other set of primes will produce that number.

Thus, every number other than 1 is either prime or composed (by multiplication) from a unique set of primes. This prompts another definition:

DEFINITION 8.3.4

A number is **composite** if it is the product of two or more primes.

In conjunction with Definition 8.3.2 and Theorem 8.3.1, we can characterize composite numbers as follows:

THEOREM 8.3.3

A number is composite if and only if it has three or more divisors.

Simply put, a prime number is primitive in terms of factors. It cannot be factored as the product of smaller numbers. A composite number is composed of other, smaller numbers, ultimately of (a unique set of) prime numbers. An interesting analogy can be drawn between numbers and chemical compounds, with primes corresponding to atoms and composites to compounds. The analogy is far from perfect, as there are an infinite number of primes that can be combined in an infinite number of ways.

The Fundamental Theorem provides the key to questions about the divisors of the square of a prime, and indeed about $D(n)$ in general, by yielding the following characterization of divisibility:

THEOREM 8.3.4

If $m > 1$, then $m \mid n$ if and only if the prime factorization of m consists of primes that form a subset of the primes in the prime factorization of n.

The proof of this theorem is fairly easy if we utilize the definition of divisibility (8.2.1) and the Fundamental Theorem. If $m \mid n$, then there is some number k such that $m \cdot k = n$. If we now replace each of m, k, and n by its prime factorization, the uniqueness of the prime factorization of n establishes that the same primes must occur on either side of the equals sign. So every prime in the factorization of m is in the factorization of n. Conversely, if the primes in the factorization of m form a subset of the primes in the factorization of n, the product of all the "excess" primes in the factorization of n form a number k such that $m \cdot k = n$ (if there are no excess primes, $m = n$ and $k = 1$), so $m \mid n$.

This theorem, applied to $n = p^2$, shows that the only divisors greater than 1 are p and p^2, since p and p^2 are the only "subsets" of p^2. Thus, there are exactly three divisors, as we suspected earlier.

In general, Theorem 8.3.4 tells us that the number of divisors of n is just the number of distinct subsets of primes that can be formed from the prime factorization of n. We can even count the divisor 1 as corresponding to the empty subset of primes.

EXAMPLE 8.3.3

The prime factorization of 126 is $2 \cdot 3^2 \cdot 7$. Therefore, the divisors of 126 are 1, 2, 3, 7, $2 \cdot 3$, $2 \cdot 7$, 3^2, $3 \cdot 7$, $2 \cdot 3^2$, $2 \cdot 3 \cdot 7$, $3^2 \cdot 7$, and $2 \cdot 3^2 \cdot 7$. Therefore, $N(126) = 12$.

Actually, it is not essential to determine the divisors of n in order to know how many divisors there are. We merely need to compute *how many* distinct combinations of primes can be formed from the prime factorization. We need not actually form them and compute what the divisors are. A process to count the number of combinations is suggested by the next example.

EXAMPLE 8.3.4

The prime factorization of 600 is $2^3 \cdot 3 \cdot 5^2$. A number will be a divisor provided that it has at most three factors of 2, at most one factor of 3, and at most two factors of 5. Each distinct choice within these constraints will correspond to a distinct divisor. Thus, we have four choices of factors of 2 (none, one, two, or three), two choices of factors of 3 (none or one) and three choices of factors of 5 (none, one, or two). All together, we have $4 \times 2 \times 3$ choices; thus, there are 24 distinct combinations of primes and 24 divisors of 600.

Our process for determining $D(n)$, then, for $n > 1$, is as follows:

1. Determine the prime factorization of n.
2. Collect like primes and write the prime factorization in exponential form.
3. Add 1 to each exponent.
4. Multiply the resulting numbers.

Step 3 determines the number of choices of factors of each prime; adding 1 accounts for the choice of using no factors of that prime. Step 4 computes the total number of combinations.

EXAMPLE 8.3.5

What is $D(9828)$?

SOLUTION   The prime factorization of 9828 is $2^2 \cdot 3^3 \cdot 7 \cdot 13$. Adding 1 to each exponent, we obtain 3, 4, 2, and 2, respectively. So $D(9828) = 3 \times 4 \times 2 \times 2 = 48$.

The problem of determining $D(n)$ is thus finally reduced to finding the prime factorization of n and applying some trivial arithmetic. The Fundamental Theorem guarantees a unique prime factorization, so any method that repeatedly factors n and avoids the trivial factor of 1 will ultimately produce *the* prime factorization. Arithmetic skill, a hand calculator, or common sense can all be used. One organized process is to factor out all factors of 2, then all factors of 3, of 5, of 7, and so on.

EXAMPLE 8.3.6

What is the prime factorization of 5280?

SOLUTION

$$
\begin{array}{r|r}
2 & 5280 \\
2 & 2640 \\
2 & 1320 \\
2 & 660 \\
2 & 330 \\
3 & 165 \\
5 & 55 \\
& 11
\end{array}
$$

So $5280 = 2^5 \cdot 3 \cdot 5 \cdot 11$.

EXAMPLE 8.3.7

What is the prime factorization of 126,000?

SOLUTION   Here, we can save some time by recognizing that each terminal "0" represents a factor of $2 \cdot 5$. Thus,

$$
\begin{array}{r|r}
2^3 \cdot 5^3 & 126000 \\
2 & 126 \\
3 & 63 \\
3 & 21 \\
& 7
\end{array}
$$

So $126{,}000 = 2^4 \cdot 3^2 \cdot 5^3 \cdot 7$.

EXAMPLE 8.3.8

What is the prime factorization of 20,889?

SOLUTION   20,889 is not divisible by 2.

$$3 \,\overline{|\,20889}$$

$$3 \,\overline{|\,6963}$$

$$2321$$

2321 is not divisible by 3, 5, or 7, but

$$11 \,\overline{|\,2321}$$

$$211$$

211 cannot be divisible by 2, 3, 5, or 7 (or 2321 would be; see Theorem 8.2.5). It is not divisible by 11, so we try the next prime, 13. It is not divisible by 13, so we try 17. 17 is not a factor, but the fact that the quotient obtained when we try is less than 17 tells us we need go no further (why?); 211 is a prime number. So $20,889 = 3^2 \cdot 11 \cdot 211$.

EXERCISES 8.3

1. Which numbers between 1 and 30 have exactly four divisors? What is their prime factorization? In general, what must be the prime factorization of a number with exactly four divisors?

2. Prove Theorem 8.3.1.

3. Give the prime factorization of each of the following.
   (a) 91
   (b) 120
   (c) 299
   (d) 313
   (e) 347
   (f) 496
   (g) 836
   (h) 1878
   (i) 4173
   (j) 100,100
   (k) 299,299
   (l) 1,000,000

4. Determine $D(n)$ for each of the numbers in Exercise 3. Which are prime? Which are composite?

5. If $n > 1$ and has no prime divisors less than or equal to $\sqrt{n}$, show that n must be prime. (See Examples 8.3.1 and 8.3.8.)

6. Prove or disprove: $D(n)$ is odd if and only if n is a square (the square of another number).

7. Prove or disprove: If p is prime, then $p^2 - 2$ is prime.

8. If $N(k) = 6$, then either $k = p^5$ or $k = p^2 \cdot q$, where p and q are distinct primes. What is the form of the prime factorization if $N(k) = 10$? 11? 12?

9. What is the smallest number that has exactly 12 divisors? Exactly 18 divisors?

10. Determine all numbers less than 200 which have exactly 12 divisors.

8.4   SUMMING DIVISORS

We now turn our attention to the problem of adding all the divisors of a given number. As with the problem of determining the number of divisors, our practical challenge is not how to get an answer, but how to do it efficiently.

Again for the sake of brevity, we introduce a notational definition:

DEFINITION
8.4.1

We shall designate the sum of all the divisors of n by $S(n)$.

Much of the necessary work is already behind us, as the development of our process to determine $D(n)$ identified that each combination of primes chosen as a subset of the prime factorization of n produces a divisor of n. Thus, we already have an efficient method of listing all the divisors of a number.

EXAMPLE 8.4.1

List all divisors of 675.

SOLUTION   The prime factorization of 675 is $3^3 \cdot 5^2$. Thus, its divisors are

$$1, 3, 5, 3 \cdot 3, 3 \cdot 5, 5 \cdot 5, 3 \cdot 3 \cdot 3, 3 \cdot 3 \cdot 5, 3 \cdot 5 \cdot 5, 3 \cdot 3 \cdot 3 \cdot 5, 3 \cdot 3 \cdot 5 \cdot 5, 3 \cdot 3 \cdot 3 \cdot 5 \cdot 5$$

The pattern of determining divisors in Example 8.4.1 should be clear. First, all divisors with no prime factors are listed (1 is the only such divisor); then all divisors with one prime factor; then all divisors with two prime factors; and so on. This is a workable process unless the factorization of n contains a large number of primes. In such a situation it is easy to overlook a few combinations. (See Exercise 8.4.1.)

A more rigid (but more foolproof) method is described next.

EXAMPLE 8.4.2

List all divisors of $1800 = 2^3 \cdot 3^2 \cdot 5^2$.

SOLUTION   Think of an odometer with three wheels (one for each distinct prime), containing, respectively, the numbers 0, 1, 2, 3; 0, 1, 2; and 0, 1, 2 (corresponding to the exponents). Each time a wheel of this odometer turns through a complete cycle, it causes the adjacent wheel to move one space. Thus, the following numbers are produced in turn: 000, 001, 002, 010, 011, 012, 020, 021, 022, 100, 101, and so on. Eventually 36 distinct numerals are produced, before the odometer returns to

"000." Those numerals correspond to the 36 divisors of 1
produced by using the three digits of a numeral as expone
and 5. For example, the numeral "212" corresponds to the

If we wish to determine $S(n)$, we can factor n, determine its
them. If $D(n)$ is small, this is reasonable and practical. But as $D(n)$ gets larger, the
process becomes tedious.

Fortunately, an observable pattern of divisors of n simplifies their addition.
Consider, for example, summing the divisors of 1800 listed in the order produced
by the "odometer" method (Example 8.4.2): $1, 5, 5^2, 3, 3 \cdot 5, 3 \cdot 5^2, 3^2, 3^2 \cdot 5,$
$3^2 \cdot 5^2, 2, 2 \cdot 5$, and so on. If we group these appropriately, we observe that the
first three sum to 31, the next three to $3 \cdot 31$, and the following three to $9 \cdot 31$.
Clearly, this is no accident, as the second three divisors are, by construction,
3 times the first three divisors and the next three are 9 times the first three. Thus, the
sum of the first nine terms is

$$31 + 3 \cdot 31 + 9 \cdot 31 = (1 + 3 + 9)(31) = (13)(31)$$

Clearly, the next nine divisors will each be double these nine, so their sum will
just be double $(13)(31)$; the next nine will sum to $2^2 \times (13)(31)$; and the last nine
will sum to $2^3 \times (13)(31)$. So the sum of all 36 divisors is

$$(13)(31) + 2(13)(31) + 4(13)(31) + 8(13)(31) = (1 + 2 + 4 + 8)(13)(31) = (15)(13)(31)$$

In conclusion, then,

$$S(1800) = S(2^3 \cdot 3^2 \cdot 5^2) = S(2^3) \cdot S(3^2) \cdot S(5^2)$$

This pattern will hold true for any number. Therefore, to compute $S(n)$:

1. Determine the prime factorization of n.
2. Collect like primes and write the prime factorization in exponential form.
3. For each power of a prime, p^k, in this factorization, compute $S(p^k)$.
4. Multiply the resulting numbers. (The strong parallel of this process with that
   discussed in Section 8.3 should be noted.)

Thus, the problem of computing $S(n)$ is practically reduced to factorization and
computation of $D(p^k)$, where p is a prime. The former has been treated earlier. For
the latter, the divisors of p^k are trivial to note: $1, p, p^2, p^3, \ldots, p^{k-1}, p^k$. If k is
small, the simplest computation is merely adding the few numbers. For large k, the
following may be used:

THEOREM
8.4.1

If $n > 1$, then

$$S(n^k) = 1 + n + n^2 + \cdots + n^k = \frac{n^{k+1} - 1}{n - 1}$$

EXAMPLE 8.4.3

Compute $S(5^3)$.

SOLUTION $S(5^3) = 1 + 5 + 25 + 125 = 156$.

EXAMPLE 8.4.4

Compute $S(2^{11})$.

SOLUTION $$S(2^{11}) = \frac{2^{12} - 1}{2 - 1} = 2^{12} - 1$$

EXERCISES 8.4

1. List all the divisors of $2^4 \cdot 3^2 \cdot 5 \cdot 7^3$ that have exactly seven prime factors. (There are 15 such divisors.)

2. Without listing them all and without counting, prove that the number of divisors of $2^4 \cdot 3^2 \cdot 5 \cdot 7^3$ which have exactly four prime factors is the same as the number which have exactly six prime factors.

3. List all the divisors of 60; of 405; of 756.

4. Determine $S(n)$ for each of the numbers in Exercise 8.3.3.

5. Prove Theorem 8.4.1. (*Hint:* Let $1 + n + n^2 + \cdots + n^{k-1} + n^k = x$. Determine a comparable equation for $n \cdot x$. Combine those two by subtraction and solve for x.)

6. Suppose you had a job that paid 1 cent the first day and doubled in pay each successive day. How much money would you earn in 30 days at this job?

8.5   PROPER DIVISORS

Numbers can be classified in many ways. They are either even or odd; they may be prime or composite; they may have exactly 12 divisors. A classification introduced by the ancient Greeks depends on the sum of the divisors of a number. The Greeks followed a slightly different way of computing from the one we have used. They considered only divisors less than n as "proper" divisors of n.

DEFINITION
8.5.1

x is a **proper divisor** of y provided that $x \mid y$ and $x < y$.

Note that y is then the only "improper" divisor of y; all other divisors are proper. We can now approach the classification introduced by the Greeks by considering sums of proper divisors.

DEFINITION
8.5.2

We shall designate the sum of all proper divisors of a number n by $P(n)$.

EXAMPLE 8.5.1

Determine $P(15)$.

SOLUTION   The proper divisors of 15 are 1, 3, and 5, so $P(15) = 1 + 3 + 5 = 9$.

EXAMPLE 8.5.2

Determine $P(1800)$.

SOLUTION   In Example 8.4.2, we noted that $D(1800) = 36$. Thus, 1800 has 35 proper divisors. It would be tedious to determine these 35 and then add them, but fortunately there is no need. In Section 8.4 we found that $S(1800) = (15)(13)(31) = 6045$. Since this is the sum of all 36 divisors, and since 1800 is the lone "improper" divisor included in the sum, $P(1800) = 6045 - 1800 = 4245$.

In general, then, $P(n) = S(n) - n$. We can thus use the techniques of Section 8.4 to compute $S(n)$, then subtract n to determine $P(n)$. As Example 8.5.1 indicates, it is not necessary to first compute $S(n)$. It would be a waste of time when there are only a few divisors. Use common sense to determine which method to use.

Finally, we can observe an interesting distinction in the results of Examples 8.5.1 and 8.5.2. That $P(1800)$ is much larger than $P(15)$ does not surprise us. What is noteworthy is that $P(1800)$ is larger than 1800, but $P(15)$ is smaller than 15. Thus, we can categorize the two numbers differently, as did the Greeks.

DEFINITION
8.5.3

If $P(n) < n$, then n is **deficient**.

DEFINITION
8.5.4

If $P(n) > n$, then n is **abundant**.

Thus, 15 is deficient and 1800 is abundant. There is one other possibility:

DEFINITION
8.5.5

If $P(n) = n$, then n is **perfect**.

The simple connection between $P(n)$ and $S(n)$ suggests the following simple characterizations:

THEOREM 8.5.1 | n is deficient if and only if $S(n) < 2n$.

THEOREM 8.5.2 | n is abundant if and only if $S(n) > 2n$.

THEOREM 8.5.3 | n is perfect if and only if $S(n) = 2n$.

EXERCISES 8.5

1. Determine $P(n)$ for each number from 1 to 30.

2. Classify each number from 1 to 30 as deficient, abundant, or perfect.

3. Determine $P(n)$ for each number in Exercise 8.3.3.

4. Classify each number in Exercise 3 as deficient, abundant, or perfect.

5. Prove that every prime is deficient. Conclude that the set of deficient numbers is infinite.

6. Prove that every multiple of an abundant number is abundant. Conclude that the set of abundant numbers is infinite.

7. Show that 2^k is deficient for every value of k.

8. Give an example of an odd, abundant number.

9. Prove: If n is composite, then $D(n) \leqslant P(n)$.

10. Prove: If $D(n)$ is prime, then n is deficient.

8.6   EVEN PERFECT NUMBERS

In doing the exercises for Section 8.5, you will have come across several perfect numbers: 6, 28, and 496. This is not a lot, but we should expect perfect numbers to be rare. There are many more ways for $P(n)$ to be unequal to n than equal.

Our goal is to identify the characteristic properties of perfect numbers. Are the perfect numbers cited above the only examples? If there are more, how many more? Is there a formula that will compute every perfect number, as the formula $[n(n + 1)]/2$ computes every triangle number?

In pursuit of answers to those questions, all we have to go on are the three perfect numbers identified so far. What do they have in common? Certainly, they are all even numbers, so we might restrict our attention to even numbers for the present. The distribution of these three perfect numbers strongly suggests that we have missed some. The numbers 6 and 28 are fairly close together, but it is a long way to 496.

We could test every even number between 28 and 496. However, before launching into such a sizable task, we should attempt some intelligent guesswork, along the following line. Since 6 and 28 differ by 22, we could follow this increment and test the number 50, or perhaps a few even numbers in the neighborhood of 50.

EXAMPLE 8.6.1

Determine whether any of 46, 48, 50, 52, and 54 is perfect.

SOLUTION $P(46) = 26$, $P(48) = 76$, $P(50) = 46$, $P(52) = 46$, and $P(54) = 66$. So 46, 50, and 52 are deficient, whereas 48 and 54 are abundant.

Another approach might be to try a multiplicative pattern. Since 28 is between 4 times 6 and 5 times 6, and 496 is between 17 times 28 and 18 times 28, we might look for a "missing" perfect number about 4 times 28, expecting 496 to be about 4 times that. Since $4 \times 28 = 112$ and $496 \div 4 = 124$, we might reasonably look at even numbers in that range.

EXAMPLE 8.6.2

Determine whether any even number from 112 to 124 is perfect.

SOLUTION $P(112) = 136$, $P(114) = 126$, $P(116) = 94$, $P(118) = 62$, $P(120) = 240$, $P(122) = 64$, and $P(124) = 100$. So 116, 118, 122, and 124 are deficient, whereas 112, 114, and 120 are abundant.

Neither of these ideas has worked. Perhaps we did not look far enough. In any event, negative results can be just as valuable as positive.

If we reflect on the things we have learned in the earlier sections, we must note that determining $D(n)$, $S(n)$, and $P(n)$ depends on first finding the prime factorization of n. In fact, the Fundamental Theorem of Arithmetic provides the proof that the prime factorization of a number determines all the properties of a number definable in terms of its divisors. So we should search for the key in the prime factorization of perfect numbers.

Table 8.1 shows a very strong pattern, or more properly, several patterns. We have a perfect number with one factor of 2, with two factors of 2, and with four factors of 2. Perhaps we should look for a perfect number with three factors of 2 (one may be missing from the list) and then one with five, six, and so on. Perhaps

TABLE 8.1   Prime Factorization of Perfect Numbers

n	Prime Factorization
6	$2 \cdot 3$
28	$2^2 \cdot 7$
496	$2^4 \cdot 31$

we should follow the pattern of exponents to seek a perfect number with seven or eight factors of 2 (1, 1 + 1, 1 + 1 + 2, 1 + 1 + 2 + 3, and so on, or 1, 1 · 2, 1 · 2 · 2, 1 · 2 · 2 · 2, and so on).

To pursue the first conjecture, we should examine numbers of the form $2^3 \cdot p$, where p is some odd prime between 7 and 31. This information is given in Table 8.2. Although we have not found a perfect number, we have some interesting information. When multiplied by 2^3, smaller odd primes produce abundant numbers; larger primes produce deficient numbers. If we consider the amount of discrepancy between n and $P(n)$, we find it smallest at 13 and 17, and increasing as we pick odd primes farther away in either direction.

TABLE 8.2

p	$2^3 \cdot p$	$P(2^3 \cdot p)$	Classification
11	88	92	abundant
13	104	106	abundant
17	136	134	deficient
19	152	148	deficient
23	184	176	deficient
29	232	218	deficient

This all suggests trying $2^3 \cdot 15 = 120$, as 15 ought to be "just right." However, as we saw in Example 8.6.1, $P(120) = 240$ and 120 is "very" abundant. This is to be expected, of course, since 15 is composite, causing 120 to have 16 divisors, while all the numbers in Table 8.2 have only eight.

What is called for is a *prime* number halfway between 13 and 17. No such number exists, but if it did, we would have the makings of another perfect number. For if 15 were prime, $P(2^3 \cdot 15)$ would equal $1 + 2 + 4 + 8 + 15 + 30 + 60 = 120$.

Another way to put all this is: The perfect numbers in Table 8.1 consist of a power of 2 times an odd prime number in "the right place," but there happens to be no prime number in the right place to go with 2^3.

We are ahead, however, in that we now know "good" odd numbers to be 3, 7, 15, and 31. You should see a pattern in these numbers. Each is twice the previous number, plus 1. Another way to describe the pattern is to note that each number is 1 less than a power of 2: $3 = 2^2 - 1$, $7 = 2^3 - 1$, $15 = 2^4 - 1$, $31 = 2^5 - 1$. The former observation is good for successively computing terms of the sequence, but the latter is preferable for a formula for a typical term: $2^k - 1$.

The next number in this sequence of odd numbers is 63. It should be multiplied by 2^5, the next power of 2. Since 63 is composite, we should expect a situation similar to that encountered with $2^3 \cdot 15$. This is exactly the case.

If we go to the next step, however, we find 127, a prime number. And when we examine $2^6 \cdot 127 = 8128$, we find that $S(8128) = S(2^6) \cdot S(127) = (2^7 - 1)(128) = (127)(128) = 16,246$. So $P(8128) = 8128$; 8128 is a perfect number!

We should now have enough information to formulate a precise conjecture and, hopefully, to prove it.

CONJECTURE
8.6.1

If $2^k - 1$ is prime, then $(2^{k-1})(2^k - 1)$ is perfect.

Attacking the conjecture head-on, let us try to calculate $S(n)$, where $n = (2^{k-1})(2^k - 1)$.

$$S(n) = S(2^{k-1}) \cdot S(2^k - 1) \qquad \text{as we saw in Example 8.4.2}$$

$$S(2^{k-1}) = 2^k - 1 \qquad \text{by Theorem 8.4.1}$$

$$S(2^k - 1) = (2^k - 1) + 1 = 2^k \qquad \text{since } 2^k - 1 \text{ is prime}$$

So $S(n) = (2^k - 1)(2^k)$. This is almost the prime factorization of n (with the factors reversed); the only difference is a factor of 2^k rather than 2^{k-1}. But 2^k is just $2 \cdot 2^{k-1}$. So we have

$$S(n) = (2^k - 1)(2^k) = (2^k - 1)(2^{k-1}) \cdot 2 = 2n$$

and, by Theorem 8.5.3, n is perfect. We have proved our conjecture and label it a theorem.

This theorem was known to the ancient Greeks, and recorded in Euclid's *Elements*. Perfect numbers of that form are thus known as Euclidean perfect numbers.

DEFINITION
8.6.1

A number of the form $(2^{k-1})(2^k - 1)$ is called a **Euclidean number**.

Our theorem could thus be reworded:

THEOREM
8.6.1

If $2^k - 1$ is prime, the corresponding Euclidean number, $(2^{k-1})(2^k - 1)$, is perfect.

Euclidean perfect numbers contain a power of 2 and a single odd prime. Might there not be even perfect numbers with two or more odd primes? Surprisingly, the answer is no, as the following theorem states.

THEOREM
8.6.2

If n is an even perfect number, it is of the form $(2^{k-1})(2^k - 1)$, where $2^k - 1$ is prime.

The proof of this theorem is a bit more involved than the others presented here, but is included for the sake of completeness.

PROOF OF THEOREM 8.6.2 (optional)   Since n is even, $n = 2^m \cdot q$, where q is odd and $m \geqslant 1$. From our observations in Section 8.4, $S(n) = S(2^m) \cdot S(q) = (2^{m+1} - 1) \cdot S(q)$, by Theorem 8.4.1. Since n is perfect, we know by Theorem 8.5.3 that $S(n) = 2n = 2(2^m \cdot q) = 2^{m+1} \cdot q$. Equating these two expressions, we have

$$2^{m+1} \cdot q = (2^{m+1} - 1) \cdot S(q) \tag{8.1}$$

This equation shows that $(2^{m+1} - 1)$ is a factor of $2^{m+1} \cdot q$. But $(2^{m+1} - 1)$ and 2^{m+1} are consecutive integers and can have no common factors other than 1. So, by Theorem 8.3.2, it follows that $(2^{m+1} - 1)$ must be a factor of q:

$$q = (2^{m+1} - 1) \cdot d \tag{8.2}$$

where d is some integer.

Substitution of this in (8.1) and cancellation of the common factor of $(2^{m+1} - 1)$ leads to

$$2^{m+1} \cdot d = S(q) \tag{8.3}$$

Suppose now that $d > 1$. Then by (8.2), q has at least three distinct divisors: 1, d, and q itself. (Why can't we claim at least four divisors?) Thus, $S(q) \geqslant 1 + d + q > d + q = d + (2^{m+1} - 1) \cdot d = 2^{m+1} \cdot d$, from the definition of $S(q)$, (8.2) and a bit of algebra. But, by (8.3), the final expression is equal to $S(q)$, so we conclude that $S(q) > S(q)$! Since this is impossible, our supposition that $d > 1$ must be false, and therefore $d = 1$.

Equation (8.2) then gives us that $q = 2^{m+1} - 1$, and (8.3) implies that $S(q) = 2^{m+1} = q + 1$. This can only be true if q is prime, which concludes the proof.

EXERCISES 8.6

1. Verify that $2^5 \cdot 63$ is abundant.

2. Show that $2^5 \cdot p$ is abundant if p is any prime less than 63 and deficient if p is any prime greater than 63. (See Table 8.2.)

3. Prove or disprove: If an even number has two or more odd prime factors, it is abundant.

4. Prove or disprove: The sum of the reciprocals of the proper divisors of a perfect number is 1.

5. Prove or disprove: Every Euclidean number is a triangle number. (See Exercise 8.1.7.)

6. Show that any multiple of a perfect number other than the perfect number itself is abundant.

8.7 MERSENNE PRIMES

In Section 8.6 we found that every even perfect number is Euclidean, with exactly one odd prime factor, of the form $2^k - 1$. The search for additional even perfect numbers, then, boils down to an examination of numbers of the form $2^k - 1$, to determine which are prime.

DEFINITION 8.7.1

A number of the form $2^k - 1$ is called a **Mersenne* number** and is denoted M_k.

In terms of this notation, we could rephrase Theorem 8.6.1 as follows:

THEOREM 8.7.1

If M_k is prime, $2^{k-1} \cdot M_k$ is perfect.

To determine if M_k is prime, we could always resort to the definition. Given sufficient time, M_k can be tested (for a specific value of k) to determine if it is prime. But, as we have said so often, there ought to be a better way.

Using what we have already observed, we can organize our information as shown in Table 8.3.

TABLE 8.3

k	$M_k = 2^k - 1$	M_k
1	1	
2	3	prime
3	7	prime
4	15	composite
5	31	prime
6	63	composite
7	127	prime

$M_8 = 255$ is clearly composite, which suggests the possibility that the Mersenne numbers are alternately prime and composite, after an initial anomaly. Unfortunately, this is disproved (or another anomaly occurs) by $M_9 = 511 = 7 \cdot 73$, which is composite.

A different conjecture, with no noticeable anomaly is readily suggested by comparing the subscript, k, with the classification of M_k. So far, M_k is prime when $k = 2, 3, 5, 7$ and composite when $k = 4, 6, 8, 9$. The pattern is striking in its

---

* Marin Mersenne (1588–1648) was a Franciscan friar. He spent a great deal of time corresponding with prominent intellectuals of his day, fostering communication among them. His interest in number theory is honored by the numbers that bear his name.

simplicity. (We even have $M_1 = 1$, which is neither prime nor composite.) Thus, the following conjecture is formed:

CONJECTURE
8.7.1

M_k is prime or composite as k is prime or composite, respectively.

This would suggest that M_{10} should be composite, M_{11} prime, M_{12} composite, and so on. Taking the "composite part" of the conjecture first, we can obtain a lead to a more precise connection by observing some of the divisors of the composite occurrences of M_k. $M_4 = 15 = 3 \cdot 5$; $M_6 = 63 = 3^2 \cdot 7$; $M_8 = 255 = 3 \cdot 5 \cdot 13$; $M_9 = 511 = 7 \cdot 73$; $M_{10} = 1023 = 3 \cdot 11 \cdot 31$. We note that a Mersenne prime is always among the prime factors. Specifically, $3 (= M_2)$ is a factor of M_4, M_6, M_8, M_{10}; $7 (= M_3)$ is a factor of M_6 and M_9; and $31 (= M_5)$ is a factor of M_{10}.

Again, there is a striking correlation among the Mersenne numbers and their subscripts, suggesting the following:

CONJECTURE
8.7.2

If $c \mid d$, $M_c \mid M_d$.

This turns out to be fairly easy to prove, using the formula of Theorem 8.7.1 and some laws of exponents.

We utilize Theorem 8.2.4, whereby we consider the fraction

$$\frac{M_d}{M_c} = \frac{2^d - 1}{2^c - 1}$$

Since $c \mid d$, we know there is some integer b such that $d = b \cdot c$, so our fraction becomes

$$\frac{M_d}{M_c} = \frac{2^{bc} - 1}{2^c - 1} = \frac{(2^c)^b - 1}{2^c - 1}$$

By Theorem 8.4.1, this fraction is the "answer" to the problem

$$1 + 2^c + (2^c)^2 + (2^c)^3 + \cdots + (2^c)^{b-1}$$

Thus, M_d / M_c equals the sum of a series of whole numbers, which must be a whole number, so $M_c \mid M_d$.

As a simple corollary, if d is composite, it has a divisor c such that $1 < c < d$. Then $1 < M_c < M_d$ and $M_c \mid M_d$, establishing that M_d is composite. This is the proof of:

THEOREM
8.7.2

If k is composite, M_k is composite.

This is half of Conjecture 8.7.1. The other half, asserting M_k prime when k is prime, turns out to be false, as it breaks down for $k = 11$ (and for many other primes). (See Exercise 8.7.1.)

This is one of the classic examples of a persuasive pattern turning out to be misleading. Many mathematicians believed in the truth of this conjecture, and M_{11} was claimed to be prime as recently as this century.* The incorrectness of such a seemingly obvious conjecture is a paramount example of why mathematics insists on rigorous proofs for every assertion, even the most obvious. Simplicity of form, successful prediction of examples, and majority belief can all be wrong.

So where do we stand in our search for Mersenne primes, the key factor to even perfect numbers? Theorem 8.7.1 narrows the search to Mersenne numbers with prime subscripts, but each such number must be tested. Although some additional narrowing can be effected by the form of the prime, it is basically true that, to date, no pattern has been found to characterize Mersenne primes. Improved algorithms for testing primes and high-speed computers have extended the list of known Mersenne primes (see Table 8.4), but no theorem that characterizes Mersenne

TABLE 8.4

k	M_k (Number of Digits)	Discoverer (date)
2	3	
3	7	unknown (before 300 B.C.)
5	31	
7	127	
13	8,191	Regius (1536)
17	131,071	Cataldi (1603)
19	524,287	Euler (1772)
31	(10)	
61	(19)	Seelhoff and Pervusin (1886)
89	(27)	Powers and Cunningham (1912)
107	(33)	Uhler (1952)
127	(39)	
521	(157)	
607	(183)	
1,279	(386)	Robinson (1952)
2,203	(664)	
2,281	(687)	
3,217	(969)	
4,253	(1,281)	
4,423	(1,332)	Lucas and Lehmer (1962)
9,689	(2,917)	
9,941	(2,993)	
11,213	(3,376)	Gillies (1964)
19,937	(6,002)	Tuckerman (1971)
21,701	(6,533)	Nickel and Noll (1978)

---

* This is true despite the fact that M_{11} was shown to be composite by Cataldi in 1603.

primes has been proved. As you can see from Table 8.4, the occurrence of Mersenne primes is sporadic, and the magnitude of the numbers involved is truly monumental. The twenty-fifth, and largest known to date, perfect number contains 13,065 digits; this number, if printed in ordinary newsprint, would require a line of print $67\frac{1}{2}$ feet long.

EXERCISES 8.7

1. What is the prime factorization of M_{11}?

2. What number should be multiplied by M_{89} to produce a perfect number?

3. Determine a prime factor of M_{25}; of M_{121}.

4. What is the prime factorization of $M_{15} = 32,767$?

8.8   RELATED TOPICS

A number of topics in number theory are closely related to perfect numbers. But before we look at these, let's discuss odd perfect numbers.

Briefly, nothing definitive is known about odd perfect numbers. More precisely, no such numbers are known, and theoretical methods have shown that any possible odd perfect numbers must be extremely large.* It is commonly believed by many mathematicians that no odd perfect numbers exist, but this conjecture has defied proof.

Another classification of numbers suggested by extending the characterization of perfect numbers (Theorem 8.5.3) is that of multiply perfect numbers.

DEFINITION
8.8.1

A number n is called **k-ply perfect** if $S(n) = k \cdot n$.

In this notation, perfect numbers are renamed **doubly perfect** numbers. In Example 8.6.2, we stumbled across the fact that $P(120) = 240$ or, equivalently, $S(120) = 360$; thus, 120 is triply perfect. Some other triply perfect numbers are 672 and 523,776.

Descartes, in 1638, gave the triply perfect number 1,476,304,896. He also gave several quadruply perfect numbers (30,240, 32,760, 23,569,920, 142,990,848, 66,433,720,320, and 403,031,236,608) and one quintuply perfect number (14,182,439,040).

---

* It can be proved, for example, that an odd perfect number must have at least six distinct prime factors, and it has been shown that there are no odd perfect numbers less than 100,000,000,000,000,000,000.

Another generalization on perfect numbers that has captured the interest of mathematicians (not to mention numerologists)* is that of amicable numbers.

DEFINITION 8.8.2

A pair of numbers, m and n, is called an **amicable pair** provided that $P(m) = n$ and $P(n) = m$.

In other words, an amicable pair of numbers has the property that each is the sum of the proper divisor of the other. The only pair of amicable numbers known to the ancient Greeks was 220 and 284. Fermat discovered the amicable pair 17,296 and 18,416. Descartes added more than 60 pairs to the list. Today approximately 400 amicable pairs are known.

Another way to view amicable pairs is to note that $P(P(m)) = m$; $n = P(m)$ has the same property, of course. We might naturally look at numbers such that $P(P(P(k))) = k$, or $P(P(P(P(k)))) = k$, and so on. It has been conjectured that if one starts with any integer k and considers $P(k)$, $P(P(k))$, and so on, a closed cycle eventually occurs.†

A topic of more modern vintage suggested by perfect numbers is that of semiperfect numbers.

DEFINITION 8.8.3

A number n is **semiperfect** if the sum of some subset of the proper divisors of n equals n.

For example, 18 is semiperfect, because $3 + 6 + 9 = 18$. Clearly, no deficient number is semiperfect and every perfect number is. Thus, only among abundant numbers is there a real question as to which are semiperfect and which are not. It seems to be the case that "most" abundant numbers are semiperfect, and that the occurrence of abundant numbers that are not semiperfect is quite rare. This has prompted a whimsical mathematician to propose the following definition:

DEFINITION 8.8.4

An abundant number that is not semiperfect is said to be **weird**.

Very little is known of semiperfect and weird numbers.‡

Another recent spin-off of perfect numbers are superperfect numbers:

---

* See E. Dickson, *History of the Theory of Numbers* (New York: Chelsea Publishing co., Inc., 1952).

† For a further discussion, see W. Sierpinski, *Elementary Theory of Numbers* (Warsaw: Pantswowe Wyclawnictwo Naukowe, 1964).

‡ See "Are All Weird Numbers Even?" *The American Mathematical Monthly*, Vol. 179, No. 7, Aug.–Sept., 1972, p. 774.

| DEFINITION 8.8.5 | A number n is **superperfect** provided that $S(S(n)) = 2n$. |

It is not difficult to show:

| THEOREM 8.8.1 | If M_k is prime, then 2^{k-1} is superperfect. |

For example, $M_5 (= 31)$ is prime; and $2^4 (= 16)$ is superperfect since $S(S(16)) = S(31) = 32 = 2 \cdot 16$. By a process similar to that employed in the proof of Theorem 8.6.2, one can prove:

| THEOREM 8.8.2 | If n is an even superperfect number, it is of the form 2^{k-1}, where M_k is prime. |

Thus, there is a one-to-one correspondence between even superperfect numbers and Mersenne primes, and hence between even superperfect and perfect numbers.

The correspondence with perfect numbers is also known to extend to odd numbers. No odd superperfect numbers are known, and a few theorems have been proved (analogous to those for odd perfect numbers) indicating that odd superperfect numbers, if any, must be very large.*

EXERCISES 8.8

1. Verify that 672 is a triply perfect number.

2. Verify Descartes's quintuply perfect number. (*Hint:* Its prime factorization is $2^7 \cdot 3^4 \cdot 5 \cdot 7 \cdot 11^2 \cdot 17 \cdot 19$. Work with the prime factorizations, and avoid multiplying.)

3. Prove that m and n are an amicable pair if and only if $S(m) = S(n) = m + n$.

4. Verify that 220 and 284 are an amicable pair.

5. Starting with 24, find $P(24)$, $P(P(24))$, and so on, until a pattern becomes clear.

6. Follow the process of Exercise 5 for each of the following numbers.
   (a) $1184 = 2^5 \cdot 37$
   (b) $14{,}264 = 2^3 \cdot 1783$ (You should come across the prime number 967 at some point.)

7. Show that every multiple of a semiperfect number is semiperfect.

---

* For more information, see D. Soryanarayana's "Super Perfect Numbers," *Elemente der Mathematik*, Vol. 24, No. 1, 1969, or H. J. Kanold, "Über 'Super Perfect Numbers,'" *Elemente der Mathematik*, Vol. 25, No. 2, 1970.

8. Show that a number n is semiperfect if the sum of some subset of the proper divisors of n equals $P(n) - n$.

9. Determine which of the following are semiperfect. Which are weird?
   (a) 12   (b) 20
   (c) 28   (d) 70
   (e) 84   (f) 244
   (g) 572   (h) 836

10. Prove Theorem 8.8.1.

8.9 CONCLUSIONS

As indicated in the introduction to this chapter, you should reflect on what you have learned in these pages not merely for their content and significance, but as a pale mirror of number theory and mathematics as a whole.

In this vein, we saw that simple questions can lead to profound techniques and demand difficult forays into uncharted regions of knowledge. We also saw a microcosm of the interdependence of mathematics in the ways in which methods and concepts were called on from one section to another.

Perhaps the greatest mathematical lessons can be learned from the failures we observed. The search for perfect numbers led us down a long and sometimes difficult road, even though we could build on earlier formulas. The pattern of the prime factorization was finally decoded in Section 8.6, with the observation of Euclidean form and Mersenne primes. But then the last piece of the puzzle, the apparent pattern of the Mersenne primes, proved false and left a disturbing hole.

Such "failures" and incompleteness are familiar in mathematics. In every branch of mathematics, unanswered questions have defied solution, sometimes for millennia.

We should observe, again, that such misleading patterns are a major reason for mathematicians' demand for rigorous proof. The fact that available techniques and all known examples fit a proposed conjecture is not considered sufficient to assert the truth of the conjecture.

In much of our discussion, especially in Section 8.8, we saw another aspect of mathematics—that one idea always suggests another. If one can learn $D(n)$ from the prime factorization of n, what can one learn of the prime factorization from $D(n)$? From perfect numbers, we generalize to k-ply perfect, amicable pairs, triples, then to semiperfect and superperfect numbers. Mathematics organizes, generalizes, interrelates, and—after centuries of searching for answers—finds new questions.

Finally, we would be remiss to suggest that this brief encounter with number theory has conveyed any sizable portion of that subject, or that number theory is representative of all that is contained in mathematics. For the wealth of other topics in number theory, investigate the references. For alternative vistas in the panorama of mathematics, turn to other chapters.

For Further Reading

"Are All Weird Numbers Even?" *The American Mathematical Monthly*, Vol. 79, No. 7, Aug.–Sept. 1972, p. 774.

Beck, A., M. Bleicher, and D. Crowe. *Excursions into Mathematics*. New York: Worth Publishers, Inc., 1969.

Bell, E. T. "The Queen of Mathematics." In *The World of Mathematics*, Vol. I, Part III. New York: Simon and Schuster, 1956.

Boyer, C. *A History of Mathematics*. New York: John Wiley & Sons, Inc., 1968.

Dickson, E. *History of the Theory of Numbers*. New York: Chelsea Publishing Co., Inc., 1952.

Gardner, M. "*Perfect Numbers.*" *Scientific American*, March 1968.

——. "Unsolved Problems in Elementary Number Theory." *Scientific American*, December 1973.

Giola, A. A., and A. M. Vaidya. "Amicable Numbers with Opposite Parity." *The American Mathematical Monthly*, Vol. 74, No. 8, Oct. 1967, pp. 969–973.

Kanold, H. J., "Über 'Super Perfect Numbers.' "*Elemente der Mathemetik*, Vol. 25, No. 2, 1970.

Niven, I., and H. S. Zuckerman. *An Introduction to the Theory of Numbers, Second Edition*. New York: John Wiley and Sons, Inc., 1966.

Ore, O. *Number Theory and Its History*. New York: McGraw-Hill Book Company, 1948.

Sierpinski, W. *Elementary Theory of Numbers*. Warsaw: Panstwowe Wyclawnictwo Naukowe, 1964.

——. *A Selection of Problems in the Theory of Numbers*. Elmsford, N.Y.: Pergamon Press, 1964.

Soryanarayana, D. "Super Perfect Numbers." *Elemente der Mathematik*, Vol. 24, No. 1, 1969.

Stein, S. *Mathematics, the Man-Made Universe*. San Francisco: W. H. Freeman and Company, Publishers, 1963.

9

FINITE GROUPS

9.1   INTRODUCTION

This chapter is intended to provide a glimpse of part of a vast, growing area of modern mathematics called "abstract algebra." Each of these two words has been known to cause discomfort, and together they can be positively terrifying, so let's begin by examining exactly what this term means. Many people consider the word "abstract" to be a synonym for "vague" or "hard to understand," but in fact its meaning is very different. Its literal meaning is derived from Latin and is better reflected by our use of the verb "to abstract," which means "to draw out of" or "to separate from." Thus, the adjective "abstract" is used to designate some property of an object which is considered separately from that thing's many other characteristics. Hence, the abstraction process is a way of *simplifying* a situation by focusing directly on a specific aspect of that situation, separating its essential features from extraneous facts that might tend to confuse the issue.

Let us turn to arithmetic for an example. Among the many, many familiar elementary arithmetic facts at our disposal, we know that the sums $1 + 7, 2 + 6$, $3 + 5$, and $4 + 4$ all equal 8, that $8^2 = 64$, and that the sums $1 + 14 + 49$, $4 + 24 + 36, 9 + 30 + 25$, and $16 + 32 + 16$ all equal 64. This collection of data has many characteristics, some interesting from one point of view, some from another. For instance, one might observe that every digit from 0 through 9 was used in the examples, or that every three-summand expression contains an even number, or that all numbers are printed with the same size of type. All these observations are true, but they may serve to distract us from more significant observations. In this example, let us rearrange the information given:

$$64 = 8^2 = (1 + 7)^2 = 1 + 14 + 49$$
$$= (2 + 6)^2 = 4 + 24 + 36$$
$$= (3 + 5)^2 = 9 + 30 + 25$$
$$= (4 + 4)^2 = 16 + 32 + 16$$

A closer look at the facts in this form should suggest that they illustrate a general fact about the behavior of numbers which can be written in the following *abstract* form:

> *If a and b are numbers such that a + b = 8, then 64 = (a + b)² = a² + 2ab + b².*

The use of the letters a and b here in place of the various pairs of numbers we started with is the essential abstraction step, unifying all four specific instances of the one pattern and bringing the pattern itself into clear focus. Moreover, this form suggests a further step in generality—the observation that the numbers 8 and 64 are irrelevant to the general behavior of such sums; for *any* numbers a and b, it is true that $(a + b)^2 = a^2 + 2ab + b^2$, as you may know from high school.

This example typifies much of what was called "algebra" in high school. High school algebra, and indeed the algebra studied and practiced by mathematicians until fairly modern times, is little more than symbolized arithmetic. Various problems and techniques involving numbers and the basic arithmetic operations $+$, $-$, $\times$, and $\div$ have been abstracted by using letters to stand for numbers, thus making it possible to focus on general rules that describe the behavior of our number system. This area of mathematics is sometimes called "classical algebra" to distinguish it from its modern counterpart.

Abstract algebra takes a further step in generality (and hence in simplicity). Rather than confining itself to the numbers and operations of our usual number system, this form of algebra looks at objects in any set* whatsoever and discusses operations on them. Of course, since the objects we consider need not be numbers (for instance, they could be letters, motions, or elephants), the usual operations of arithmetic generally will not apply to them, and hence the fundamental characteristics of the idea "operation" must be abstracted from its particular numerical setting and put in a general form that admits application to other collections of things. It is this shift of attention from the concept of number to the concept of operation that distinguishes modern algebra from its classical counterpart; in fact, abstract algebra may be *defined* as the study of operations on sets.

9.2   OPERATIONS

As suggested in Section 9.1, the first step in our investigation will be to isolate the idea of "operation" in some general form. If we look at the familiar operations of addition, subtraction, multiplication, and division, we can observe that they all

---

* A **set** is any collection of things of any sort.

share a common general characteristic: They all are ways of combining two (or more) numbers to form another number. Abstracting this idea from its numerical setting, we get a useful general definition of operation.

DEFINITION
9.2.1

An **operation** on a set S is any rule or method that assigns to each ordered pair of elements of S exactly one element of S.*

[*Note:* An "ordered pair" of elements is just a pair of elements in which the first element can be distinguished from the second. We usually write (x, y) to denote the ordered pair in which x is the first element and y is the second. (x, y) and (y, x) are different ordered pairs. By way of contrast, the set $\{x, y\}$ is considered to be indistinguishable from the set $\{y, x\}$.]

EXAMPLES 9.2

1. The four arithmetic operations are obvious first examples of operations on the set of all real numbers. Addition assigns to any ordered pair (x, y) of numbers the sum $x + y$; subtraction assigns to (x, y) the difference $x - y$; multiplication assigns to (x, y) the product $x \cdot y$; and division assigns to (x, y) the quotient $x \div y$. These examples also suggest a convenient notational shorthand:

NOTATION: An operation on a set S will generally be denoted by a symbol such as $*$. In particular, the element of S that $*$ assigns to the ordered pair (x, y) will be written $x * y$.

MORE EXAMPLES 9.2

2. If S is any set, we can define an operation $*$ on it by assigning to each ordered pair the first element of that pair. For instance, if $S = \{?, \$, \%\}$, then $\$ * ? = \$$, $\% * \$ = \%$, and so forth.

3. Let $S = \{n, d, q\}$, and define an operation $*$ on S by listing all possible ordered pairs and elements assigned to them at random, as given in Figure 9.1.

FIGURE 9.1

$$n * n = q \qquad n * d = d \qquad n * q = n$$
$$d * n = d \qquad d * d = n \qquad d * q = n$$
$$q * n = n \qquad q * d = n \qquad q * q = d$$

---

* Strictly speaking, such operations are usually called "binary operations" because they combine two elements at a time. Since this is the only kind of operation we shall consider, the word "binary" has been omitted for simplicity.

Since every ordered pair of S is assigned exactly one element of S, this listing specifies an operation on S.

Listing the assignment for each ordered pair is sometimes the only way a particular operation can be written down, because it is not always possible to give a succinct general rule for it. However, we can make the job shorter and the result clearer by organizing the information in an **operation table**. An operation table works much like the grid used for reading a road map, and is perhaps most easily explained by example. To put the operation of Example 9.2.3 into table form, we list the elements of S in a row and in a column as shown in Figure 9.2. This arrangement creates a natural array of nine "boxes" to be filled, as indicated by the dashed lines. If we "name" each box by the letter indicating its row and then the letter for its column, we have a box for each ordered pair of elements of S, and can fill each one with the element assigned to that pair by the operation. Thus, the box filled in Figure 9.2 shows the entry that corresponds to $d * q$.

FIGURE 9.2

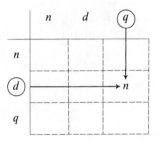

Figure 9.3 gives the complete table for this operation. The information supplied is the same as that given in Figure 9.1, but is displayed much more compactly. The symbol for the operation is noted for convenience in the upper left corner.

FIGURE 9.3

$*$	n	d	q
n	q	d	n
d	d	n	n
q	n	n	d

STILL MORE EXAMPLES 9.2

4. Let $S = \{a, b, c\}$, and define the operation $*$ on S by the table in Figure 9.4. This table says $a * a = b$, $a * b = a$, $a * c = c$, and so on.

5. The tables shown in Figure 9.5 define two operations, $*$ and $\circ$, on the set $\{0, 1\}$.

FIGURE 9.4

*	a	b	c
a	b	a	c
b	c	b	a
c	a	c	b

FIGURE 9.5

*	0	1
0	0	1
1	1	0

○	0	1
0	0	0
1	0	1

Let us denote the set $\{\ldots, -3, -2, -1, 0, 1, 2, 3, \ldots\}$ of all integers by I and the set $\{1, 2, 3, \ldots\}$ of all positive integers by I^+. Ordinary addition and multiplication are operations on both I and I^+. However, subtraction is an operation on I but not on I^+, since, for example, subtraction assigns to the pair of positive integers $(3, 4)$ the number -1, which is in I but not in I^+. (Remember that the definition of operation requires the assigned elements to come from the original set.) In general, an operation defined on a set S is not necessarily an operation on every subset* of S, since the operation on S may assign to an ordered pair of subset elements some element of S that is not in the subset. If, however, the operation assigns to each pair of subset elements an element from that subset, we say that the subset is **closed** under the operation (or sometimes that the operation is closed on the set).

AND MORE EXAMPLES 9.2

6. The even integers are closed under addition but the odd integers are not.

7. The table in Figure 9.6 defines an operation $*$ on the set $A = \{a, b, c, d\}$. The subset $\{a, b\}$ is closed under $*$, but the subsets $\{c, d\}$ and $\{a, b, c\}$ are not.

FIGURE 9.6

*	a	b	c	d
a	a	b	c	d
b	b	a	d	c
c	c	d	a	b
d	d	c	b	a

8. We can define a new operation $*$ on the set I of integers by the statement $x * y = x + 2y$ for all integers x and y. The set of all even integers is closed under $*$, and so is the set of all odd integers. (Can you find a subset of I that is not closed under $*$?)

---

* A **subset** of a set S is any set whose elements are also elements of S.

EXERCISES 9.2

1. Referring to the operation $*$ defined by Figure 9.6, find $a * b$, $b * c$, $d * d$, and $c * a$.

2. Let $S = \{w, x, y, z\}$ and define an operation $*$ on S as follows: Any pair containing w is assigned w; any pair containing x is assigned the other element in that pair; $x * x = x$, $y * z = z * y = z$, $y * y = z * z = y$. Describe $*$ by means of an operation table.

3. Let $S = \{1, 2\}$. Write tables describing *all* possible operations on S.

4. Let $T = \{x, y, z\}$.
   (a) Write five different operation tables for T.
   (b) How many operations are possible on T? Why?
   (c) Which of the operations you defined in part (a) is closed on the set $\{y, z\}$?

5. (a) Define two new operations on the set I of integers.
   (b) Which of the operations you defined in part (a) are closed on the set I^+?
   (c) Which of the operations you defined in part (a) are closed on the set of all even integers?

9.3   SOME PROPERTIES OF OPERATIONS

The abstract definition of operation is so general that the list of all possible operations even on a small set would be far too long to permit a meaningful examination of each one individually. For instance, on any three-element set we can define 19,683 different operations!* Hence, to organize this bewildering variety of things in some manageable way, we shall look at several special properties that make some operations "nicer" than others. The properties discussed here should already be familiar to you, since they are merely generalizations of elementary arithmetic concepts.

The most obvious "nice" properties are suggested by a close look at the definition of operation itself. That definition specifies that the operation must deal with *ordered* pairs, implying that the element assigned to a pair in one order might differ from the one assigned to that pair taken in reverse order. Indeed, this is what happens with the usual arithmetic operations of subtraction and division:

$$5 - 3 = 2 \quad \text{but} \quad 3 - 5 = -2$$
$$10 \div 2 = 5 \quad \text{but} \quad 2 \div 10 = \tfrac{1}{5}$$

However, addition and multiplication combine two numbers independently of their order—

---

* Since any operation on a three-element set can be written as a 3-by-3 table, and since each of the nine entries can be any of the three elements in the set, there are a total of 3^9 such operation tables. $3^9 = 19,683$.

$$5 + 3 = 3 + 5 = 8$$

$$10 \times 2 = 2 \times 10 = 20$$

and thus are easier to work with. This suggests the following general definition:

**DEFINITION
9.3.1**

An operation $*$ on a set S is **commutative** if

$$a * b = b * a$$

for all elements a and b in S.

EXAMPLE 9.3.1

The operation defined in Figure 9.4 is not commutative since, for example, $a * b = a$ but $b * a = c$. However, the operations given in Figures 9.3, 9.5, and 9.6 are all commutative. An easy way to see that a table defines a commutative operation is to imagine folding it along the diagonal from its upper left corner to its lower right corner. If the table entries that match by this folding are actually the same, then the operation is commutative. Figure 9.7 illustrates this.

FIGURE 9.7

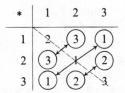

If an operation is not expressible as a table, then a proof that it is commutative might be difficult, since a general argument is needed to show that *every* pair is assigned the same element in both orders. However, since the definition of commutativity requires such reversibility to hold in every case, to prove that an operation is not commutative it suffices to find an example of one pair of elements which is assigned different elements in its two orders. For instance, we can prove that the operation $*$ on the integers defined in Example 9.2.8 is not commutative merely by observing that $3 * 5 = 3 + 2 \cdot 5 = 13$ but that $5 * 3 = 5 + 2 \cdot 3 = 11$.

Returning to the definition of operation, let us now focus on the requirement that operations assign elements to *pairs* of elements. It is natural to ask how three or more elements can be combined. Any attempt to apply an operation to three elements must proceed two at a time, and, since not all operations are commutative, the order in which the elements are considered is important. However, even if the order of the elements is fixed, there are still two ways to apply the operation. If a, b, and c are three elements (taken in that order), we could combine a and b and then combine c with $a * b$, or we could combine b and c and then combine a with $b * c$;

in symbols, we could consider either $(a * b) * c$ or $a * (b * c)$. For instance, if asked to subtract the numbers 9, 5, and 2, we might compute

$$(9 - 5) - 2 = 4 - 2 = 2 \quad \text{or} \quad 9 - (5 - 2) = 9 - 3 = 6.$$

On the other hand, adding 9, 5, and 2 is independent of this grouping process:

$$(9 + 5) + 2 = 14 + 2 = 16 \quad \text{and} \quad 9 + (5 + 2) = 9 + 7 = 16$$

This suggests another general property which some operations possess.

DEFINITION 9.3.2

An operation $*$ on a set S is **associative** if

$$(a * b) * c = a * (b * c)$$

for all elements a, b, and c in S.

EXAMPLE 9.3.2

The operation defined by Figure 9.4 is not associative, because, for example,

$$(b * b) * c = b * c = a \quad \text{but} \quad b * (b * c) = b * a = c$$

Both operations in Figure 9.5 are associative, as is the operation defined by Figure 9.6. Unfortunately, there is no easy way to observe this from the tables (as there is for commutativity), so a proof of these facts would require checking all possible cases. Even for an operation on a two-element set this is a somewhat tedious task, because there are eight possible three-element arrangements to be checked, so we shall give here just one example for each operation of Figure 9.5; you may verify the other cases at your leisure.

$$(1 * 1) * 0 = 0 * 0 = 0 \quad \text{and} \quad 1 * (1 * 0) = 1 * 1 = 0$$
$$(1 \circ 1) \circ 0 = 1 \circ 0 = 0 \quad \text{and} \quad 1 \circ (1 \circ 0) = 1 \circ 0 = 0$$

Ordinary multiplication is associative, but a proof of that fact requires an understanding of number systems which is beyond the scope of this unit. We give here instead a general proof of associativity for a simpler operation on the integers. Let $*$ be the operation defined by $x * y = y$ for all integers x and y. To prove that $*$ is associative, we observe that for any integers a, b, and c, $(a * b) * c = b * c = c$ and $a * (b * c) = a * c = c$, by the definition of $*$.

EXERCISES 9.3

1. (a) Give an example which proves that ordinary division is not associative.
   (b) Give an example which proves that the operation defined in Example 9.2.8 is not associative.

2. In all the examples we have examined so far, operations either have been both associative and commutative or have had neither of these properties. It is natural to ask whether these properties always exist together. Show that such is not the case by finding among the 16 operations on the set $\{1, 2\}$ (see Exercise 9.2.3) an example of
   (a) an operation that is commutative but not associative.
   (b) an operation that is associative but not commutative.

3. Construct an operation on the set $\{p, q, r\}$ which is both commutative and associative.

4. Give a general argument to prove that the table in Figure 9.8 defines an associative operation on the set $\{1, 2, 3, 4, 5\}$. (A case-by-case verification is a permissible but unattractive strategy, because there are 125 cases to be checked.)

FIGURE 9.8

*	1	2	3	4	5
1	1	1	1	1	1
2	2	2	2	2	2
3	3	3	3	3	3
4	4	4	4	4	4
5	5	5	5	5	5

9.4 THE DEFINITION OF GROUP

Of the four elementary arithmetic operations, the two "nicer" ones seem to be addition and multiplication, since, as we saw in Section 9.3, they are both commutative and associative. They are also well behaved in other ways. For instance, if a and b are integers, then any equation of the form $a + x = b$ has an integer solution. Similarly, if c and d are positive rational numbers (fractions), any equation of the form $c \cdot x = d$ has a rational solution. Now, if we examine the ways in which these kinds of equations are solved, we find a common pattern that can be abstracted and made into a definition with surprisingly far-reaching consequences. Let us look at two examples, solving each equation in painfully inefficient but enlightening detail.

To solve $3 + x = 5$, we add -3 to both sides of the equation, regroup by associativity, and let -3 "cancel" the 3 on the left, leaving us with the desired solution:

$$-3 + (3 + x) = -3 + 5$$
$$(-3 + 3) + x = 2$$
$$0 + x = 2$$
$$x = 2$$

Similarly, to solve $3 \cdot x = 5$, we multiply both sides by $\frac{1}{3}$, regroup by associativity, and let $\frac{1}{3}$ "cancel" the 3 on the left, leaving the desired solution:

$$\frac{1}{3} \cdot (3 \cdot x) = \frac{1}{3} \cdot 5$$

$$(\frac{1}{3} \cdot 3) \cdot x = \frac{5}{3}$$

$$1 \cdot x = \frac{5}{3}$$

$$x = \frac{5}{3}$$

In both cases we needed a number to "cancel" the number that was paired with x, where "cancel" meant that we got a number that left x unchanged when they were combined. We also needed associativity to get the numbers combined before having to deal with x. (Notice that commutativity was not needed.) These observations can be translated into specific properties of an operation on a set to guarantee that simple equations will always be solvable in the system.

DEFINITION 9.4.1

Let $*$ be an operation on a set G with the following properties:

1. $*$ is associative on G.
2. There is an element z in G such that $z * a = a$ and $a * z = a$ for any a in G.
3. For each a in G, there is an element a' in G such that $a * a' = z$ and $a' * a = z$.

Then G with the operation $*$, which we shall sometimes abbreviate as $(G, *)$, is called a **group**. z is called an **identity** element, and a' is called an **inverse** of a.

EXAMPLES 9.4

1. As we have already implied, the system $(I, +)$ of integers with addition is a group; the identity element is 0, and the inverse of any integer a is $-a$. Similarly, the set of positive fractions with multiplication is a group; the identity is 1 and the inverse of any a is $1/a$.

2. The first table in Figure 9.5 represents a group; 0 is the identity, $0' = 0$, and $1' = 1$. The second table does not represent a group, however; although there is an identity, 1, there is no inverse for 0.

3. Figure 9.4 is not a group table, because it has no identity element.

4. Figure 9.6 represents a group. Its identity element is a, and each element is its own inverse. (Notice that an identity element is easy to recognize in a table, because its row and column must be identical with the reference row and column at the top and side of the table, respectively.)

Many other examples of groups will turn up in the course of our work, but for now the foregoing examples should suffice to illustrate the definition. Notice that a group operation need *not* be commutative. A group whose operation is commutative will be specifically called a **commutative group**.* So far, all the examples we have seen are commutative groups, but noncommutative ones will begin to appear shortly.

Groups are of fundamental importance in many areas of modern mathematics. The simplicity of their defining properties belies the wealth of information that mathematicians have unearthed in their exploration of group theory since its beginnings in the early nineteenth century. This chapter is too brief to do more than scratch the surface of such a vast field of knowledge, but even here we shall have an opportunity to see some of the power of the group axioms as we investigate a few specific problems. The first of these problems can be stated immediately and will be the focus of our attention in the next section: We know from our discussion at the beginning of Section 9.3 that there are 19,683 possible operations on a three-element set. Which of these are groups?

EXERCISES 9.4

1. Consider the following two operation tables:

TABLE I

$\circ$	p	q	r	s	t
p	s	r	t	p	q
q	t	s	p	q	r
r	q	t	s	r	p
s	p	q	r	s	t
t	r	p	q	t	s

TABLE II

$\square$	0	1	2	3	4
0	0	1	2	3	4
1	1	2	3	4	0
2	2	3	4	0	1
3	3	4	0	1	2
4	4	0	1	2	3

(a) $t \circ r =$ _____

(b) $3 \square 4 =$ _____

(c) $(q \circ r) \circ p =$ _____

(d) $1 \square (3 \square 2) =$ _____

(e) Is there an identity element in Table I? If so, what is it?

(f) Is there an identity element in Table II? If so, what is it?

(g) Does q have an inverse in Table I? If so, what is it?

(h) Does 1 have an inverse in Table II? If so, what is it?

(i) Solve for x: $r \circ x = t$.

(j) Solve for x: $2 \square x = 3$.

(k) Give an example to show that one of these operations is not commutative.

(l) Give an example to show that one of these operations is not associative.

(m) One of these tables is not a group. Which is it, and why?

---

* Commutative groups are often called **Abelian groups,** in recognition of the pioneering work in commutative operations done by the Norwegian mathematician Niels Henrich Abel (1802–1829).

2. Let $S = \{a, b, c\}$. For each of the following statements, make an operation table for S that satisfies the condition given or give reasons to show why such an operation is not possible.
   (a) S is commutative but has no identity element.
   (b) S has an identity element but not every element has an inverse.
   (c) S has an identity element but no element has an inverse.
   (d) S is a commutative group.
   (e) S is a group that is not commutative.

3. Let $*$ be the operation on the set I of integers defined by $x * y = x - 3 + y$ for all integers x and y.
   (a) Find an identity element for $(I, *)$.
   (b) Find $7'$ and $2'$.
   (c) Find a' for an arbitrary integer a.

4. Give an example of a commutative group containing six elements. (Call this group A for future reference.)

9.5   SOME BASIC PROPERTIES OF GROUPS

The problem stated at the end of the previous section can be approached in many ways. The most obvious one is also the most tedious and the least enlightening: We could check each of the 19,683 different three-element tables to see if they possessed the three properties required by the definition of group. "There must be an easier way!" you say disconsolately, and you are right. Instead of taking the direct, dull approach, let us see if we can first establish some easily recognizable properties that a group table must have, and thereby eliminate many ineligible candidates at once.

As Example 9.4.4 suggests, an easily recognizable group property in a table is the existence of an identity, so let us begin there. Every group we have seen so far has had only one identity element, but the definition does not explicitly require such uniqueness. We answer the obvious question as follows:

THEOREM 9.5.1

Any group has only one identity.

PROOF   (We shall prove this theorem by showing that, if y and z both represent identity elements in a group G, then they cannot be different.) Let $(G, *)$ be a group, and suppose that y and z both represent identity elements in G. Consider the element $y * z$. Since y is an identity, $y * z = z$; on the other hand, since z is an identity, $y * z = y$. Hence, we have $y = y * z = z$, so y must equal z, and the theorem is proved.

A useful general property is suggested by our experience with addition and multiplication of numbers. We know, for example, that $3 + x = 3 + y$ implies that $x = y$, and also $3x = 3y$ implies that $x = y$; in both cases the 3 can be "canceled." We abstract from such numerical examples a general property shared by all groups.

THEOREM 9.5.2 (Cancellation Laws)

Let a, b, and c be any elements of a group $(G, *)$.

1. If $a * b = a * c$, then $b = c$.
2. If $b * a = c * a$, then $b = c$.

PROOF   [We shall prove (1) here; the proof of (2) is almost exactly the same and is left as an exercise.] Since $(G, *)$ is a group, the element a has an inverse a' in G. Combining a' with both sides of the given equation, we get

$$a' * (a * b) = a' * (a * c)$$

so $$(a' * a) * b = (a' * a) * c \qquad \text{by associativity}$$

Denoting the identity of $(G, *)$ by z, we then have

$$z * b = z * c \qquad \text{by definition of inverse}$$

implying $$b = c \qquad \text{by definition of identity}$$

Cancellation is an important property of groups, which in turn allows us to prove other basic properties. For example:

THEOREM 9.5.3

Any element of a group has only one inverse.

PROOF   (As in the proof of Theorem 9.5.1, the approach here is to show that any two inverses of the same element must be equal.) Let $(G, *)$ be a group with identity element z, and let g be an arbitrary element of G. Suppose that x and y both represent inverses of g. Then, by definition of inverse,

$$g * x = z \qquad \text{and} \qquad g * y = z$$

so $$g * x = g * y$$

Applying the first cancellation law, we get $x = y$, and the theorem is proved.

Besides providing a powerful tool for proving other general facts about groups, the cancellation laws are easily recognizable properties of any operation expressed in table form. Suppose, for example, that Figure 9.9 represents an operation $*$ on

FIGURE 9.9

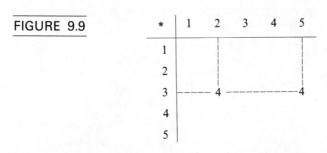

the set $\{1, 2, 3, 4, 5\}$, and that two entries in the third row are the same, as shown. In this case, we have $3 * 2 = 4$ and $3 * 5 = 4$. Hence, $3 * 2 = 3 * 5$, but $2 \neq 5$, violating cancellation law (1). Thus, cancellation law (1) implies that there can be no repetitions in any row. Similarly, cancellation law (2) precludes repetitions in any column. Coupling these observations with the definition of operation, we may conclude that in each row of any table that represents a group, every element of the set must appear exactly once!

Now we are ready to solve the problem of the three-element tables. As you recall, the problem is to decide which of the 19,683 three-element tables are groups. Let us use the general machinery just developed to try to construct some group tables for the set $\{1, 2, 3\}$.

Notice first that, since every group must have an identity element, the choice of a single element to play that role immediately determines five of the nine entries of a potential group table. For instance, the choice of 1 as identity requires that the row and the column for 1 be identical with the reference row and column, as shown in Figure 9.10. There are then only 81 $(= 3^4)$ ways to fill in the rest of the table (since there are only four remaining entries, each of which can be chosen in three ways).

FIGURE 9.10

	1	2	3
1	1	2	3
2	2		
3	3		

Hence, the identity requirement alone reduces the number of possible three-element group tables from 19,683 to a mere 243 (since there are only 81 for each choice of identity and only three possible identities). But the cancellation laws allow us to go even further. Referring to Figure 9.10, it is not hard to see that the nonrepetition restriction on the rows and columns *requires* that the middle entry of the table be 3, since a 2 there would give us an obvious repetition in the second row and column, and a 1 there would lead to repeated 3's in the third row and column. This entry in turn requires 1's at the end of the second row and column, and finally a 2 in the lower right corner. That is, the *only* way to complete the table in Figure 9.10 without row or column repetitions is as shown in Figure 9.11. Hence, there is *at most one* three-element group which has 1 as its identity element. Reasoning in precisely the same way, we can verify that there is at most one three-element group with identity 2 and at most one with identity 3. We have reduced the

FIGURE 9.11

	1	2	3
1	1	2	3
2	2	3	1
3	3	1	2

previously mentioned 243 possibilities of three-element group tables to only three.

Note that we have *not* yet proved that these tables actually represent groups; there are operations satisfying cancellation which are not associative (see Exercise 9.5.8), so we still must check these tables directly for associativity. However, we do know that no other three-element tables with identity elements need be checked, because none of the others satisfy cancellation. Thus, we now check each of the three tables directly for the group properties. For instance, referring to Figure 9.11, it is easy to observe that the operation is closed, 1 is the identity, $1' = 1$, $2' = 3$, and $3' = 2$. Checking associativity is somewhat tedious (27 cases), but it, too, can be verified. Hence, the table in Figure 9.11 is actually a group table. The group properties for the other two tables can be verified in exactly the same way. So the problem has been solved. Of the 19,683 different three-element tables, *exactly three* are groups!

EXERCISES 9.5

1. Write an operation table for the set $\{1, 2, 3, 4\}$ which satisfies
   (a) both cancellation laws.
   (b) cancellation law (1) but not (2).
   (c) cancellation law (2) but not (1).
   (d) neither cancellation law.

2. Check that the operation shown in Figure 9.11 is associative. (Work out all necessary cases.)

3. The table begun in Figure 9.12 starts to describe an operation on the set $\{1, 2, 3, 4\}$ with identity 1. Show that there are two different ways to complete this table so that the cancellation laws hold. Are they both group operations? Why?

FIGURE 9.12

	1	2	3	4
1	1	2	3	4
2	2			
3	3			
4	4			

4. Prove cancellation law (2).

5. Prove: If a and b are any elements of a group $(G, *)$, then the equation $a * x = b$ has exactly one solution in G.

6. Prove: For *any* element g of *any* group, $(g')'$ must be g. (That is, the inverse of the inverse of an element is the original element.) Give an elementary arithmetic example of this situation.

7. Let G be a finite group. Prove that, if the number of elements contained in G is even, then at least one element besides the identity is its own inverse.

8. The table given in Figure 9.13 clearly satisfies the cancellation laws. (Why?) Moreover, there is an identity element (what is it?) and each element has an inverse (what are they?). Prove that this is *not* a group operation.

FIGURE 9.13

	a	*b*	*c*	*d*	*e*
a	*c*	*d*	*a*	*e*	*b*
b	*e*	*c*	*b*	*a*	*d*
c	*a*	*b*	*c*	*d*	*e*
d	*b*	*e*	*d*	*c*	*a*
e	*d*	*a*	*e*	*b*	*c*

9. For the operation $*$ defined on the integers by $x * y = 2y$, does either cancellation law hold? Justify your answer.

Research Problem (optional): Sets with operations that have an identity and an inverse for each element and satisfy the cancellation laws (but not necessarily associativity) are called **loops**. Investigate the similarities and differences between loops and groups.

Specifically, it is natural to ask:

Do the analogs of Theorems 9.5.1 and 9.5.3 hold for loops?
Is every commutative loop a group?
Do the analogs of Exercises 5, 6, and 7 hold for loops?

Similar questions for loops may be asked by analogy as we develop some more elementary group properties in the sections to come.

9.6   SUBGROUPS

One way to proceed with our investigation of finite groups would be to continue the approach of Section 9.5, trying to examine all four-element operation tables, then all five-element tables, and so on. But this route is tiresome and inefficient, since the factorial growth of the number of possibilities supplies too many cases for even the powerful cancellation laws to dispose of. Checking associativity in a five-element table can be tedious indeed, and by the time tables with 10 elements hove into view we would be well beyond the limits of human patience. (There are 1000 associativity cases to be checked for a 10-element table!) We choose instead to seek another general insight into group theory which, like the cancellation laws, will allow us to dispose of many particular cases at once. This path will lead us to a major result in group theory, Lagrange's Theorem. We need one more definition:

DEFINITION
9.6.1

Let G be a group with operation $*$. If S is a subset of G and S is itself a group with respect to the same operation $*$, then we call $(S, *)$ a **subgroup** of $(G, *)$.

EXAMPLES 9.6

1. The set $\{a, c\}$ forms a subgroup of the group in Figure 9.6 since the operation $*$ applied to $\{a, c\}$ yields the table given here, which conforms to the pattern of a

$*$	a	c
a	a	c
c	c	a

two-element group, as we have seen. Similarly, $\{a, b\}$ and $\{a, d\}$ are subgroups, but $\{b, c\}$ is not, since the set is not even closed under $*$. In fact, no other two- or three-element subsets are subgroups of this group.

$*$	b	c
b	a	d
c	d	a

2. The set of even integers is a subgroup of the set of all integers under addition, but the set of odd integers is not. (Why?)

It is important to note that the definition of subgroup requires that the same operation be used for the subset as was used in the original group. Otherwise, any subset could be made into a group, making the subgroup concept worthless. Besides, since a group operation is necessarily associative when applied to *any* three elements of the group, it is automatically associative on any subset; hence, in checking to see if a subset is a subgroup we may ignore the troublesome associativity property. In other words, to verify that a subset S of a group is a subgroup of $(G, *)$ we need only check three properties:

1. S is closed under $*$.
2. S contains the identity element of $(G, *)$.
3. Each element of S has its inverse contained in S.

EXERCISES 9.6

1. Define one operation on the set $\{b, c\}$ that makes it a group, and one that does not.

2. Find all the subgroups of your six-element group A of Exercise 9.4.4.

3. The table given in Figure 9.14 defines a group; call it $(B, \circ)$.

FIGURE 9.14

$\circ$	j	f	g	h	i	k
j	j	f	g	h	i	k
f	f	j	i	k	g	h
g	g	h	j	f	k	i
h	h	g	k	i	j	f
i	i	k	f	j	h	g
k	k	i	h	g	f	j

(a) Find all subgroups of B.

(b) Count the number of elements in each of the subgroups. How do these numbers relate to the total number of elements in B?

(c) How does this group differ from the group A of problem 2?

4. Define an operation $\oplus$ on the set $C = \{0, 1, 2, 3, 4, 5, 6, 7, 8, 9, 10, 11\}$ by $x \oplus y =$ the remainder left when $x + y$ is divided by 12.

   (a) $9 \oplus 7 = $ _____

   (b) $(6 \oplus 8) \oplus 11 = $ _____

   (c) $6 \oplus (8 \oplus 11) = $ _____

   (d) $\oplus$ is an associative operation. Verify the rest of the properties needed to make $(C, \oplus)$ a group.

   (e) Find all the subgroups of C.

   (f) Count the number of elements in each of the subgroups. How do these numbers relate to the total number of elements in C?

5. Prove that any subset of a *finite* group which satisfies properties (1) and (2) preceding this group of exercises must also satisfy property (3).

6. Prove that the identity element of any subgroup must be the same as the identity element of the group containing it.

7. (a) Let G be a commutative group with operation $*$. Prove that, for any elements a and b in G, $(a * b)' = a' * b'$.

   (b) In the noncommutative group $(B, \circ)$ defined in Exercise 3, find an example of two elements for which the property stated in part (a) does *not* hold.

   (c) If $(G, *)$ is a group but is not commutative, how can you write $(a * b)'$ in terms of a' and b'? Justify your answer.

9.7   SUBGROUPS (continued)

To illustrate the existence and variety of subgroups within a group, we shall consider the set $D = \{0, 1, 2, 3, 4, 5, 6, 7, 8, 9\}$ of single digits and define an operation $\oplus$ on D by

$$x \oplus y = \text{the remainder left when } x + y \text{ is divided by 10}$$

It is not hard to verify that $(D, \oplus)$ is a group and that its identity element is 0. Moreover, the inverses are as follows (why?):

$$
\begin{array}{ll}
0' = 0 & 5' = 5 \\
1' = 9 & 6' = 4 \\
2' = 8 & 7' = 3 \\
3' = 7 & 8' = 2 \\
4' = 6 & 9' = 1
\end{array}
$$

Let us try to find all possible subgroups of $(D, \oplus)$. We know from subgroup property (2) preceding Exercises 9.6 and Exercise 9.6.6 that any subgroup of D must contain 0; in fact, $\{0\}$ itself is a subgroup. We can locate others by trial and error, but a little ingenuity can save a lot of effort. For example, any subgroup containing 1 must also contain $1 \oplus 1$, which is 2. Continuing this closure argument, we see that such a subgroup must then contain $2 \oplus 1 = 3$, and hence $3 \oplus 1 = 4$, $4 \oplus 1 = 5$, and so on. That is, any subgroup of D containing 1 also contains every other element of D and hence must be D itself. So far we have located a one-element subgroup and a ten-element subgroup (D itself). Clearly, $\{0\}$ is the only possible one-element subgroup. (Why?) Are there any two-element subgroups? If so, each must contain 0, which is its own inverse, and hence the other element must be its own inverse, as well. Thus, the only two-element subgroup of D is $\{0, 5\}$. How about three-element subgroups? In this case, subgroup property (3) requires that the two nonzero elements be inverses of each other. Let us try the various combinations: $\{0, 1, 9\}$ is not a subgroup because it does not contain $1 + 1 = 2$; $\{0, 2, 8\}$ does not contain $2 \oplus 2 = 4$; $\{0, 3, 7\}$ does not contain $3 \oplus 3 = 6$; $\{0, 4, 6\}$ does not contain $4 \oplus 4 = 8$. Hence, there are no three-element subgroups of D. These computations also ensure that there are no four-element subgroups, since any such subgroup must contain 0 and 5 (by Exercise 9.5.7) and some other element with its inverse; but adding 5 to each of the three-element sets above does not correct any of the closure problems shown above. A five-element subgroup must contain 0 and two pairs of elements that are inverses of each other (why?), and cannot contain 1, as we have seen. Moreover, it cannot contain the pair 3 and 7, because $7 \oplus 7 \oplus 7 = 1$. The only possibility, then, is $\{0, 2, 4, 6, 8\}$; since the sum of even numbers is even and division by 10 preserves evenness, this is indeed a subgroup. Similar arguments may be used to rule out the possibility of six-, seven-, eight-, and nine-element subgroups, so the only subgroups of D are $\{0\}$, $\{0, 5\}$, $\{0, 2, 4, 6, 8\}$, and D itself.

Notice that the number of elements in each subgroup of this ten-element group D is a divisor of the number 10. Looking back at the results of Exercises 2, 3, and 4 of Exercises 9.6, we can observe the same phenomenon: the six-element groups in Exercises 2 and 3 only have one-, two-, three-, and six-element subgroups, and the twelve-element group C in Exercise 4 only has subgroups containing one, two, three, four, six, or twelve elements. This is no accident; in fact, the assertion that subgroups are always related to their "parent" group in this way is the first major theorem of group theory:

| LAGRANGE'S THEOREM* | The number of elements in any subgroup of a finite group must be a divisor of the number of elements in the whole group. |

(Recall that a whole number a is a **divisor** of a whole number b if there is a whole number c such that $a \cdot c = b$.)

The proof of Lagrange's Theorem is not trivial, but it is well within the scope of the mathematical tools we now have at our disposal. Before examining its proof, however, we shall look at a few consequences of the theorem itself. These examples will serve to illustrate the power of Lagrange's Theorem, both in handling specific examples and in proving other general properties of groups.

For a first instance of Lagrange's Theorem at work, we turn to an example just like the one that began this section. Consider the set $E = \{0, 1, 2, 3, 4, 5, 6, 7\}$ and define an operation $\oplus$ on E by

$$x \oplus y = \text{the remainder left when } x + y \text{ is divided by 8}$$

$(E, \oplus)$ is a group with identity element 0 and inverses as follows:

$$0' = 0 \qquad 4' = 4$$
$$1' = 7 \qquad 5' = 3$$
$$2' = 6 \qquad 6' = 2$$
$$3' = 5 \qquad 7' = 1$$

(Can you justify these statements about the identity and the inverses?)

PROBLEM   Find all possible subgroups of $(E, \oplus)$.

SOLUTION   Since E contains eight elements, Lagrange's Theorem tells us that it is only possible to have subgroups containing one, two, four, or eight elements. Since every subgroup must contain 0, $\{0\}$ is the only one-element subgroup. Clearly, the only eight-element subgroup is E itself. Now, since a two-element subgroup must contain an element besides 0 which is its own inverse (see Exercise

---

* Named after Joseph-Louis Lagrange (1736–1813), outstanding French mathematician and friend of Napoleon, but proved by Augustin-Louis Cauchy (1789–1857), who pioneered in group theory.

9.5.7), we see from the list of inverses above that the only two-element subgroup is $\{0, 14\}$. Finally, any four-element subgroup must contain 0, 4, and two elements that are inverses of each other. (Why?) Checking the available inverse pairs, we must discard 1 and 7 because $1 \oplus 1 = 2$, and hence closure would require a fifth element in the subgroup. Similarly, the pair 3 and 5 will not work, because $3 \oplus 3 = 6$. However, the pair 2 and 6 fit with 0 and 4 to form the only possible four-element subgroup of $(E, \oplus)$, as shown by the table in Figure 9.15. Thus, we have found all the subgroups of $(E, \oplus)$.

FIGURE 9.15

$\oplus$	0	2	4	6
0	0	2	4	6
2	2	4	6	0
4	4	6	0	2
6	6	0	2	4

Compare this solution with the long closure arguments used in finding all subgroups of the group $(D, \oplus)$ at the beginning of this section; Lagrange's Theorem allowed us to avoid a lot of tedious work! Here is an even more striking example of the same sort: Define an operation $\oplus$ on the set $F = \{0, 1, 2, \ldots, 96\}$ by

$$x \oplus y = \text{the remainder left when } x + y \text{ is divided by } 97$$

PROBLEM   Find all subgroups of this 97-element group $(F, \oplus)$.

SOLUTION   Since the only divisors of 97 are 1 and 97, Lagrange's Theorem tells us that the only possible subgroups are of these two sizes. But $\{0\}$ is the only possible one-element subgroup, and $(F, \oplus)$ itself is the only possible 97-element subgroup. Hence, we have found *all* subgroups of the group $(F, \oplus)$.

This last problem is a particular case of an important general consequence of Lagrange's Theorem:

THEOREM 9.7.1   If the number of elements in a group is prime,* then there are only two subgroups— the one-element subgroup consisting of the identity alone, and the entire group itself.

PROOF   Obvious.

---

* Recall that a **prime number** is a number whose only divisors are 1 and the number itself.

We close the section with one more interesting corollary of Lagrange's Theorem. This result is a twin to the often-used Exercise 9.5.7.

THEOREM 9.7.2

Let G be a finite group. If the number of elements contained in G is odd, then *no* element besides the identity can be its own inverse.

PROOF Let z denote the identity element of G, and suppose there were some other element g in G which is its own inverse. Then $\{z, g\}$ would be a two-element subgroup of G (why?), and so, by Lagrange's Theorem, 2 must divide the number of elements in G. But this is impossible, because the number of elements in G is odd. Hence, there cannot be an element besides z which is its own inverse.

EXERCISES 9.7

1. Let $S = \{0, 1, 2, 3, 4, 5, 6, 7, 8\}$ and define $\oplus$ on S by

    $$x \oplus y = \text{the remainder left when } x + y \text{ is divided by 9}$$

    $(S, \oplus)$ is a group. Find all its subgroups.

2. The two tables in Figure 9.16 represent groups. Find all their subgroups.

FIGURE 9.16

	p	q	r	s	t	u	v
p	p	q	r	s	t	u	v
q	q	r	s	t	u	v	p
r	r	s	t	u	v	p	q
s	s	t	u	v	p	q	r
t	t	u	v	p	q	r	s
u	u	v	p	q	r	s	t
v	v	p	q	r	s	t	u

	1	2	3	4	5	6	7	8
1	1	2	3	4	5	6	7	8
2	2	3	4	1	8	7	5	6
3	3	4	1	2	6	5	8	7
4	4	1	2	3	7	8	6	5
5	5	7	6	8	1	3	2	4
6	6	8	5	7	3	1	4	2
7	7	6	8	5	4	2	1	3
8	8	5	7	6	2	4	3	1

3. (a) If a group has a two-element subgroup, a five-element subgroup, and a nine-element subgroup (and perhaps others), what is the smallest number of elements that it can possibly contain? Why?

   (b) If a group has a two-element subgroup, a five-element subgroup, and a fifteen-element subgroup (and perhaps others), what is the smallest number of elements it can possibly contain? Why?

4. Let G be a finite group. Prove that the only subgroup of G which contains more than half of the total number of elements in G is G itself.

5. Let $T = \{a, b, c, \ldots, t\}$, the set of the first 20 letters of the alphabet, and let $U = \{b, d, g, j\}$.
   (a) Split T into a collection of subsets such that
      (i) Every element of T is in exactly one subset.
      (ii) U is one of the subsets in the collection.
      (iii) All subsets in the collection contain the same number of elements.
   (b) Is there more than one way to choose a collection of subsets as described in part (a)? Why?
   (c) Give an example of a subset V of T such that the splitting-up process described in part (a) *cannot* be carried out if V is substituted for U.
   (d) Is it possible to find a set W containing a different number of elements than U such that the same kind of splitting-up process as described in part (a) can be carried out on T if W is substituted for U? If such a set W can be found, what are the possible numbers of elements that it could contain? If such a set W cannot be found, why not?

9.8 LAGRANGE'S THEOREM

The proof of Lagrange's Theorem provides an opportunity for us to see all the basic machinery developed so far interacting to produce a major mathematical result. It is hardly surprising that the argument used is neither short nor trivial; rather, it is remarkable that such an elegant and far-reaching result can be proved at all from the few facts that we have established so far! For a preliminary idea of how the proof will proceed, we begin by examining the general version of Exercise 9.7.5.

Suppose that a set G contains n elements and a subset H of G contains r elements. Suppose further that G can be split up into a collection of subsets such that:

1. Every element of G is in exactly one subset.
2. H is one of the subsets in the collection.
3. All subsets in the collection contain the same number of elements.

In such a case, (2) and (3) imply that each subset contains r elements. Then, since each of the n elements of G is in exactly one subset, the total number of elements in G can be found by counting the number of subsets in the collection and multiplying that number by r. That is, r must be a divisor of n. To prove Lagrange's Theorem, then, we need only show that, for any group G and *subgroup H*, there is a way of splitting G up into a collection of subsets satisfying conditions (1), (2), and (3) listed above. To do this, we need one new concept.

DEFINITION 9.8.1   Let H be a subgroup of a group G with operation $*$. If g is some fixed element of G, then the set of all elements $g * h$ for each element h of H is called a **coset** of H in G and is denoted by $g * H$.

(Note that the fixed element g is always on the left when combined with the elements of H. It is important to keep this in mind when forming cosets, since $*$ may not be a commutative operation.*)

EXAMPLES 9.8

1. As we have seen before, one of the subgroups of the group $(D, \oplus)$ defined at the beginning of Section 9.7 is $\{0, 2, 4, 6, 8\}$; call this subgroup S. Then the coset of S determined by 3 is

$$3 \oplus S = \{3 \oplus 0, 3 \oplus 2, 3 \oplus 4, 3 \oplus 6, 3 \oplus 8\} = \{3, 5, 7, 9, 1\}$$

Similarly, the coset of S determined by 4 is

$$4 \oplus S = \{4 \oplus 0, 4 \oplus 2, 4 \oplus 4, 4 \oplus 6, 4 \oplus 8\} = \{4, 6, 8, 0, 2\}$$

A complete list of cosets of S in D is as follows:

$$0 \oplus S = \{0, 2, 4, 6, 8\} \qquad 1 \oplus S = \{1, 3, 5, 7, 9\}$$
$$2 \oplus S = \{2, 4, 6, 8, 0\} \qquad 3 \oplus S = \{3, 5, 7, 9, 1\}$$
$$4 \oplus S = \{4, 6, 8, 0, 2\} \qquad 5 \oplus S = \{5, 7, 9, 1, 3\}$$
$$6 \oplus S = \{6, 8, 0, 2, 4\} \qquad 7 \oplus S = \{7, 9, 1, 3, 5\}$$
$$8 \oplus S = \{8, 0, 2, 4, 6\} \qquad 9 \oplus S = \{9, 1, 3, 5, 7\}$$

Notice that all the sets in the left column are the same, as are all the sets in the right column (since two sets are equal if they contain the same elements, regardless of order). Thus, there are *only two different cosets* of S in D.

2. One of the subgroups of the group $(B, \circ)$ defined by Figure 9.14 is $\{j, f\}$; call this subgroup J. Then the coset of J in B that is determined by the element h is

$$h \circ J = \{h \circ j, h \circ f\} = \{h, g\}$$

A complete list of cosets of J in B is as follows:

$$j \circ J = \{j, f\} \qquad g \circ J = \{g, h\} \qquad i \circ J = \{i, k\}$$
$$f \circ J = \{f, j\} \qquad h \circ J = \{h, g\} \qquad h \circ J = \{k, i\}$$

Notice that there are only three different cosets of J in B.

A closer look at these two examples can provide even more insight into the arguments at the beginning of this section. In Example 9.8.2, notice that the three

---

* Such cosets are usually called "left cosets," but since they are the only kind of coset we shall use, we may omit the word "left" without ambiguity.

different cosets $\{j, f\}$, $\{g, h\}$, and $\{i, k\}$, are a collection of subsets of B which satisfy conditions (1), (2), and (3) at the beginning of this section with respect to the subgroup J; that is:

1. Every element of B is in exactly one subset.
2. J is one of the subsets in the collection.
3. All three of the subsets contain the same number of elements.

Notice further that the number of elements in the subgroup J (two) multiplied by the number of different cosets (three) gives us the total number of elements in the original group B (six)!

The same observations can be made about Example 9.8.1: The two different cosets $\{0, 2, 4, 6, 8\}$ and $\{1, 3, 5, 7, 9\}$ form a collection of subsets of D satisfying conditions (1), (2), and (3) above with respect to the subgroup S, and the number of elements in S (five) multiplied by the number of cosets (two) yields the total number of elements in D (ten). It seems, then, that a reasonable way to proceed with the proof of Lagrange's Theorem is by trying to show that these three conditions hold for the collection of different cosets of *any* subgroup in *any* finite group.

PROOF   Suppose that $(G, *)$ is a finite group containing n elements, and H is a subgroup of G containing r elements. We want to show that the collection of all different cosets of H in G satisfies conditions (1), (2), and (3) at the beginning of this section.

1. Since the identity element, z, must be in the subgroup H, each element g of G is in *at least* one coset, namely, the one it determines. (That is, g is in $g * H$, because z is in H and $g * z = g$.) To show that no element of G can be in two different cosets, we prove that any two cosets which contain a common element must be equal: Suppose that two cosets $x * H$ and $y * H$ both contain some particular element g. Then g in $x * H$ implies that $g = x * h_1$ for some h_1 in H, and g in $y * H$ implies that $g = y * h_2$ for some h_2 in H. Thus,

$$x * h_1 = y * h_2 \tag{a}$$

Since H is a group, the inverse of h_1 is in H, so we may combine h_1' with both sides of equation (a) to get

$$(x * h_1) * h_1' = (y * h_2) * h_1' \tag{b}$$

Now by associativity this becomes

$$x * (h_1 * h_1') = y * (h_2 * h_1') \tag{c}$$

and, since $h_1 * h_1'$ is the identity element, we get

$$x = y * (h_2 * h_1') \tag{d}$$

We use (d) to show that *every* element of $x * H$ must also be in $y * H$. Suppose that $x * h$ is an arbitrary element of $x * H$. By (d), we have

$$x * h = (y * (h_2 * h_1')) * h \tag{e}$$

and by associativity this can be rewritten as

$$x * h = y * ((h_2 * h_1') * h) \qquad \text{(f)}$$

But h_2, h_1', and h are all in H, and since H is a subgroup it must be closed under the operation $*$; therefore, $(h_2 * h_1') * h$ is in H, so (f) tells us that the element $x * h$ can be written as $y *$ (some element of H) and hence $x * h$ is in $y * H$. Thus, $x * H$ is a subset of $y * H$. By an argument just like this one, except for the use of h_2' instead of h_1' in passing from equation (a) to an equation like (b), we can show that $y = x * (h_1 * h_2')$ and then that $y * H$ is a subset of $x * H$ (see Exercise 9.8.5). Since each of the cosets $x * H$ and $y * H$ is a subset of the other, they must be equal, and hence condition (1) has been verified. That is, every element of G is in exactly one of the different cosets of H in G. (Whew!)

2. This part is easy. If we denote the identity element of G by z, then the coset $z * H$ is H itself (since $z * h = h$ for each element h in H). Therefore, H is one of the cosets, satisfying condition (2).

3. To show that all cosets have the same number of elements, we must prove that any coset of H in G contains the same number of elements as H itself. Consider any coset $g * H$. If we write out the elements of H and $g * H$, there appears to be a natural matching between the two sets which shows they are the same size:

$$H = \{h_1, \quad h_2, \quad h_3, \quad \ldots, \quad h_r\}$$
$$g * H = \{g * h_1, g * h_2, g * h_3, \ldots, g * h_r\}$$

One thing must be checked, however, before this becomes a valid argument. We must be sure that different h's are paired with *different* coset elements. That is, whenever we have two different elements of H, say $h_i \neq h_j$, then we must be sure that $g * h_i \neq g * h_j$. But this is easy to verify; for if $g * h_i = g * h_j$, then the cancellation law would imply $h_i = h_j$, contrary to our hypothesis. Hence, each coset contains r elements, so condition (3) is satisfied.

At this point of the proof, all the hard work has been done. All that remains is to put the pieces together as suggested earlier in the section. There are only a finite number of different cosets; say there are k of them. Since each coset contains r elements and no two different cosets contain any common elements, the total number of elements in the cosets (that is, in their union) is $r \cdot k$. But each element of G is in some coset, so $r \cdot k = n$. Therefore, r is a divisor of n, and Lagrange's Theorem is proved!

EXERCISES 9.8

1. Recall the twelve-element group C defined in Exercise 9.6.4. Two of its subgroups are $S = \{0, 3, 6, 9\}$ and $T = \{0, 4, 8\}$. Write out the following cosets of these subgroups in C.

(a) $5 + S$ (b) $5 + T$
(c) $2 + S$ (d) $2 + T$
(e) $0 + S$ (f) $9 + S$
(g) $4 + T$ (h) $4 + S$
(i) $3 + T$

2. Write out all the cosets of the subgroup $\{a, c\}$ in the group defined by Figure 9.6.

3. Let K denote the system described by the operation table in Figure 9.17 and let L denote the subset $\{p, q, r\}$.

FIGURE 9.17

$\circ$	p	q	r	s	t	u
p	p	q	r	s	t	u
q	q	p	s	s	u	t
r	r	s	p	s	t	q
s	s	s	s	p	q	u
t	t	u	t	q	p	r
u	u	t	q	u	r	p

(a) For each element x in K, write out the set $x \circ L = \{x \circ p, x \circ q, x \circ r\}$.
(b) How many different sets did you get in part (a)? Do all these sets contain the same number of elements?
(c) Is $(K, \circ)$ a group? Justify your answer in terms of your answer to part (b).

4. Let $(G, *)$ be defined by the table in Figure 9.18.

FIGURE 9.18

$*$	1	2	3	4	5	6
1	1	2	3	4	5	6
2	2	5	4	3	6	1
3	3	4	5	6	1	2
4	4	3	6	1	2	5
5	5	6	1	2	3	4
6	6	1	2	5	4	3

(a) Write a table describing this same operation $*$ on the subset $H = \{1, 4\}$.
(b) Is $(H, *)$ a group? Why?
(c) Write out the sets $1 * H$, $2 * H$, $3 * H$, $4 * H$, $5 * H$, and $6 * H$.
(d) What information do your results give about whether or not G is a group? Why?

5. In our proof of Lagrange's Theorem, condition (1) was established by showing that two cosets $x * H$ and $y * H$ were equal. The argument which proved that $x * H$ must be a subset of $y * H$ was written out in detail; mimic that argument to show that $y * H$ must be a subset of $x * H$.

Review Exercise: Decide whether or not each of the following tables represents a group. If it does, give the identity element and the inverse of 2. If it does not, give a reason *other than* "*not associative*" to justify your answer. (*Hint:* Exactly two are groups.)

(a)

	1	2	3
1	3	2	1
2	2	1	3
3	1	3	2

(b)

	1	2	3
1	1	2	3
2	2	3	1
3	3	1	2

(c)

	1	2	3	4
1	0	1	2	3
2	1	2	3	4
3	2	3	0	1
4	3	4	1	2

(d)

	1	2	3	4	5
1	1	3	5	1	4
2	3	5	4	2	3
3	5	4	2	3	1
4	1	2	3	4	5
5	4	3	1	5	1

(e)

	1	2	3	4	5
1	3	4	1	5	2
2	5	3	2	1	4
3	1	2	3	4	5
4	2	5	4	3	1
5	4	1	5	2	3

(f)

	1	2	3	4	5
1	1	2	3	4	5
2	2	5	1	3	4
3	3	4	2	5	1
4	4	1	5	2	3
5	5	3	4	1	2

(g)

	1	2	3	4	5	6
1	2	3	4	5	6	1
2	3	4	5	6	1	2
3	4	5	6	1	2	3
4	5	6	1	2	3	4
5	6	1	2	3	4	5
6	1	2	3	4	5	6

(h)

	1	2	3	4	5	6	7	8
1	1	2	3	4	5	6	7	8
2	2	3	1	5	6	7	8	4
3	3	1	2	6	7	8	4	5
4	4	5	6	1	8	2	3	7
5	5	6	7	8	1	4	2	3
6	6	7	8	2	4	3	5	1
7	7	8	4	3	2	5	1	6
8	8	4	5	7	3	1	6	2

9.9   GROUPS OF SYMMETRIES (optional)

An interesting instance of finite groups in mathematics arises from an examination of some simple planar geometric figures. For example, consider a straight split log lying flat side down in sand on a beach. If the log is picked up, it is easy to see that there are exactly two ways to replace it so that it coincides with its imprint in the sand (flat side down again, of course)—it can either be returned to its original position, or be reversed and then put down (see Figure 9.19). The rigidity of the

FIGURE 9.19

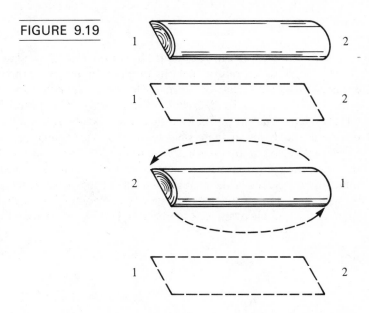

log ensures that we can describe each of these movements by the initial and final positions of its endpoints. Thus, if the ends are numbered 1 and 2, the log movements may be symbolized by

$$1 \to 1 \qquad 1 \underset{2}{\overset{1}{\times}} 1$$
$$2 \to 2 \qquad 2 \;\; 2$$

If we call these two movements of the log *l* (for "leave alone") and *s* (for "switch"), respectively, and look at what happens when we follow one of these movements by another, we can see that:

> *l* followed by *s* is the same as just doing *s*
>
> *s* followed by *l* is the same as just doing *s*
>
> *s* followed by *s* is the same as just doing *l*

and   *l* followed by *l* is the same as just doing *l*

This information can be summarized by the operation table in Figure 9.20, where ∘ denotes "followed by." Notice that this is a group table. The log movements l and s are examples of **rigid motions**, a geometric concept of figure movement which allows no distortion of the figure. With this understanding, we could easily replace our log model by a strictly geometric one using a line segment.

FIGURE 9.20

∘	l	s
l	l	s
s	s	l

A more interesting group can be obtained by considering the rigid motions of an equilateral triangle which returns it to its original "imprint" in the plane. Such motions must take each vertex (or "corner") to some original vertex position (not necessarily its own), and since the triangle cannot be distorted, the positions of the three vertices determine the position of the entire triangle. It is not hard to verify that there are exactly six of these rigid motions for an equilateral triangle (see Exercise 9.9.1). Three are rotations in the plane of 120°, 240°, and 360° (or 0°), and the other three are reflections with respect to the three medians. A **median** is a line determined by the center of the triangle and one of the vertices; **reflection** of the triangle with respect to a median can be pictured as holding the median fixed and then flipping the triangle over (see Figure 9.21). If the vertices are numbered 1, 2,

FIGURE 9.21

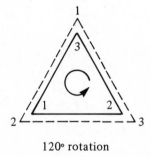
120° rotation

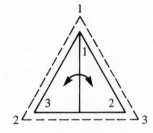
"Flip," fixing the median through vertex 1

and 3 (counterclockwise), any rigid motion can be represented by a chart indicating how each vertex relates to the original vertex positions. For example, if the triangle is rotated 120° (counterclockwise), vertex 1 goes to the place originally occupied by vertex 2, vertex 2 goes to the place originally held by 3, and vertex 3 goes to the place originally held by 1, as shown in Figure 9.22.

FIGURE 9.22

Vertex	Place

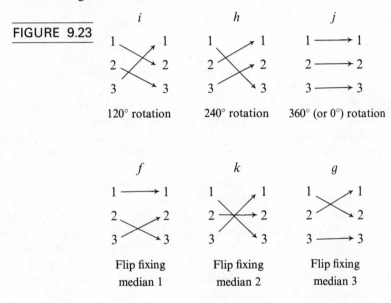

Using this representation scheme, the six rigid motions of the triangle are as shown in Figure 9.23.

FIGURE 9.23

Now let us consider what happens when we follow one of these rigid motions by another. For example, look at the 120° rotation followed by the flip fixing median 3. The vertices are moved around as follows: Vertex 1 is moved to position 2 by the rotation, and, since the flip fixing (the original) median 3 takes that position 2 to position 1, vertex 1 ends up back where it started, in position 1. Similarly, the rotation takes vertex 2 to position 3, and the flip leaves position 3 alone, so vertex 2 ends up in position 3. Finally, vertex 3 is rotated to position 1, and position 1 is then flipped over to position 2, so vertex 3 ends up in position 2. Looking at the result of all this, we can see that the final position of the triangle is the same as if it had only been moved by the flip fixing median 1. The process of following one motion by another is actually much easier than the preceding description indicates, provided that we adopt some convenient notation. Let us name the six rigid motions by single letters as shown in Figure 9.23,* and denote the "followed by"

---

* A reason for this apparently random lettering will appear shortly.

process by "∘", as we did in Figure 9.20. Then, using the diagrams of Figure 9.23, it is easy to see the result described above just by following arrows, as shown in Figure 9.24. Following the arrows from the left column to the right, we see that $1 \to 1$, $2 \to 3$, and $3 \to 2$; that is, $i \circ g = f$.

FIGURE 9.24

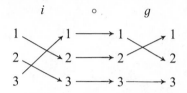

Since all the possible rigid motions of a triangle are listed in Figure 9.23, any one of them followed by another will necessarily yield the same result as one of the original six motions; that is, the set $\{j, f, g, h, i, k\}$ of rigid motions is closed under the operation ∘. In fact, if we compute all the possible combinations of rigid motions (using the method described above) and arrange the results in an operation table, we actually get the *group* table shown in Figure 9.14. Thus, the set of all rigid motions of an equilateral triangle forms a noncommutative group of six elements.

Similar groups may be formed by considering the rigid motions of various geometric figures. For example, there are eight rigid motions of a square—four rotations (90°, 180°, 270°, 360°) and four reflections, or "flips," which hold fixed one of the four dashed lines in Figure 9.25. The resulting eight-element group is noncommutative and is often referred to as "the octic group" (see Exercise 9.9.3). A few such groups are explored in the exercises. Some of these so-called "groups of symmetries" have become useful in physics and chemistry because of their relationships to problems of crystalline structure.

FIGURE 9.25

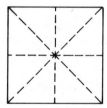

EXERCISES 9.9

1. Cut an equilateral triangle out of cardboard and number the corners 1, 2, and 3 on both sides. Trace the outline of your triangle on a piece of paper and number the corners (outside the traced figure). Then by moving your triangle around, verify that there are only six different ways to replace the triangle in its tracing, corresponding to the six rigid motions listed in Figure 9.23.

2. (a) Compute the following combinations of rigid motions of an equilateral triangle according to the process illustrated by Figure 9.24:

$$f \circ k, \quad g \circ i, \quad i \circ h, \quad j \circ f, \quad g \circ g, \quad h \circ h$$

   (b) Compute the six combinations listed in part (a) by actually moving around the triangle you made for Exercise 1.

3. Construct an operation table for the octic group (the group of rigid motions of a square), and verify that all the defining properties of a group are satisfied. Is it commutative?

4. (a) Find all rigid motions of a rectangle that is not a square (there are four) and construct an operation table for them.
   (b) Verify that this table satisfies all the defining properties of a group. Is it commutative?
   (c) Relabel these four rigid motions using the letters a, b, c, and d so that your operation table matches the table in Figure 9.6.
   (d) Find a subgroup of the octic group (Exercise 3) that matches this rectangle group.

5. (a) Find all rigid motions of an isosceles right triangle and construct an operation table for them.
   (b) Verify that this table satisfies all the defining properties of a group. Is it commutative?
   (c) Relabel these rigid motions so that your operation table matches one of the tables in Section 9.2.
   (d) Can you find a subgroup of the equilateral triangle group (described in this section) which matches this isosceles right triangle group? Why or why not?

For Further Reading

Berlinghoff, William. *Mathematics: The Art of Reason.* Lexington, Mass.: D. C. Heath and Company, 1968, Chap. IV.

Fraleigh, John B. *A First Course in Abstract Algebra.* Reading, Mass.: Addison-Wesley Publishing Company, Inc., 1967, Sects. 1–13.

Herstein, I. N., and I. Kaplansky. *Matters Mathematical.* New York: Harper & Row, Publishers, 1974, Chap. 4.

Newman, James R., ed. *The World of Mathematics.* New York: Simon and Schuster, Inc., 1956, Vol. 3, Part IX.

Penney, David E. *Perspectives in Mathematics.* Menlo Park, Calif.: W. A. Benjamin, Inc., 1972, Chap. 4.

10

AXIOMATIC SYSTEMS

10.1 INTRODUCTION

In this chapter we shall investigate the structure of an axiomatic system. In some sense, we shall thus study *the* fundamental structure of all mathematics, for all of mathematics can in principle be cast into the axiomatic structure. In practice, few mathematicians spend time relating the mathematics they do to some axiomatic foundation, and those who do encounter some of the most difficult questions of mathematics and frequently disagree with each other over the answer.

We shall see some of these fundamental questions, although we cannot develop all the material to fully explain them, let alone the possible answers. We can and will see the basic structure of an axiomatic system together with some of the more fundamental properties such a system may possess.

To oversimplify somewhat, an axiomatic system is a collection of statements logically interrelated so that some, called theorems, are logical consequences of others, called axioms. Those, and a few other details, will be discussed at length in the next section. But first let us briefly review the history of the genesis of the axiomatic method.

The **axiomatic method**, or the organization of information into an axiomatic system, can be traced to the Greeks. Mathematics prior to the Greek era was a collection of useful facts about numbers and shapes. The Greeks, from approximately 600 B.C., began to study and record logical interconnections among mathematical facts, proving some as consequences of others. They used logic and proof to discover new facts and to reject others as being inconsistent with accepted facts. About 300 B.C., Euclid added a new dimension by taking most of the extant mathematics of his time and organizing it so that virtually all the statements were proved from a relatively small beginning collection of definitions and axioms— and hence the axiomatic method was born.

The product of Euclid's genius was some 400 theorems, most on geometry, in the 13 books of the *Elements*. Although some of the theorems and most of the proofs in the *Elements* are original to Euclid, much was a compilation of previously known propositions. What has proved to be of greatest import, though, is not the content of the *Elements*, but its form. So profound, in fact, was the impact of this work that it became the paradigm of mathematics, and to a certain extent of all science, for the next 2000 years. With some slight modifications and clarification, this axiomatic method has become identified as *the* structure and the distinguishing mark of mathematics.

A tremendous amount of material has been added to mathematics in the intervening centuries since Euclid. But no mathematical claim has been accepted unless it has met the test of being provable from basic axioms. And where sizable bodies of intuitively persuasive mathematical propositions exist, mathematicians have followed the pattern of Euclid in working to identify an axiomatic basis.

Today, the concepts of axiom, theorem, proof, definition, and so on, have themselves undergone rigorous scrutiny, resulting in a more precise understanding of axiomatic systems. By modern standards, Euclid's *Elements* is somewhat flawed—some of his definitions are circular or meaningless, and some proofs erroneous. Ironically, this modern rigor, which sets standards too high for Euclid, is a direct result of the failure of a centuries-long attempt to improve on Euclid.

It was felt by many mathematicians that one of Euclid's axioms, the parallel postulate, was actually a statement provable from the other axioms. If so, this statement should not be identified as an axiom, and the whole structure of geometry would rest on a smaller, simpler, more elegant set of axioms.

The dozens of attempts to prove this all failed, although much interesting geometry was discovered as a "spin-off." But the prestige of the problem, the temptation of besting the master, kept professionals and amateurs coming back to the challenge. The final result was more earth-shaking than could have been imagined.

In the early nineteenth century, publications by Nikolai Lobachevsky and Janos Bolyai showed that a geometry radically different from Euclid's could be deduced from Euclid's axioms with the controversial parallel postulate replaced by an alternative axiom. Although at first rejected as erroneous, or at best a curiosity, the new non-Euclidean geometry was slowly accepted by mathematicians. But this acceptance necessitated a profound reassessment of geometry.

It became clear that Euclidean geometry could no longer be thought of as the "true" geometry of the real world, but only as one of several possible, mutually exclusive but logically consistent geometries.* With the shattering of the long-

---

* One of the first to recognize the impact was Karl Friedrich Gauss, who realized the consistency of non-Euclidean geometry earlier than Lobachevsky or Bolyai. In a letter in 1817, he wrote, "I am becoming more and more convinced that the necessity of our Euclidean geometry cannot be proved. . . . We must place geometry not in the same class with arithmetic, which is purely a priori, but with mechanics."

assumed connection between Euclidean geometry and the physical world, it was discovered that much of that geometry actually depended on the erroneous link to the world rather than on logical deduction from the axioms.

A period of close analysis and criticism of the existing mathematics and of the axiomatic method itself ensued. The result was a freeing of mathematics from artificial ties to the physical world, followed by an explosion of new mathematics. It also resulted in an increased understanding of and commitment to the axiomatic method.

10.2 CONSTITUENTS

An **axiom system** is generally considered to have four essential constituents: undefined terms, defined terms, axioms, and theorems. As we shall see, this is somewhat oversimplified, but essentially correct.

In any system intended to communicate information from one individual to another, there must be some common understanding of the meaning of terms. In English a word may have many meanings, controlled by the context of use; and one idea may be represented by many different words, differing almost imperceptibly. Such variety and nuance provides the wealth of the language.

But in mathematics, variation or nuance of meaning is not desirable. Rather, we seek precision and simplicity in terminology, and require a word to have a single meaning which does not vary from one context to another.

Simply put, when using a word in a mathematical system, we must agree to use it consistently with one meaning. A definition, then, must identify a single, unambiguous idea which the defined word, phrase, or symbol represents from that point on—no more and no less.

This imposes the first criterion of a good mathematical definition—that it be **characteristic**. For example, to define a square as a four-sided figure would fail this criterion; for although this describes an important property of a square, it does not characterize a square among all geometric figures.

The second criterion of a good definition is that it be **noncircular**. In other words, we cannot define a word by using the word itself in its definition or by using words which are in turn defined in terms of the word being defined. For example, if we identify an odd number as one that is not even and an even number as one that is not odd, we have defined nothing, but have involved ourselves in a circular loop of words.

This requirement of noncircularity, so seemingly obvious, leads to the necessity of some words being undefined. Otherwise, we either have circular definitions or an infinite regression. If words are to be defined in terms of more basic words, in turn defined in terms of those more basic, and so on, there must be some "most basic" words, not defined.

The fact that a term is undefined does not mean that it is meaningless. For example, the term "set" is undefined in most mathemeticians' lexicons. But they have a common understanding that the intuitive meaning of the term is "bunch," "collection," and so on, just as it is in ordinary English. They also agree that the term is not formally defined. Its formal meaning, which closely parallels its intuitive meaning, is acquired not by definition but by constraints imposed by the axioms of the system that use the term. Thus, undefined terms acquire meaning by context.

EXAMPLE 10.2.1

Assuming that the terms "angle," "acute angle," "right angle," "obtuse angle," "triangle," and "side" have been introduced, define *right triangle* (as normally understood in Euclidean geometry).

Trial A   A right triangle is a triangle that is right.

This statement is circular; as a simple grammatical statement, it is persuasive, but it does not define anything.

Trial B   A right triangle is a triangle containing exactly two acute angles.

This statement is not characteristic of right triangles. It does define something, but not what we would want to call a right triangle. (Both right and obtuse triangles fit this definition.)

Trial C   A right triangle is a triangle in which the product of the lengths of some two of the sides is twice the area.

This statement, as a statement of Euclidean geometry, is characteristic of right triangles; but it fails as a definition because it depends on terms not yet introduced (length of sides, area) according to the statement of the problem.

Trial D   A right triangle is a triangle containing one right angle.

This is a good definition.

The namesake of an axiom system is analogous to an undefined term. An **axiom** is a statement that is accepted without proof; it is the logical starting point for developing the system, just as undefined terms are the linguistic starting point.

It is sometimes said that an axiom is "accepted as true." Actually, mathematics has nothing to do with truth. The development of an axiom system produces theorems that are logical consequences of the axioms. Thus, the axiomatic method, which is the foundation of all mathematics, deals with validity—logical necessity—not with truth. Truth is a concern of philosophy, not mathematics.

The fourth element of an axiom system—in some sense its goal—is the theorem. A **theorem** is any statement that is logically derivable from the axioms. Of course, deriving theorems from axioms is not the same as adding a column of figures to produce an answer. Frequently, the axioms are actually created after the theorems; a number of theorems may be of interest, suggesting some intuitively simpler, logically prior statements which can be identified as axioms. Even when we start temporally and logically with axioms, there is no fixed procedure to determine theorems, or to prove whether or not a given statement is a theorem. The creative talent to do these things is the mark of a mathematician.

Before proceding to examine some particular axiom systems, a few ancillary topics should be considered. In one sense, all these may be lumped under the heading of "level of formality."

In a highly formal axiomatic development, not only are undefined and defined terms and axioms given, but also there are explicit rules of logic, usually quite restricted. Frequently, such systems are highly symbolic, as opposed to verbal, because symbols have less danger of leading to error or oversight due to an assumed, but unjustified, connotation in a term or an axiom. Words, even in mathematics, can be deceiving.

In a less formal development, rules of logic are not specified, but instead we allow ourselves a sort of "undefined" logic. This is the formalized (and sometimes symbolized) mathematical logic which is merely a more precise form of common sense. This is the logic used in everyday discourse and practiced in arithmetic, algebra, science, and, especially, geometry.

Even in this sort of development, a considerable spread of precision and formalism is found. In some, nothing is accepted as "understood" about the system except what is contained in the axioms and definitions. Even slight variations, such as tense or mood of verbs, is not allowed unless formally introduced as a definition or language rule. In others, familiar variations of grammar are accepted without question, and results of other familiar parts of mathematics are used without formal justification.

We shall adopt this last, most informal mode. This is almost certainly the style you have seen in a high school geometry course. There is an obvious comfort in allowing ourselves to use "common sense" and grammatical *laissez faire*, but we must beware of the concomitant danger. We can grow so casual that we believe we have proved a theorem, when we have actually been duped by a persuasive but flawed argument. We must be careful not to argue conversely, by example, or merely by analogy. A brief review of the terminology of logic and sets contained in Appendixes A and B is appropriate for the remainder of the unit.

EXERCISES 10.2

1. What are the four components of an axiomatic system?

2. Why must some terms be undefined in an axiom system?

3. Are the definitions of words in a dictionary definitions in the sense of an axiomatic system? What are the undefined terms of English?

4. Given that the terms "angle," "side," "quadrilateral," "parallel," "perpendicular," and "congruent" have been introduced (as undefined or defined terms) with their usual meanings, give a definition of:
   (a) square   (b) parallelogram
   (c) trapezoid   (d) rectangle

5. For the terms in Exercise 4, is there any advantage in defining certain ones before others? Explain. In what order should they be defined?

6. Assuming that the terms "point," "line," "segment," "angle," "right angle," "acute angle," "obtuse angle," "parallel," "perpendicular," "congruent," "triangle," "quadrilateral," and "opposite" have been introduced, which of the following are good definitions? ("Good" means noncircular, characteristic, and based on previously introduced terms.) Identify the flaw(s) of those which are not.

   (a) *M* is the midpoint of segment *AB* if it is midway from *A* to *B*.
   (b) *M* is the midpoint of segment *AB* if segment *AM* is half of segment *AB*.
   (c) *M* is the midpoint of segment *AB* provided that some line through *M* bisects *AB*.
   (d) *M* is the midpoint of segment *AB* provided that segment *AM* is congruent to segment *MB*.
   (e) A right triangle is a triangle in which one side is perpendicular to another.
   (f) A right triangle is one containing a right angle.
   (g) An obtuse triangle is a triangle that is larger than a right triangle.
   (h) An obtuse triangle is one containing an obtuse angle.
   (i) A scalene triangle is one that is not isosceles.
   (j) Two triangles are similar if they have the same shape but different size.
   (k) A parallelogram is a figure with opposite sides parallel.
   (l) A parallelogram is a quadrilateral with opposite sides equidistant from each other.
   (m) A rectangle is a quadrilateral that is rectangular.
   (n) A rectangle is a parallelogram with right angles.
   (o) A trapezoid is a quadrilateral with one pair of opposite sides parallel.
   (p) A semisquare is a quadrilateral with one pair of opposite sides congruent.
   (q) A circle is the set of points at a given distance from a fixed point, called the center.
   (r) A circle is a closed curve that has constant curvature everywhere.
   (s) A circle is the set of all points, *Q*, such that segment *PQ* is congruent to segment *AB*, where *P* is some point and *AB* is some segment.

10.3   AN AXIOMATIC SYSTEM

Let us now examine one very simple axiom system to exemplify the features identified in the previous section. We shall take our undefined terms to be "point" and "line," and shall freely borrow the terminology of elementary set theory.

   A1: A line is a set of points.
   A2: Given two distinct points, there is one and only one line containing them.
   A3: Every line contains at least two points.
   A4: Every point is contained in at least two lines.
   A5: A point exists.

Before proceeding to develop some theorems, let us examine the axioms. The first axiom seems to be a definition; it tells us exactly what a line is, does it not? The answer is, "no." If this statement were a definition, it would identify the abbreviation "line" as a shorthand for "a set of points." Thus, every set of points would be a line. This is not what axiom A1 says. Rather, it identifies the *undefined* term "line" as a set of points, but allows that there may be sets of points that are not lines.

Axiom A2 is a familiar one from geometry. We could paraphrase this axiom to state that two points determine exactly one line. Although this would probably be understood, that understanding would depend on a preconditioning to the word "determine" in a geometric setting. The language chosen relies only on terms of set theory, and, in a sense, makes precise what "determining" is.

You may think at first that axiom A3 is unnecessary. You may well be thinking: "Why bother to say that a line has at least two points, when everyone knows that lines have an infinite number of points?" This reaction is natural but indicates that you have not accepted "line" or "points" as undefined. In some other context (Euclidean geometry, for example), lines and points may derive certain properties from those axioms. But those axioms and the corresponding intuition you may have built up should not be transferred to *this* axiom system.

Don't assume anything about undefined terms (or for that matter, defined terms) which cannot be justified by the axioms. Note that any two words could be used as the undefined terms, not just "point" and "line." If axiom A3 said, "Every banoogy contains at least two nerds," we would not presume that a "banoogy" contained an infinite number of "nerds." Just as we have no preconceptions about banoogies and nerds, we should have no preconceptions about points and lines.

Similar comments could be made about axiom A4. Axiom A5 is quite unlike the others. Whereas axioms A1 through A4 state conditions interconnecting lines and points, axiom A5 simply states the existence of a point. The first thing to note is that this axiom is not superfluous. Axiom A3, for example, does not guarantee the existence of any point; it only guarantees that *if* there is a line, then there are points. Axioms A1, A2, and A4 are similarly conditional in nature. Axiom A5, in this sense, is stronger than the first four in that it states the existence of a point unconditionally. Without this axiom, we could not claim, within the axiom system, that there were any points or lines.

Let us now consider some theorems which can be derived from these axioms.

T1 A line exists.

PROOF   This follows immediately from axioms A5 and A4. A5 states that a point exists, and A4 claims that given a point there is a line containing it. (Actually, A4 yields more; namely two lines.)

The parenthetical remark indicates that we could have stated a slightly stronger theorem. But we are not required to state the "best" theorem. Often, we need to prove a weaker theorem in order to use it as a tool to prove other, stronger, theorems. Sometimes, a theorem is stated in a form weaker than necessary for aesthetic reasons. Here, for example, we might note that T1 is an analog to A5.

T2 | There are at least three points.

PROOF   By A5, there is a point; let us label it p. Then by A4 there are two lines, say L and M, which contain p. Now by A3, L must contain some point $q \neq p$. Similarly, M must contain a second point $r \neq p$. (We have now established three distinct *labels* of points. We have not established that these labels name three distinct points.) Suppose that $q = r$; then by A2 there is only one line containing p and q, which must be the one line containing p and r. But this would imply that $L = M$, contradicting A4. Therefore, $q \neq r$, and the three points are distinct.

T3 | There are at least three lines.

PROOF   By the same steps listed in the proof of T2, we obtain two distinct lines, L and M, containing (at least) the three distinct points p, q, and r, with $p, q \in L$ and $p, r \in M$.

Since $q \neq r$, A2 implies that there is a line, N, containing q and r. (As in the previous proof, we are not finished until we establish that N is distinct from L and M. To do this, we again use a proof by contradiction.) Suppose that $N = L$. Then $p, q, r \in L$, and, by A2, $L = M$. This contradiction shows that the supposition must be false; $N \neq L$. Similarly, $N \neq M$.

It is tempting to shortcut the proof above by using T2 to claim the existence of three points, and then claim the existence of three lines. Unfortunately, with T2 as a starting point, we do not have enough information to establish the distinctness of the three lines. We have in T2 an example of a theorem which states less than what was proved. What was proved, but not "claimed," in the statement of T2 is: "There exist three points, no two of which are on the same line."

Can we prove that there are more than three points or three lines? A careful look at the axioms, and especially at the proofs of theorems T2 and T3, shows that we cannot. Does that mean, then, that there are exactly three lines and three points in this axiom system? The answer to this, too, is no! It is *possible* that there are more points and lines; the axioms do not allow us to conclude that there are only three points and lines, nor that there are more than three.

The statement, "There are more than three points," is said to be independent of the axioms. We cannot prove it, so it is not a theorem; nor can we prove it false. In Section 10.5 we shall look more carefully at the notion of independence. For now, it is sufficient to note that an axiom system does not always provide all the logical foundations to make a judgment on the validity of every possible statement.

There are several more theorems which we could prove in this axiom system, and some of these are given in the exercises. Of course, this set of axioms is relatively simple; and a more complicated axiom set can produce a much greater variety of theorems.

Let us consider some "natural" concepts which we could define, even in this simple system. Recall the criteria for a good definition, and also note that the purpose of a definition is to identify or abbreviate a recurring or important concept.

D1   A set of points is said to **colline**, or be **collinear**, if there is some line that contains all the points.

With this definition, we could state succinctly the observation made earlier as: "There exist three noncollinear points."

A familiar geometric concept that can easily be defined in terms of the simple ideas already identified is parallelism.

D2   Two lines are **parallel** if they have no points in common.

Notice, in keeping with our earlier discussion on undefined terms that we should not read too much into this definition. It does fit our intuitive idea of parallel lines and is a good definition. But we should not impose our Euclidean preconception of parallelism. We might just as well be defining that two banoogies are larapep if they have no nerds in common. Parallel lines (whatever lines are) are certain disjoint sets of points (whatever they are).

Can we prove that there are parallel lines, or that given a line there is a line parallel to it? No; the situation is like that of the existence of more than three points. We should note, then, that definitions are independent of existence. We can define parallel lines without being able to prove that there are any. Of course, there are conditional theorems involving parallelism. For example:

T4   If there exist two parallel lines, then there exist at least four points.

The proof is left as an exercise.

EXERCISES 10.3

All exercises refer to the axiom system introduced in this section.

1. Prove the following: Given a line, there exists a point not on it.

2. Prove: Given a point, there exists a line not through it.

3. Prove theorem T4.

4. Prove: If there exist two parallel lines, then there exist at least six lines.

5. If axiom A2 is removed from the axiom set, show that theorem T2 can still be proved. Can theorem T3 be proved under these conditions?

10.4   MODELS

The word "model" is used in mathematics in much the same way as in ordinary English. Frequently, one speaks of a model boat, car, and so on, meaning a copy, usually much smaller, which attempts to mimic many of the features of the real object. The accuracy of the copying and the number of features represented is in some sense a measure of the value of the model.

Mathematics has a modeling value like this, and this is the use of models most frequently associated with mathematics. For example, one might construct a mathematical model of the weather patterns over North America in order to make long-range predictions. The more features that affect the weather which can be identified and represented in the model, the better will be the predictions, and the greater the value of the model.

Another kind of model is used in everyday life. The best example is an architect's or builder's model of a proposed building. In this instance, the model-builder is not copying an existing object, but is working from a set of blueprints or other description of the proposed building. Conceivably, two different model-builders might construct quite different models from the same set of blueprints, for the choice of materials, scale of the model, landscaping, and so on, would not be prescribed by the blueprints, and considerable variance might appear in the two models.

Again, mathematics mirrors this kind of modeling, and it is this type of model that we are interested in. Our blueprints are the axioms of a system. Just as an architect's model makes design ideas more concrete and visible, a model of an axiom system makes its content more concrete. And, as there may be many things *not* specified by the axioms, considerable variety in models may be possible.

What, exactly, is a model for an axiom system? It is merely the resulting structure, physical or mathematical, when concrete entities, again physical or mathematical, are assigned to or identified with the undefined terms of the axiom system in such a way that each axiom becomes a true statement about those concrete entities.

For example, using the axiom set introduced in the previous section, if we let "point" be represented by "a number in the set $\{1, 3, 5\}$," and "line" by "one of the sets $\{1, 3\}, \{1, 5\}, \{3, 5\}$," we can verify each of the axioms. Axiom A1, for example, becomes: "Each set $\{1, 3\}, \{1, 5\}$, and $\{3, 5\}$ is a set of numbers chosen from the set $\{1, 3, 5\}$." This is, of course, a true statement. You should verify that the remaining four axioms of that system also become true statements when interpreted according to the foregoing assignments.

You should also note that an arbitrary assignment of concrete "points" and "lines" does not automatically produce a model. If we chose $\{1, 2, 3, 4\}$ to be the set of points, and the sets $\{1, 2, 3\}, \{1, 2, 4\}, \{1, 3, 4\}, \{2, 3, 4\}$ to be lines, then we have not constructed a model, for axiom A2 is not a true statement with this assignment of terms: for example, the points 1 and 2 are contained in two distinct lines.

Let us formally define the terminology we have introduced:

| DEFINITION 10.4.1 | An **interpretation** of an axiom system is any assignment of concrete meaning to the undefined terms of that system. |

| DEFINITION 10.4.2 | If an axiom becomes a true statement when its undefined terms are interpreted, then the interpretation **satisfies** the axiom. |

| DEFINITION 10.4.3 | A **model** for an axiom system is an interpretation that satisfies all the axioms. |

Thus, both of the examples given above are interpretations of the axiom system of Section 10.3, but only the first is a model.

As suggested earlier, there are other models of our axiom system which are quite different from the one given. For the sake of brevity, in the examples below, we shall not use the usual set theoretic notation, but shall list the "lines" as vertical columns of "points." When not stated otherwise, we shall understand that the set of points is just the collection of all those listed. With this notation, the first example would be listed as in Example 10.4.1.

EXAMPLES 10.4

1. 1 1 2
   2 3 3

2. 1 1 1 2 2 3
   2 3 4 3 4 4

3. 1 2 3 4 5 6 7
   2 3 4 5 6 7 1
   4 5 6 7 1 2 3

Let us examine how we would verify that the interpretation of Example 10.4.1 is actually a model:

1. With the agreed-upon variance from usual set *notation*, each line is a set of points (numbers), and A1 is satisfied.
2. For the distinct points (numbers) 1 and 2, there is one and only one line (the set $\{1, 2\}$) containing them. A similar statement is true of the distinct points 1 and 3 and of the pair 2 and 3. These are the only pairs of distinct points in this

interpretation, so A2 is satisfied. (Note that *each* pair of distinct points must be checked.)

3. Each line in the interpretation is a set of at least two points. (Exactly two is certainly at least two.) Therefore, A3 is satisfied.

Similarly, A4 and A5 are satisfied, so this interpretation is a model of the axiom system.

In the exercises you will be asked to go through the details to show that the interpretations of Examples 10.4.2 and 10.4.3 are also models. For now, let us take that as an established fact and observe what information models give us about an axiom system.

The utility of models arises from logic and their definition. By definition, every axiom becomes a true statement in a model. And by logic, from true statements we can deduce only other true statements. Thus, every theorem of the axiom system, which is just a statement logically derived from the axioms, must become a true statement in every model. This fact has a dual utility.

On the one hand, a fact that is observed to be true in a model *may* be a theorem; the possibility is stronger if the fact is true in every known model. Ultimately, of course, a theorem must be proved from the axioms. The persistence of some fact in many models is not a proof; it may only indicate that we tend to create models with similarities.

On the other hand, a fact that is observed to be false in one model cannot be a theorem in the axiom system. Thus, the fact that the model of Example 10.4.2 contains four points establishes with finality that the statement "There are exactly three points" is not a theorem.

If we consider the statements "There are parallel lines" and "There are no parallel lines," Example 10.4.1 shows that the first is not a theorem, and Example 10.4.2 shows that the second is not. As discussed in Section 10.3, these statements on the existence of parallels are independent of the axiom system. Models provide the method to demonstrate that a statement cannot be proved, and hence to prove that a statement is independent.

Note that not every unprovable statement is independent. For example, the statement that there are only two points is not independent; it is contradictory to the axioms.

Your use of models is not restricted to observing properties of models someone else has constructed. Usually, you have some statement to test which you have reason to believe is not a theorem (cannot be proved). Your challenge is to construct a model which, by design, fails to satisfy the statement in question. This can often be as challenging as finding a proof of a theorem.

For example, consider the following statement:

If one line contains three points, then every line contains three points.(10.4.1)

All the models of our axiom system that we have seen so far satisfy this statement, suggesting that it is a theorem. (The models of Examples 10.4.1 and 10.4.2 do not contradict this statement, but rather satisfy it **vacuously** since they fail to satisfy the antecedent of the conditional. The model of Example 10.4.3 satisfies the statement nonvacuously since it has three-point lines.)

If you try to prove this statement, though, you will fail, since it cannot be proved. To demonstrate this, we need to construct a model that fails to satisfy this statement. Equivalently, this model must satisfy the negation of the statement: One line contains three points, but not every line contains three points (some contain fewer).

This suggests how to start constructing our interpretation. Using numbers as points, we form one three-point line, say {1, 2, 3}, and one with fewer than three points. Since we must construct a model, satisfying in particular axiom A3, our lines must contain at least two points.

Could we, then, use {1, 2} as our second line? No, for then we would have two lines containing the same pair of distinct points, and we would violate A2. We could choose a two-point line parallel to the three-point line, but there is no purpose in doing more than we have to—no need to introduce a fourth *and* a fifth point. So let us make the second line {1, 4}.

Of course, the interpretation so far is not a model, since now A2 is not satisfied for the pairs of points 2 and 4 and 3 and 4. This is most easily remedied by making these doubletons into two-point lines, {2, 4} and {3, 4}.

It is now fairly easy to see that this interpretation, with one three-point line and three two-point lines, is a model that fails to satisfy statement (10.4.1), thus showing it to be independent of the axioms.

EXERCISES 10.4

1. Verify that Example 10.4.2 is a model.

2. Verify that Example 10.4.3 is a model.

3. Verify that the interpretation testing statement (10.4.1) is a model.

4. Determine which of the following interpretations are models of the system of Section 10.3. For those which are not models, identify which axioms are satisfied and which are not.

   (a) 2
       4
       6

   (b) 5 5 8
       8 8 9
          9

   (c) Points = {a, b, c, d}

             a   c   a   b

             b   d   d   c

(d) Points $= \{x, y, z\}$

w	w	w	x	x	y
x	y	z	y	z	z

5. For each of the following statements, give a model which shows that the statement is not a theorem. (Take care that the interpretation you give really is a model.)
   (a) There are no more than seven points.
   (b) If one line has a parallel, then every line has a parallel.
   (c) If there are four points, then there are six lines.

10.5   METAMATHEMATICS

In this section we shall continue to study the simple axiom system introduced in Section 10.3. But rather than look at further definitions or theorems or proofs, all of which are part of the system, we shall step back one logical step and seek information about the system as a whole. We shall look at some of the logic, or prelogic, which makes the axiomatic method what it is. This study of the system, from outside the system, is one part of *metamathematics* (above or outside of mathematics).

The first metamathematical property to consider is consistency. It is easier to define its opposite first.

DEFINITION
10.5.1

An axiomatic system is **inconsistent** if it is possible to prove both a statement and its negation from the axioms. A system is **consistent** if it is not inconsistent.

At first, you might think that it would be impossible to prove a statement and its negation. How can a statement and its negation both be true? The answer is that they cannot, but the problem is not specious, for mathematics deals not with truth, but with validity. Of course, mathematics with internal contradictions is thoroughly useless, so an inconsistent axiom system is worthless—but it is not impossible.

Curiously, the problem with an inconsistent system is not that nothing can be proved, but that everything can be proved. (If a statement and its negation are both theorems, then their conjunction is a theorem. But this conjunction, necessarily false, used as an antecedent, will produce a true implication, regardless of the conclusion. Thus, any statement is, formally, a theorem.) The goal of an axiomatic system, then, is certainly not to prove everything. A consistent system allows us to discriminate between provable and unprovable statements.

We can easily construct an inconsistent system, by taking two contradictory statements as axioms. We would not do so, except to exemplify the concept of inconsistency. But we might accidentally devise an inconsistent system, one where the contradiction is not obvious. Since consistency requires that no pair of contradictory statements both be theorems, and since proofs are often elusive, how are we to know that any system is consistent?

The answer lies in models, and the feature of models noted in the previous section: namely, every theorem of an axiom system becomes a true statement about the model when interpreted. Thus, a model of an inconsistent system would have to exemplify contradictory true statements—an impossibility. In other words, the existence of a model implies consistency. In particular, any of the models of the axiom system of Section 10.3 demonstrates its consistency.

Lest we breathe an unwarranted sigh of relief, it should be noted that we have not removed the problem of consistency; we have translated it into a problem of models. In most branches of mathematics, the axiom systems involved implicitly or explicitly infer the existence of infinite sets of entities (number, points, lines, and so on). Models, then, cannot be tested completely without relying on some general mathematical principle, itself a theorem of some axiom system with "infinite-type" axioms. The result is that such systems have "models" via the entities of other systems, whose consistency is just as questionable. In the end, we can only judge these systems to be "relatively consistent." This is the situation for all branches of mathematics.

A metamathematical property closely related to consistency is completeness. Consistency requires that no statement and its negation both be theorems. Completeness requires that at least one be a theorem. To formally define the concept, let us first formalize an idea introduced in the previous section.

DEFINITION 10.5.2	If a statement and its negation both fail to be theorems in an axiom system, then the statement is **independent** of the axioms.

DEFINITION 10.5.3	A **complete** axiom system is a consistent one that contains no independent statements.

Looking at the situation from another viewpoint, a theorem is clearly dependent on the axioms. Similarly, the negation of a theorem can be thought of as dependent upon the axioms. In any model of the axiom system, the truth of the former and the falseness of the latter follow automatically. But a statement which is independent is true in some models and false in others, and thus is not logically dependent on the axioms. (Its negation, of course, has the same property.)

In the system of Section 10.3, we noted several independent statements. Thus, this system is not complete. We observed this through the use of models. We can "explain" the incompleteness by virtue of the simplicity of the system.

It should be noted that completeness, unlike consistency, is not essential to an axiom system. It is best seen as an aesthetic property, one which may in some instances be undesirable, depending on the purpose of the axiom system.

About the only way to show that a system is complete is to show that essentially only one model can exist (aside from labeling in the model). In this situation, validity of theorems and truth of statements in the model becomes a two-way, rather than a one-way street, and completeness is guaranteed. Of course, this is only possible in those relatively simple systems where consistency is demonstrable.

An independent statement cannot be contradictory to the axiom and theorems of a system. If it were, its negation would be a theorem, proved by contradiction. Thus, an independent statement can be added to the set of axioms, and the enlarged set will be consistent (assuming the original set was). Alternatively, the negation of the statement could be used, the result being a very different, although consistent, system. This is precisely the kind of theory described in the introduction relative to the development of non-Euclidean geometry.

The process of enlarging an axiom set by adding an independent statement suggests a way of "completing" an axiom system. If a system we are dealing with is shown to be incomplete, we add on the independent statement and search for more theorems. If another independent statement is found, we add it on as another new axiom. Eventually, it would seem, we must produce a system that is complete. For simple systems, this is the case; but, surprisingly, for systems of any complexity (including all the familiar branches of mathematics), it has been proved that an infinite regression of independent statements exists, and the system is **essentially incomplete.***

A third metamathematical property is that of independence. We have already encountered independent statements, and this same notion provides the terminology for this system property.

DEFINITION 10.5.4   If a statement that is an axiom can be proved from the other axioms in an axiom set, it is said to be **dependent.** Otherwise, it is **independent**. The axiom set, if consistent, is **independent** if all the axioms are independent; otherwise, it is dependent.

We are not creating a contradictory definition of independent or dependent here; we are merely extending it consistently to refer meaningfully to axioms, and then to the axiom set.

An independent statement is one that is unprovable from the axioms. An axiom is trivially provable from the axioms, since it is one. But if we consider the axiom set formed by deleting the one in question, we may think of the deleted axiom as no longer an axiom. It is then quite reasonable to question whether this statement is provable (dependent) or not (independent) in the new system.

This property of an axiom set being independent is, as with completeness, not essential. Most mathematicians, though, find it aesthetically desirable. For if some

---

* Kurt Godel, in a paper in 1931, established this surprising fact, since called Godel's incompleteness theorem. It was a shock to mathematicians, because it implies that no axiomatization of mathematics can be complete, a fact contrary to the intuitive beliefs of mathematicians of the nineteenth century.

axiom is dependent on the others, it could be (permanently) deleted from the set, be proved as a theorem, and all the theorems of the former system could be proved in the new system, with one fewer axiom. Since one goal of an axiom system is to establish a sizable set of theorems on a minimal set of axioms, any dependent axioms should be eliminated as superfluous.

On the other hand, it is not illogical to have a dependent axiom set. There are several reasons why one might knowingly construct a dependent system. For heuristic purposes, it might be much easier to prove a number of elementary theorems with a dependent set of axioms. Then, when the student has gained some familiarity with the system and skill in proof, the dependent axiom, whose proof is difficult, could be pointed out. (The first part of this scenario is not unusual in textbooks. The second is less commonly seen.)

Another heuristic, or structural, reason for a dependent system is as follows. Frequently, an axiom system is best studied in parts, with some axioms and theorems presented first, then more axioms added, more theorems proved, and so on, until the entire axiom set is finally identified. In such a development, it is not unusual for some of the early axioms to be dependent in the final set of axioms, but their omission at the outset would have made the early development of theorems impossible.

Probably the most common reason for an axiom set being dependent is that the designer of the set did not notice the dependence. Frequently, dependence is not demonstrated until much time and effort have been expended just getting to know the system.

The method of establishing independence or dependence of axioms has already been identified. If a proof of an axiom from the others is exhibited, it is dependent. If an interpretation is given that satisfies all the other axioms, but fails to satisfy the axiom being tested—or, in other words, satisfies its negation—then the axiom is independent. For this reason, such an interpretation is called an **independence model** for the axiom.

EXAMPLE 10.5.1

In the axiom set of Section 10.3, determine the dependence/independence of axiom A1.

SOLUTION   The following is an independence model for A1:

$$\text{Points: } \{1, 2, 3\}$$

$$
\begin{array}{cccc}
\text{Lines:} & 1 & 1 & 2 \\
 & 2 & 3 & 3 \\
 & 4 & 4 &
\end{array}
$$

Note that A1 is not satisfied since some lines are *not* sets of points (4 is not a point). Note also that A2 is satisfied: we do not have two lines containing the same two distinct points. 1 and 4 are indeed distinct, but 4 is not a point.

EXAMPLE   10.5.2

Give an independence model for axiom A3.

SOLUTION   Points: $\{1, 2\}$

Lines: 1   1

2

You should verify that axioms A1, A2, A4, and A5 are satisfied and that A3 is not.

There are several other metamathematical properties of interest, but these three, the most important, should suffice to convey the idea. For a deeper treatment of these topics, the reader is directed to the references at the end of the chapter.

EXERCISES 10.5

1. Verify the remaining parts of Example 10.5.1 to show that the given interpretation is an independence model.

2. Verify the parts of Example 10.5.2 to show that the given interpretation is an independence model.

3. Suppose that the following statement is added to the axiom set of Section 10.3: Every line contains at most two points. Is the resulting system consistent? Is it complete?

4. Construct an axiom set containing at least three axioms in which every axiom is dependent but the deletion of any axiom leaves an independent set. The undefined terms need not be geometric. Alternatively, prove that such a set of axioms cannot exist.

The remaining questions refer to the following axiom set.

B1: A line is a set of points.
B2: Given two distinct points, there is at least one line containing them.
B3: Given two distinct points, there is at most one line containing them.
B4: Every line contains at least two points.
B5: Every line contains at most two points.
B6: Given a line, there is a point not on it.
B7: Given a line and a point not on it, there is at least one line through that point parallel to the given line.
B8: Given a line and a point not on it, there is at most one line through that point parallel to the given line.
B9: A line exists.

5. Is axiom system B consistent?

6. Is axiom system B complete?

7. Determine the independence/dependence of each axiom in this system.

For Further Reading

Golos, Ellery B. *Foundations of Euclidean and Non-Euclidean Geometry.* New York: Holt, Rinehart and Winston, 1968, Chaps. 1–3.

Kline, Morris. *Mathematical Thought from Ancient to Modern Times.* New York: Oxford University Press, 1972, Chaps. 36, 42, 43, 51.

Mathematics in the Modern World: Readings from Scientific American. San Francisco: W. H. Freeman and Company, Publishers, 1968, Chaps. 27, 31.

Nagel, E., and J. R. Newman. *Godel's Proof.* New York: New York University Press, 1958.

Stoll, Robert R. *Sets, Logic and Axiomatic Theories*, 2nd ed. San Francisco: W. H. Freeman and Company, Publishers, 1974, Chap. 3.

Witter, George. *Mathematics, The Study of Axiom Systems.* Waltham, Mass.: Blaisdell Publishing Company, 1964.

BASIC LOGIC

If one were required to single out the dominant characteristic of mathematics, it would have to be "reasonableness." In fact, mathematics is utterly dependent upon the innate rationality of humans, and hence if we discuss mathematics at all, we soon find ourselves confronted with a consideration of that elusive concept called reason. But since every intellectual consideration is by nature rational, we are forced to use reason to analyze reason, an awkward situation at best, leading to intricacies better left to courses in philosophy. We shall limit this section to a brief exposition of those basic logical principles which are essential to mathematical discourse.

The notion underlying all discourse is that of a meaningful statement (which we shall not attempt to define rigorously here). Each meaningful statement has a **truth value**, which in our logical system is either **true** or **false**, depending upon the context in which the statement appears. The assertion that there is no category between "true" and "false" in which we may place a statement is usually called the **Law of the Excluded Middle**. We also assert that a statement may not simultaneously possess both truth values; this is the **Law of Contradiction**.

From any statement we may derive several related statements:

1. The **negation** or **contradictory** of the statement, which is characterized by the fact that it always has the opposite truth value from the original statement.

2. **Contrary** statements, which are false whenever the original statement is true, but may still be false when the original statement is false.

EXAMPLE A.1

Consider the statement "The wall is red." Its negation is "The wall is not red." Several contrary statements are "The wall is green," "The wall is yellow," and "The wall is blue." The statement "The wall is concrete" is neither the negation nor a contrary of the original statement.

NOTATION: If a statement is denoted by s, its negation is often denoted by $\sim s$.

Statements may also be categorized with regard to their scope, and thus are either universal or existential. A **universal** statement is an assertion about all things of a certain kind, whereas an **existential** statement merely asserts the existence of at least one thing that satisfies the statement. The negation of a universal statement is existential, and the negation of an existential statement is universal.

EXAMPLES A

2. "All buildings have flat roofs" is a universal statement; its negation is "There is a building that does not have a flat roof."

3. "There are pink elephants" is an existential statement; its negation is "No elephants are pink." (That is, "All elephants possess the property of not being pink.")

Having considered statements singly, we move on to combinations of statements. It suffices to consider pairs of statements here, because more complex combinations can be analyzed two at a time. The truth value of a combination can be deduced from the truth value of its component statements, according to several fairly straightforward rules. For convenience, let us represent two statements by s and t; then their basic combinations can be described as follows:

1. The **conjunction** of these two statements is "s and t." It is true whenever s and t are both true, and is false otherwise.
2. The **disjunction** of the two statements is "s or t." It is false if s and t are both false, and is true otherwise.

A slightly more complicated but very important combination is the assertion of a "cause-and-effect" kind of relationship between two statements (in a particular order). It is called a **conditional** statement, and is usually in the form "If s, then t," or "s implies t." In this case, s is called the **hypothesis** of the conditional, and t is called the **conclusion**. A conditional statement is false whenever its hypothesis is true and its conclusion is false; in all other cases the conditional is considered to be true. If we take a conditional statement and interchange its hypothesis and conclusion, we obtain the **converse** of that conditional.

EXAMPLES A

Consider the statements "It is spring" and "The grass is green." Then:

4. The conjunction of the two statements is "It is spring and the grass is green."

5. The disjunction of the two statements is "Either it is spring or the grass is green."

6. "If it is spring, then the grass is green" is a conditional in which "It is spring" is the hypothesis and "The grass is green" is the conclusion. The converse is "If the grass is green, then it is spring."

The truth value of a conditional statement is not related to that of its converse; that is, a true conditional may or may not have a true converse. If it does, then the hypothesis and conclusion are said to be **equivalent** statements (in that each implies the other). The two most common ways of expressing the equivalence of statements s and t is by saying "s is necessary and sufficient for t" or "s is true if and only if t is true." This type of statement is called a **biconditional** or an **equivalence**.

EXAMPLE A.7

"$x + 3 = 5$" and "$x = 2$" are equivalent statements.

NOTATION: "If and only if" is abbreviated as **iff**.

The truth of a conditional statement is usually referred to as **validity**; that is, a conditional is said to be valid if its conclusion follows from its hypothesis. A valid conditional is also called an **implication**. The fact that a conditional is valid does not ensure the truth of its conclusion; for that we also need the truth of the hypothesis. A **deductive argument** is simply a chain of implications in which the conclusion of each implication is at least part of the hypothesis of the next, and the argument is valid if and only if each implication is valid. The truth of its conclusion, however, also requires the truth of the initial hypotheses. Thus, a deductive argument is simply a process for guaranteeing that the truth of a certain statement (conclusion) follows from the truth of one or more other statements (hypotheses). The establishment of the truth of a statement by making it the conclusion of a deductive argument whose initial hypotheses are taken as true is called a **direct proof** of the statement.

There is another type of proof, based on both deduction and the negation of a statement. Since a true hypothesis and a valid argument together must yield a true conclusion, a valid argument that yields a false conclusion must proceed from a false hypothesis. Hence, we may also prove the truth of a statement s by forming its negation, $\sim s$, and using $\sim s$ as the hypothesis of a valid argument whose conclusion is known to be false. This implies that $\sim s$ must itself be false, and hence s must be true, by the way we defined negation. This procedure is known as **proof by contradiction**, or **indirect proof**. Both the direct and indirect methods of proof are exemplified by the deductive arguments used throughout this book.

EXERCISES A

1. Give the negation and a contrary of each of the following propositions.
   (a) The flower is purple.
   (b) All freshmen are flowers.
   (c) Some birds are fire engines.
   (d) The development of the number systems is motivated exclusively by a consideration of algebraic equations.
   (e) Several meetings have been held to discuss the renovation of the schedule.
   (f) All dogs have fleas.
   (g) There are three trees in the front yard.
   (h) $8 + 3 = 11$.
   (i) No textbooks are useful.
   (j) For each x, there exists a y such that y is less than x.

2. In each of the following conditionals, state the hypothesis, the conclusion, and the converse.
   (a) If all flying things are airplanes, then birds are airplanes.
   (b) p is true if q is true.
   (c) The existence of a propositional function implies the existence of an infinity of propositions.
   (d) Truth implies truth.
   (e) We will be able to sleep through this lecture if nobody asks a question.
   (f) If there are sufficient funds and if the carpenters do not strike, the entire building will be renovated by September.
   (g) If rabbits eat carrots, then they have good eyesight.
   (h) I will be lucky if I pass.
   (i) Equality implies equivalence.
   (j) A conditional is true if its hypothesis is false.

3. Let the propositions p and q be "Roses are red" and "Snowmen like carrots," respectively. Translate into acceptable English:
   (a) p implies q (b) q implies p
   (c) p if and only if q (d) $(\sim p)$ implies q
   (e) p implies $(\sim q)$ (f) $(\sim q)$ implies $(\sim p)$

4. Follow the directions of Exercise 3 for the propositions

$$p: \quad \text{"Several errors have been made"}$$

and $\qquad\qquad q: \quad$ "All the answers are incorrect"

BASIC SET THEORY

B.1 THE LANGUAGE OF SETS

A **set** is a collection of objects called the **elements** or **members** of the set. There are many examples of sets with which the reader is familiar. For example, the reader is an element of the set of students in a class. A student not in the class is not a member of this set.

A set must be **well-defined** in the sense that we can determine whether or not an element belongs to a given collection. Some examples of well-defined sets are:

1. The set of letters in the English alphabet.
2. The set of presidents of colleges in the state of Connecticut.
3. The set of mathematics teachers in Cheshire High School.
4. The set of books published by Allyn and Bacon.
5. The set of points on a given line.
6. The set of positive numbers.

Each of the foregoing sets is so clearly defined that we can easily determine whether or not an element belongs to a given collection. On the other hand, the set of smart students in a given class is not well-defined, because different persons may include different students in the set.

Sets are designated customarily by capital letters A, B, C, . . . and members of a set are denoted by lowercase letters a, b, c, ... If a set A consists of the numbers 2, 4, 6, 8, 10, 12, we write

$$A = \{2, 4, 6, 8, 10, 12\}$$

If a set has a large number of elements, we may abbreviate in listing its members. For example, we may denote the set of letters of the alphabet by

$$\{a, b, c, \ldots, y, z\}$$

and interpret the dots to mean that the unlisted letters d through x, inclusive, are also members of the set.

Essentially, a set may be specified in two different ways. One way is to list all the elements of a set as we have illustrated; another way is to enclose in braces a defining property of membership by which it can be determined whether or not a given member belongs to that set. Consider the set

$$\{\text{Monday, Tuesday, Wednesday, Thursday, Friday, Saturday, Sunday}\}$$

To belong to this set an element must be one of the days of the week. We may denote this set, by the symbol, called the **set-builder notation**, as follows:

$$\{x \,|\, x \text{ is a day of a week}\}$$

and read this symbol: "the set of all x such that x is a day of a week." The vertical line in the notation is read "such that." The letter x of the alphabet in the notation is called a *variable*.

EXAMPLE B.1.1

List the elements of the set

$$A = \{x \,|\, x \text{ is a vowel in the alphabet}\}$$

SOLUTION

$$A = \{a, e, i, o, u\}$$

EXAMPLE B.1.2

List the elements of the set

$$B = \{x \,|\, x \text{ is a number on a die}\}$$

SOLUTION

$$B = \{1, 2, 3, 4, 5, 6\}$$

It is important not to confuse the concept of an *element* with the concept of a *set*. If A is a set whose only element is 7, we write $A = \{7\}$ so as not to confuse the counting number 7 with the set $\{7\}$ consisting of one element.

To express the fact that a number 6 belongs to the set A, we write

$$6 \in A$$

where the symbol $\in$ is interpreted to mean "is an element of," "is a member of," or "belongs to." Thus, if $A = \{5, 6, 7, 8\}$, then

$$5 \in A$$

$$6 \in A$$

$$7 \in A$$

$$8 \in A$$

but

$$9 \notin A$$

The notation $9 \notin A$ means "the number 9 does not belong to the set A."

DEFINITION B.1.1

A set with no elements is called the **null set**; it is denoted by $\{ \ \}$ or $\varnothing$.

EXAMPLE B.1.3

The set $\{x \mid x^2 + 1 = 0, x \text{ is a counting number}\}$ is an empty set, because there is no counting number x that satisfies the equation $x^2 + 1 = 0$.

The set of all secretaries who can type 1000 words per minute, the set of sales executives in Sears and Roebuck over 300 years old, the set of all men who weigh more than 1500 pounds, and the set of all counting numbers greater than 3 and less than 4 are all examples of an empty set.

Universal Set

The set of *all* logical possibilities $\mathscr{U}$ in a given collection constitutes the basis for a discussion and must be agreed upon before the discussion proceeds in an unambiguous manner. For example, if we wish to talk about some of the incorporated business firms in Illinois, the **universal set** would be the set of all incorporated business firms in that state, and firms incorporated in the state of New York or any other state are then excluded from the discussion.

The universal set may be the set of the days of a week or the set of counting numbers less than 8. It could also be a very large set, such as the heights of all males in the United States, or it could be a relatively small set, such as the number of females in a given household. Nevertheless, the concept of a universal set means the set of all possibilities under discussion.

Relation between Sets—Subsets

Next, consider the sets

$$\{1, 2, 3\} \quad \text{and} \quad \{1, 2, 3, 4, 5\}$$

How do these sets compare? Perhaps an obvious statement is that every element of the first set is also an element of the second set. Such a relation between sets is common. For example, each element of the set of vowels in the alphabet is also an element of the set of all letters of the alphabet.

DEFINITION B.1.2
A set is a **subset** of set B if and only if every element of set A is also an element of set B.

The notation $A \subseteq B$ or $B \supseteq A$ means "A is a subset of B." Thus, we say that $A \subseteq B$ if $x \in A$ implies that $x \in B$, but if there exists some element $x \in A$ that does not belong to B, then A is not a subset of set B.

EXAMPLE B.1.4

In the universe of counting numbers,

$$\{3, 4, 5\} \subseteq \{2, 3, 4, 5\}$$
$$\{2, 4, 6, 8\} \subseteq \{2, 4, 6, 8, 10\}$$
$$\{1, 2, 3\} \subseteq \{1, 2, 3, 7\}$$
$$\{5, 6, 7\} \subseteq \{5, 6, 7\}$$

Set A is a subset of itself, $A \subseteq A$, since every element of A is an element of A and the definition of a subset is satisfied. We also consider the null set as a subset of every set A, because there is no element in $\varnothing$ that does not belong to A; therefore, $\varnothing \subseteq A$.

DEFINITION B.1.3
Let A and B be two sets. Set A is a **proper subset** of B, written $A \subset B$, if $A \subseteq B$ and set B contains at least one element that does not belong to set A.

The null set is a proper subset of every nonempty set, whereas a set A is not a proper subset of itself. All subsets of $\mathscr{U}$, except $\mathscr{U}$ itself, are proper subsets of the universal set $\mathscr{U}$. Here are some examples.

EXAMPLE B.1.5

In the universe of counting numbers,

$$\{2, 3, 4\} \subset \{2, 3, 4, 6\}$$

$$\{2, 4, 6, 8\} \subset \{2, 4, 6, 8, 12\}$$

but

$$\{1, 2, 3\} \not\subset \{1, 2, 3\}$$

How many subsets does a given set have? If the set is empty, it has one subset, itself. If the set has one element, then it has two subsets, itself and the empty set $\varnothing$. Before we generalize, we consider a few more examples.

EXAMPLE B.1.6

(a) $A = \{1, 2\}$. The subsets of A are

$$\{ \}, \quad \{1\}, \quad \{1, 2\}$$
$$\{2\}$$

(b) $B = \{1, 2, 3\}$. The subsets of B are

$$\{ \}, \quad \{1\}, \quad \{1, 2\}, \quad \{1, 2, 3\}$$
$$\{2\}, \quad \{1, 3\}$$
$$\{3\}, \quad \{2, 3\}$$

Notice that we systematically wrote the subset containing no element, then all subsets that have one element, two elements, and so on, finally including the original set itself. Table B.1 points out a relationship between the number of possible subsets and the number of elements in a given set. Because the set $\{1, 2, 3\}$ is not a proper subset of itself, it follows that there are $2^3 - 1 = 7$ proper subsets of the set containing three elements 1, 2, and 3. This leads us to a conjecture that if n is a nonnegative integer, then a set with n elements has 2^n subsets, of which $2^n - 1$

TABLE B.1

Number of Elements in a Set	Number of Subsets
0	$1 = 2^0$
1	$2 = 2^1$
2	$4 = 2^2$
3	$8 = 2^3$

are proper subsets. Readers should convince themselves by considering several examples that this conjecture is correct.

Remember that the two symbols $\in$ and $\subseteq$ are distinct notions; one denotes the set membership and other the set inclusion. The difference between these concepts becomes evident if one compares the statement

$$x \in A \qquad \text{and} \qquad \{x\} \subseteq A$$

where $x \in A$ states that x is an element of a set A, and $\{x\} \subseteq A$ means that the set containing one element x is a subset of set A.

DEFINITION B.1.4

Two sets A and B are said to be equal if A is a subset of B and B is a subset of A. In symbols, we write $A = B$.

This means that two sets A and B are equal if and only if every element of A is an element of B and every element of B is an element of the set A.

EXAMPLES B.1

7. The set $A = \{1, 2, 3, 4, 5\}$ and the set $B = \{2, 4, 5, 1, 3\}$ are equal because each set is a subset of the other. This example illustrates the fact that equal sets have exactly the same members and the order in which the elements of the set are listed is immaterial.

8. The set $A = \{a, b, c, d, e\}$ and the set $B = \{2, 3, 4, 5, 6\}$ are not equal even though they have the same number of elements. Why?

DEFINITION B.1.5

Two sets A and B are said to be in **one-to-one correspondence** if every element of set A can be paired with exactly one element of set B, and every element of set B can be paired with exactly one element of set A.

EXAMPLE B.1.9

The set $A = \{a, b, c, d\}$ and the set $B = \{1, 2, 3, 4\}$ can be placed in one-to-one correspondence in several different ways, three of which are shown as follows:

$$
\begin{array}{ccc}
\{a, b, c, d\}, & \{a, b, c, d\}, & \{a, b, c, d\} \\
\updownarrow\,\updownarrow\,\updownarrow\,\updownarrow & \updownarrow\,\updownarrow\,\updownarrow\,\updownarrow & \updownarrow\,\updownarrow\,\updownarrow\,\updownarrow \\
\{1, 2, 3, 4\}, & \{2, 3, 4, 1\}, & \{3, 4, 1, 2\}
\end{array}
$$

DEFINITION
B.1.6

Two sets A and B are said to be **equivalent** if they can be placed in one-to-one correspondence with each other. In symbols, we write $A \sim B$.

Note that if two sets are equal, then they are equivalent, but the converse is not true. The equality of the sets require that their members be identical. The equivalent sets, on the other hand, consist of the same number of elements, and it is not necessary that their members be identical.

EXERCISES B.1

1. Determine which of the following sets are well-defined sets.
   (a) good secretaries in Wayne High School
   (b) beautiful girls on the campus
   (c) older folks from Rhode Island
   (d) senators from the state of Iowa
   (e) wheat farmers in the United States
   (f) chemistry teachers in Hamden High School
   (g) counting numbers less than 10
   (h) residents of the state of Rhode Island
   (i) new automobiles in Albany, New York
   (j) numbers substituted for x so that $3x + 2 = 8$

2. Use the set-builder notation to describe each of the following sets.
   (a) $\{1, 2, 3, 4, 5, 6\}$
   (b) $\{2, 4, 6, 8, 10, 12, 14, 16\}$
   (c) the set of months of the year
   (d) the set of vice-presidents in an insurance company
   (e) the set of odd positive counting numbers

3. List or indicate within braces the elements or members of the following sets.
   (a) the set of senators from your state
   (b) the set of English teachers in your high school
   (c) the set of females in your household
   (d) the set of states of the United States that border Canada
   (e) the set of counting numbers between 15 and 20, inclusive

4. Given that $A = \{1, 2, 3, 4, 5, 6, 7, 8\}$, which of the following statements are true and which are false?
   (a) $5 \in A$
   (b) $12 \in A$
   (c) $0 \in A$
   (d) $\{5\} \subseteq A$
   (e) $A \subseteq A$
   (f) $\varnothing \in A$
   (g) $\varnothing \subset A$
   (h) $\{1, 2, 3\} \subseteq \{1, 2, 3, 4\}$
   (i) $\{1, 2, 3\} \subset \{1, 2, 3, 4\}$
   (j) $\{2, 3, 4, 5\} \subset \{2, 3, 4, 5\}$

5. Classify the following statements as true or false.
   (a) $\{\text{Sunil, Anita}\} \subset \{\text{Sunil, Anita, Tracey}\}$
   (b) $\{\text{Judy, Susan}\} \subseteq \{\text{Judy, Susan}\}$
   (c) $\{\text{Jeanie, Judy, Tracey}\} \subseteq \{\text{Tracey, Judy, Kathy}\}$
   (d) $\{1, 2, 3, 4, 5\} \subseteq \{5, 4, 3, 2, 1\}$
   (e) $\{x \mid 3x + 4 = 13\} = \{5\}$
   (f) $\{x \mid x^2 = 4\} = \{-2, 2\}$
   (g) $\{x \mid 2x + 1 = 7\} = \{3\}$

6. Which of the following sets are equal?
   (a) $\{1, 2, 3, 4\}$ (b) $\{4, 1, 2, 3\}$
   (c) $\{2, 4, 3, 1\}$ (d) $\{3, 1, 4, 5, 2\}$
   (e) $\{3, 4, 5, 1, 2\}$ (f) $\{5, 4, 3, 2, 1\}$

7. Given that
   $A = \{x \mid x \text{ is a counting number between 3 and 8, inclusive}\}$
   $B = \{x \mid x \text{ is a counting number less than 6}\}$

   Which of the following sets are equal to A and B?
   (a) $\{4, 5, 6, 7, 8\}$ (b) $\{3, 4, 5, 6, 7, 8\}$
   (c) $\{1, 2, 3, 4, 5\}$ (d) $\{4, 3, 2, 1, 5\}$
   (e) $\{3, 4, 5, 6, 7\}$ (f) $\{3, 8, 6, 5, 7, 4\}$

8. Which of the following sets are equivalent?
   (a) $\{\text{Kathy, Elaine}\}$ (b) $\{\text{Harry, Bob, Mike}\}$
   (c) $\{1, 2, 3\}$ (d) $\{\text{apple, orange}\}$
   (e) $\{5, 6, 7, 8\}$ (f) $\{x \mid x \text{ is a counting number less than 5}\}$

9. Given that $\mathscr{U} = \{\text{Bob, Charles, Dick, Edward}\}$, list all subsets that contain
   (a) two elements
   (b) three elements

10. Given that $A = \{1, 2, 3, 4, 5\}$, list all subsets of A that have
    (a) one element   (b) two elements
    (c) three elements   (d) four elements

B.2   OPERATIONS ON SETS

We have discussed the concept of a set, notations for describing sets, and some of the important relations on sets. It seems reasonable to ask how one can form new sets from the sets under discussion. To answer this question, we consider a family of sets that are subsets of some fixed universal set $\mathscr{U}$. We now introduce the operations of union, intersection, and complement of sets.

DEFINITION B.2.1	Let A and B be subsets of the universal set $\mathcal{U}$. The **union** of A and B, denoted $A \cup B$, is the set of all elements that belong to A or to B or to both. In symbols,

$$A \cup B = \{x \mid x \in A \text{ or } x \in B\}$$

EXAMPLES B.2

1. Let $A = \{\text{Susan, Joe, Mike}\}$ and $B = \{\text{Mary, Anita, Joe, Dick}\}$. Then

$$A \cup B = \{\text{Susan, Joe, Mike, Mary, Anita, Dick}\}$$

2. Let $A = \{2, 3, 4, 5, 6\}$ and $B = \{1, 2, 3, 4, 8, 9\}$. Then

$$A \cup B = \{1, 2, 3, 4, 5, 6, 8, 9\}$$

It may be helpful to represent sets and their operations using a pictorial method known as **Venn diagrams**. In a Venn diagram, a rectangle is used to represent the universal set $\mathcal{U}$, and subsets of $\mathcal{U}$ are represented by regions bounded by closed curves within the rectangle $\mathcal{U}$. Since $A \cup B$ consists of elements that are in A or in B or in both, the striped region of Figure B.1 represents the union of two sets, A and B.

FIGURE B.1

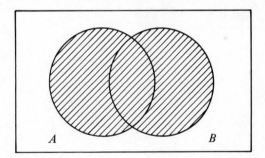

Sometimes two sets have elements in common. The striped area of Figure B.2 represents the elements common to both the sets.

FIGURE B.2

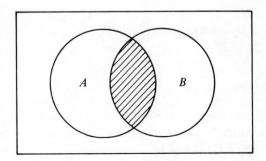

DEFINITION
B.2.2

Let A and B be the subsets of $\mathcal{U}$. The **intersection** of A and B, denoted $A \cap B$, is the set of all elements that belongs to both A and B. In symbols,

$$A \cap B = \{x \mid x \in A \text{ and } x \in B\}$$

EXAMPLE B.2.3

Let $\mathcal{U} = \{1, 2, 3, 4, 5, 6, 7, 8, 9, 10\}$, $A = \{1, 2, 3, 4, 5, 6\}$, and $B = \{4, 5, 6, 7, 8, 9, 10\}$. Then

$$A \cap B = \{4, 5, 6\}$$

Frequently, two sets do not have any element in common. For instance, if $\mathcal{U}$ is the set of all sales executives in a business firm and A and B represent, respectively, those attending a sales meeting in Philadelphia and those attending a conference at the head office in Chicago at the same time, then $A \cap B$ represents an empty set, because no person can be two places at the same time. Figure B.3 represents two sets that have no elements in common.

FIGURE B.3

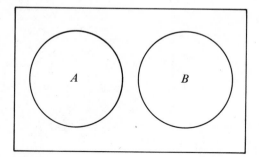

DEFINITION
B.2.3

Let A and B be subsets of $\mathcal{U}$. The sets A and B are said to be **disjoint** if and only if they have no elements in common; that is,

$$A \cap B = \varnothing$$

EXAMPLE B.2.4

Let $\mathcal{U} = \{4, 5, 6, 7, 8, 9, 10, 11, 12\}$, $A = \{4, 6, 8, 10, 12\}$, and $B = \{5, 7, 9, 11\}$. Then A and B do not have any element in common. In other words, the intersection of A and B is the null set.

A subset of the universal set $\mathscr{U}$ that is intimately related to the set A is the set called the **complement** of A. The region outside the circle but within the rectangle $\mathscr{U}$ in Figure B.4 represents the complement of a set A.

FIGURE B.4

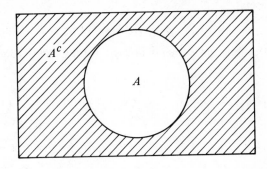

DEFINITION
B.2.4

Let A be any subset of $\mathscr{U}$. Then the set of all elements in $\mathscr{U}$ that do not belong to A is called the **complement** of A and is denoted by A^c. In symbols,

$$A^c = \{x \mid x \in \mathscr{U}, x \notin A\}$$

EXAMPLES B.2

5. Let $\mathscr{U} = \{1, 2, 3, 4, 5, 6\}$ and $A = \{1, 3, 5, 6\}$. Then $A^c = \{2, 4\}$. Observe that

    (a) $A \cup A^c = \mathscr{U}$ (b) $A \cap A^c = \varnothing$

    (c) $\mathscr{U}^c = \varnothing$ (d) $\varnothing^c = \mathscr{U}$

6. Let $\mathscr{U} = \{1, 2, 3, 4, 5, 6, 7, 8, 9, 10, 11, 12\}$, $A = \{1, 2, 3, 4, 5, 6\}$, and $B = \{4, 5, 6, 7, 8, 9\}$. Then
   (a) $A^c = \{7, 8, 9, 10, 11, 12\}$
   (b) $B^c = \{1, 2, 3, 10, 11, 12\}$
   (c) $A \cup B = \{1, 2, 3, 4, 5, 6, 7, 8, 9\}$
   (d) $A \cap B = \{4, 5, 6\}$
   (e) $(A \cup B)^c = \{10, 11, 12\}$
   (f) $(A \cap B)^c = \{1, 2, 3, 7, 8, 9, 10, 11, 12\}$
   (g) $A^c \cup B^c = \{1, 2, 3, 7, 8, 9, 10, 11, 12\}$
   (h) $A^c \cap B^c = \{10, 11, 12\}$

Clearly,

$$(A \cup B)^c = A^c \cap B^c \qquad\qquad \text{(B.1)}$$

because they consist of precisely the same elements. Similarly,

$$(A \cap B)^c = A^c \cup B^c \qquad\qquad \text{(B.2)}$$

The properties (B.1) and (B.2) are called **DeMorgan's laws**.

EXERCISES B.2

Let $\mathcal{U} = \{1, 2, 3, 4, 5, 6, 7, 8, 9, 10\}$ and suppose that we have the following subsets:

$$A = \{2, 3, 4, 5, 6\} \qquad B = \{4, 6, 7, 8, 10\}$$

Determine the elements in each of the sets in exercises 1–8.

1. A^c
2. B^c
3. $A \cup B$
4. $A \cap B$
5. $A^c \cup B^c$
6. $A^c \cap B^c$
7. $(A \cup B)^c$
8. $(A \cap B)^c$

Let $\mathcal{U} = \{1, 2, 3, 4, 5, 6, 7, 8, 9\}$ and suppose that we have the following subsets:

$$A = \{1, 3, 5, 7, 9\} \qquad B = \{2, 4, 5, 6, 8\} \qquad C = \{3, 6, 7, 9\}$$

Determine the elements in each of sets in exercises 9–20.

9. A^c
10. B^{c}
11. C^c
12. $A \cup B$
13. $B \cup C$
14. $A \cup C$
15. $A \cup B \cup C$
16. $B \cap C$
17. $A \cup (B \cap C)$
18. $(A \cup B) \cap (A \cup C)$
19. $A \cap (B \cup C)$
20. $(A \cap B) \cup (A \cap C)$

Let $\mathcal{U} = \{1, 2, 3, 4, 5, 6, 7, 8, 9, 10, 11, 12\}$ and let A, B, and C be subsets of $\mathcal{U}$ as follows:

$$A = \{1, 5, 6, 10, 11, 12\} \qquad B = \{2, 3, 7, 8, 9, 10, 12\} \qquad C = \{2, 5, 8, 9, 12\}$$

Show that

21. $(A \cup B)^c = A^c \cap B^c$

22. $(A \cap B)^c = A^c \cup B^c$

23. $A \cap (B \cup C) = (A \cap B) \cup (A \cap C)$

24. $A \cup (B \cap C) = (A \cup B) \cap (A \cup C)$

Let $\mathcal{U}$ be the set of employees in an insurance company. Suppose that we have the following subsets:

$$A = \{x \mid x \text{ is a secretary in an insurance company}\}$$
$$B = \{x \mid x \text{ is a stockholder in the insurance company}\}$$
$$C = \{x \mid x \text{ is a keypunch operator in the insurance company}\}$$

Describe each of the sets in exercises 25–32 in words.

25. $A \cup B$ 26. $A \cap B$

27. $B \cap C$ 28. $B \cup C$

29. $A \cup C$ 30. $A \cap C$

31. $A \cap (B \cup C)$ 32. $A \cap B \cap C$

Let $\mathscr{U} = \{$all students in the college$\}$. Suppose that we have the following subsets:

$$A = \{\text{all students taking a course in chemistry}\}$$
$$B = \{\text{all students taking botany}\}$$
$$C = \{\text{all students taking physics}\}$$

Describe each of the sets in exercises 33–40 in words.

33. A^c 34. B^c

35. $A \cup B$ 36. $A \cap B$

37. $A \cap B \cap C$ 38. $A \cup (B \cap C)$

39. $A \cup B \cup C$ 40. $B \cup (A \cap C)$

APPENDIX C

ON THE NATURE OF MATHEMATICS

Mathematics has its roots in the empiricism of the Babylonians and other pre-Hellenic peoples, whose trial-and-error methods of boundary demarcation and building eventually were organized into numerical rules of procedure. Thus, from its earliest stages mathematics has been a servant of humanity, aiding endeavors in other fields of activity as a single exact language applicable to all fields that require reasoning. Today the general theories and procedures of mathematics simplify and refine the treatment of countless topics, and by translation into this common tongue the methods of many different disciplines are brought to bear on each other's problems. Our most common mathematical experience is the use of arithmetic in day-to-day business transactions, but one need not look very far to find mathematics in a host of other fields. Physical scientists depend upon calculus and methods of analysis; business experts and behavioral scientists are becoming involved more and more with statistical methods, probability, linear programming, and decision theory; philosophers recognize the importance of Boolean algebra as a tool for the study of logic; even artists use geometric ideas such as symmetry and projection to aid in their creative expression. These examples of the varied uses of mathematics in other fields could be multiplied many times over, but instead let us proceed to another viewpoint.

A science is characterized by its devotion to the discovery and organization of general truths and its concern for the operation and application of general laws. In this sense we shall claim that mathematics is a science, for the very heart of the subject is the establishment of orderly procedures and the study of the logical implications of statements. Even in your own mathematical experiences you can observe that Euclidean geometry possesses the characteristics of a science apart from any application it may have to the physical world, and the same may be said of algebra or any other branch of the field. Moreover, mathematics governs the use of scientific principles and thus becomes the organizer of all science.

But this characterization of mathematics is far from sufficient. Mathematics is not merely a science. It is not even just the "queen of sciences," as some have said. There is in this field of thought a type of creativity found only in the fine arts. The men who first constructed the various systems of non-Euclidean geometry participated in a creativity quite similar to that of Rembrandt or Michelangelo. Mathematical imagination at least equals and often surpasses that required by the other fine arts, because mathematicians are neither aided nor confined by material forms of expression. Theirs is a world of pure abstraction; the mathematician "lives in 'the wildness of logic' where reason is the handmaiden and not the master."* Although many mathematical theories have arisen in response to the challenge of specific problems in the natural or behavioral sciences, creative mathematicians often generalize the original solution of a question and from it build a logical edifice, posing and investigating questions of abstract structure without regard to any connection with the physical world. They design and construct systems with a taste for order and harmony, pattern and symmetry, precision and generality. Nevertheless, some writers assert that mathematicians do not create any more than Leverrier and Adams created the planet Neptune or Admiral Peary created the North Pole. It is claimed that, because they are bound by reason, the people who design mathematical systems are merely discovering some of the many patterns of thought that already exist as logical consequences of the various initial statements used as hypotheses. One could reply that this is tantamount to asserting that Rodin *discovered* "The Thinker," since he merely shaped a piece of stone into one of the many forms it was already capable of assuming. However, the roots of this question lie deep in philosophy and we shall not attempt to resolve the controversy here.

It should be apparent by now that the nature of mathematics defies simple description. As an art mathematics creates new worlds, and as a science it explores them. It is a common unifying force present in all human intellectual endeavor, forever broadening the horizons of the mind, exploring virgin territory, and organizing new information into weapons for another assault on the unknown. "It is a language, a tool, and a game—a method of describing things conveniently and efficiently, a shorthand adapted to playing the game of common sense."* It is "the subject in which we never know what we are talking about nor whether what we say is true,"† and yet it is basic to the organization and interpretation of all truth. It demands a novelist's imagination, a poet's perception of analogy, an artist's appreciation of beauty, and a politician's flexibility of thought. Mathematics is indeed an integral and indispensable part of every truly liberal education. It is "the thinking man's liberal art."‡

* Marston Morse, "Mathematics and the Arts," *The Yale Review*, Vol. 40, No. 4 (June 1951), p. 612.
* Mario G. Salvadori, "Math's a Pleasure," *Harper's Magazine*, Vol. 209 (August 1954), p. 90.
† Bertrand Russell, *International Monthly*, Vol. 4 (1901).
‡ William E. Hartnett, "Mathematics and Humane Education," *America*, Vol. 103 (April 23, 1960), p. 113.

FOR DISCUSSION OR ESSAY

1. Compare mathematics with the various fine arts that you know, indicating similarities and differences.

2. In the last paragraph of this appendix, each sentence after the first one is a proposed description of mathematics. Choose *any one* of those sentences and comment on whether or not, in your opinion, that sentence is an adequate or appropriate description of mathematics as a whole. Give reasons (from your experience, reading, or anywhere else) to support whatever position you take.

For Further Reading

Eves, Howard. "Mathematics." In *Encyclopedia Americana* (International Ed.). Vol. 18, pp. 431–434. New York: Americana Corporation, 1965.

Hartnett, William E. "Mathematics and Humane Education." *America*, Vol. 103 (April 23, 1960), pp. 112–116.

Morse, Marston. "Mathematics and the Arts." *The Yale Review*, Vol. 40, No. 4 (June 1951), pp. 604–612.

Newman, James R., ed. *The World of Mathematics*. Vols. I–IV. New York: Simon and Schuster, Inc., 1956.
   (a) Introduction.
   (b) Bell, E. T., "The Queen of Mathematics."
   (c) Jourdain, P. E. B., "The Nature of Mathematics."
   (d) Pierce, C. S., "The Essence of Mathematics."
   (e) Sylvester, James Joseph, "The Study That Knows Nothing of Observation."
   (f) Weyl, Hermann, "The Mathematical Way of Thinking."

Rapoport, Anatol. "Mathematics and the Modern Sciences." *Journal of College Science Teaching*, Vol. 1, No. 3 (February 1972), pp. 33–42.

Rapport, Samuel, and Helen Wright, eds. *Mathematics*. New York: New York University Press, 1963.
   (a) Hardy, G. H., "A Mathematician's Apology."
   (b) Poincaré, Henri, "Mathematical Discovery."
   (c) Whitehead, Alfred North, "The Nature of Mathematics."

Salvadori, Mario G. "Math's a Pleasure." *Harper's Magazine*, Vol. 209 (August 1954), pp. 88–91.

Schaaf, William L., ed. *Our Mathematical Heritage*. New York: Collier Books, 1963.

Smith, David Eugene. *The Poetry of Mathematics and Other Essays*. New York: Scripta Mathematica, Yeshiva University Press, 1947.

THE STANDARD NORMAL DISTRIBUTION

z	0.00	0.01	0.02	0.03	0.04	0.05	0.06	0.07	0.08	0.09
0.0	0.5000	0.5040	0.5080	0.5120	0.5160	0.5199	0.5239	0.5279	0.5319	0.5359
0.1	0.5398	0.5438	0.5478	0.5517	0.5557	0.5596	0.5636	0.5675	0.5714	0.5753
0.2	0.5793	0.5832	0.5871	0.5910	0.5948	0.5987	0.6026	0.6064	0.6103	0.6141
0.3	0.6179	0.6217	0.6255	0.6293	0.6331	0.6368	0.6406	0.6443	0.6480	0.6517
0.4	0.6554	0.6591	0.6628	0.6664	0.6700	0.6736	0.6772	0.6808	0.6844	0.6879
0.5	0.6915	0.6950	0.6985	0.7019	0.7054	0.7088	0.7123	0.7157	0.7190	0.7224
0.6	0.7257	0.7291	0.7324	0.7357	0.7389	0.7422	0.7454	0.7486	0.7517	0.7549
0.7	0.7580	0.7611	0.7642	0.7673	0.7704	0.7734	0.7764	0.7794	0.7823	0.7852
0.8	0.7881	0.7910	0.7939	0.7967	0.7995	0.8023	0.8051	0.8078	0.8106	0.8133
0.9	0.8159	0.8186	0.8212	0.8238	0.8264	0.8289	0.8315	0.8340	0.8365	0.8389
1.0	0.8413	0.8438	0.8461	0.8485	0.8508	0.8531	0.8554	0.8577	0.8599	0.8621
1.1	0.8643	0.8665	0.8686	0.8708	0.8729	0.8749	0.8770	0.8790	0.8810	0.8830
1.2	0.8849	0.8869	0.8888	0.8907	0.8925	0.8944	0.8962	0.8980	0.8997	0.9015
1.3	0.9032	0.9049	0.9066	0.9082	0.9099	0.9115	0.9131	0.9147	0.9162	0.9177
1.4	0.9192	0.9207	0.9222	0.9236	0.9251	0.9265	0.9279	0.9292	0.9306	0.9319
1.5	0.9332	0.9345	0.9357	0.9370	0.9382	0.9394	0.9406	0.9418	0.9429	0.9441
1.6	0.9452	0.9463	0.9474	0.9484	0.9495	0.9505	0.9515	0.9525	0.9535	0.9545
1.7	0.9554	0.9564	0.9573	0.9582	0.9591	0.9599	0.9608	0.9616	0.9625	0.9633
1.8	0.9641	0.9649	0.9656	0.9664	0.9671	0.9678	0.9686	0.9693	0.9699	0.9706
1.9	0.9713	0.9719	0.9726	0.9732	0.9738	0.9744	0.9750	0.9756	0.9761	0.9767
2.0	0.9772	0.9778	0.9783	0.9788	0.9793	0.9798	0.9803	0.9808	0.9812	0.9817
2.1	0.9821	0.9826	0.9830	0.9834	0.9838	0.9842	0.9846	0.9850	0.9854	0.9857
2.2	0.9861	0.9864	0.9868	0.9871	0.9875	0.9878	0.9881	0.9884	0.9887	0.9890
2.3	0.9893	0.9896	0.9898	0.9901	0.9904	0.9906	0.9909	0.9911	0.9913	0.9916
2.4	0.9918	0.9920	0.9922	0.9925	0.9927	0.9929	0.9931	0.9932	0.9934	0.9936
2.5	0.9938	0.9940	0.9941	0.9943	0.9945	0.9946	0.9948	0.9949	0.9951	0.9952
2.6	0.9953	0.9955	0.9956	0.9957	0.9959	0.9960	0.9961	0.9962	0.9963	0.9964
2.7	0.9965	0.9966	0.9967	0.9968	0.9969	0.9970	0.9971	0.9972	0.9973	0.9974
2.8	0.9974	0.9975	0.9976	0.9977	0.9977	0.9978	0.9979	0.9979	0.9980	0.9981
2.9	0.9981	0.9982	0.9982	0.9983	0.9984	0.9984	0.9985	0.9985	0.9986	0.9986
3.0	0.9987	0.9987	0.9987	0.9988	0.9988	0.9989	0.9989	0.9989	0.9990	0.9990
3.1	0.9990	0.9991	0.9991	0.9991	0.9992	0.9992	0.9992	0.9992	0.9993	0.9993
3.2	0.9993	0.9993	0.9994	0.9994	0.9994	0.9994	0.9994	0.9995	0.9995	0.9995
3.3	0.9995	0.9995	0.9995	0.9996	0.9996	0.9996	0.9996	0.9996	0.9996	0.9997
3.4	0.9997	0.9997	0.9997	0.9997	0.9997	0.9997	0.9997	0.9997	0.9997	0.9998
3.6	0.9998	0.9998	0.9999	0.9999	0.9999	0.9999	0.9999	0.9999	0.9999	0.9999

ANSWERS TO ODD-NUMBERED EXERCISES

CHAPTER 2

EXERCISES 2.2

1. All except (d) and (h)   **3.** (b), (c), (g), (h), and (i)   **5.** (a) Z, WX (b) $W, ZYXW_2Z, WX$
(c) W_2ZYXW, Z (d) Six   (e) None   (f) Infinite   **7.** (a) False   (b) True   (c) True
(d) Cannot be determined   (e) True   (f) True   (g) True   (h) Cannot be determined
(i) Cannot be determined

9.

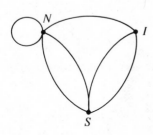

N-north bank, S-south bank, I-island

11. (a)    (b)

(c) Impossible; a vertex path guarantees a path from any vertex to any other; hence, the graph must be connected.   (d) Impossible; if the graph is connected, then there must be edges joining at least two pairs (of the three possible) of the vertices. Thus, some vertex (say it is labeled P) must be joined to each of the others (say Q and R). Then there is a vertex path QPR.   (e) See answer (b) above.   (f) Impossible; by the reasoning of answer (c) above, the existence of a vertex path guarantees that the graph is connected. With only three vertices, there must be an edge path, regardless of loops or multiple edges.   (g) See answer (a) above.

EXERCISES 2.3

1. $D, 5$; $E, 5$; $F, 4$; $G, 4$; $H, 4$; $J, 6$; $K, 4$; $L, 4$ **3.** $a, 2$; $b, 4$; $c, 0$; $d, 0$; $e, 6$; $f, 0$; $g, 2$
5. Step 1 is modified so that a path is randomly constructed starting at any vertex, P. The path must eventually terminate at P, because all vertices are even. The remaining steps in the proof are the same as in the proof of Theorem 2.3.4.
7. Start a path at any odd vertex, proceeding as described in step 1. This path must end at some other odd vertex. If there is another odd vertex, start another path there, and it will end at some distinct odd vertex. Proceeding in this fashion, odd vertices can be paired, thus proving there must be an even number of them.
9. Adding one edge to Figure 2.12 changes two vertices to even, leaving only two odd.
11. Three edges, joining distinct pairs of odd vertices. K and M, L and Q, R and S.
13. The closed edge path must pass through every vertex, because the graph is connected. If there are n vertices, this path must contain at least n edges if it passes through every vertex (at least once and returns to its start).

15.

17. Two. (See the answer to Exercise 15. A nonnull graph on one vertex must consist of one or more loops, forming an Euler graph.)

EXERCISES 2.4

1. (a)   **3.** (f)
5. Yes. For example, if the vertices (intersections) are labeled alphabetically left to right (top row: A, B, C; next row D through I; and so on), one possible path is $ABCIONTUZYXRKQWVPJDEFLMSHGA$.
7. If a vertex path omits an H-edge at a vertex, it must use the (only) other edge at that vertex; so this vertex must be one end of the path. Since there can be only two ends to the path, there can be at most two H-edges omitted.
9. Suppose that A is a terminal vertex, connected by its one edge to B, which is in turn connected to other vertices in the graph. The deletion of B, with its edges, disconnects A from the remainder of the graph, so B is critical.

EXERCISES 2.5

1.

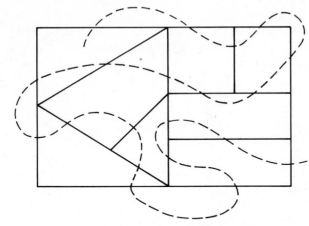

3. (c), (d), (e), and (g)

5. (a)

Face	Vertex (Dual)	Edge	Edge (Dual)
ABC	A	AB	AC
BCE	B	BE	BC
CED	E	ED	EC
ACD	D	DA	DC
ABDE	C	AC	DA
		BC	AB
		EC	BE
		DC	ED

(b)

Face	Vertex (Dual)	Edge	Edge (Dual)
FGH	J	FH	GJ
FGJ	H	FJ	GH
JGH	F	JH	FG
FJH	G	FG	JH
		GH	FJ
		GJ	FH

7.

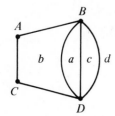

9. (a) Loop   (b) Face of degree one   (c) Face of degree zero (that is, the entire plane). Note that the only polygonal graph with an isolated vertex is the null graph on one vertex, a self-dual graph.

11. Impossible. The dual would have exactly one odd vertex. (See Exercise 2.3.7.)

EXERCISES 2.6

1. $V = 7, E = 7, F = 3, V - E + F = 3$ **3.** $V - E + F = 4$ **5.** $O = 4$ and $D = 4$; squares. $O = 6$ and $D = 3$; hexagons   **7.** (a) $V = 16, E = 32, F = 16; V - E + F = 0$ (b) $V = 9, E = 18, F = 9, V - E + F = 0$

EXERCISES 2.7

1. (a) A, 1 in, 2 out; B, 2 in, 1 out; C, 0 in, 4 out; D, 3 in, 0 out; E, 2 in, 1 out (b) P, 1 in, 2 out; Q, 3 in, 3 out; R, 1 in, 1 out; S, 1 in, 2 out; T, 3 in, 1 out   **3.** (a) A, -1; B, 1; C, -4; D, 3; E, 1 (b) P, -1; Q, 0; R, 0; S, -1; T, 2   **5.** One has net-degree -1; one has net-degree 1; all others have net-degree 0

CHAPTER 3

EXERCISES 3.4

1. $x = 3$, $y = 3$, $z = 7$ **3.** $x = \frac{5}{2}$, $y = -\frac{5}{2}$, $z = 1$ **5.** $x = 4$, $y = -3$, $z = -3$

7. $A + B = (0, 2, 4, 4, 5)$; $A - B = (4, 4, 4, -2, 9)$ **9.** $A + B = \begin{bmatrix} 4 \\ 4 \\ 4 \end{bmatrix}$; $A - B = \begin{bmatrix} -4 \\ 6 \\ 10 \end{bmatrix}$

11. $A + B = \begin{bmatrix} -1 & 1 & -1 \\ 4 & 4 & 4 \end{bmatrix}$; $A - B = \begin{bmatrix} 3 & 3 & 7 \\ 2 & -2 & 0 \end{bmatrix}$ **13.** $A + B = \begin{bmatrix} 1 & 3 & 2 & 7 \\ 7 & -2 & 4 & -2 \end{bmatrix}$;

$A - B = \begin{bmatrix} 3 & -1 & 4 & 1 \\ 1 & 0 & 0 & 8 \end{bmatrix}$ **15.** $\begin{bmatrix} 58 & 63 \\ 67 & 93 \end{bmatrix}$ **17.** $\begin{bmatrix} 13 & 15 \\ 16 & 22 \end{bmatrix}$

EXERCISES 3.5

1. 47   **3.** 67   **5.** (a) $x = 3$ (b) $x = 4$ **7.** $x = 2$, $y = 1$ **9.** $[7, 4, 1, 3, 2]$;

$\begin{bmatrix} \$0.37 \\ \$0.89 \\ \$0.99 \\ \$0.83 \\ \$1.69 \end{bmatrix}$; total bill $- \$13.01$ **11.** $A = [100, 200, 300, 500]$; $B = \begin{bmatrix} \$25.00 \\ \$34.00 \\ \$22.00 \\ \$12.50 \end{bmatrix}$; $AB = \$22,150.00$

EXERCISES 3.6

1. 3×5 **3.** 3×3 **5.** 3×4 **7.** Both AB and BA **9.** Neither   **11.** AB only

13. $\begin{bmatrix} 22 & 40 \\ 26 & 4 \end{bmatrix}$; $BA = \begin{bmatrix} 32 & 38 \\ 20 & -6 \end{bmatrix}$ **15.** $\begin{bmatrix} 8 & 15 \\ 16 & 23 \end{bmatrix}$; $BA = \begin{bmatrix} 19 & 31 & -5 \\ 8 & 13 & -2 \\ -3 & -4 & -1 \end{bmatrix}$

17. $\begin{bmatrix} 14 & 20 & 20 \\ 35 & 30 & 31 \end{bmatrix}$ **19.** $\begin{bmatrix} 13 & 9 & 19 \\ 14 & 18 & 29 \\ 15 & 9 & 30 \end{bmatrix}$ **21.** $\begin{bmatrix} 37 & 32 \\ 48 & 42 \end{bmatrix}$ **23.** $\begin{bmatrix} 24 & 29 \\ 33 & 36 \end{bmatrix}$

25. $\begin{bmatrix} 1 & 8 \\ 19 & 59 \end{bmatrix}$ **27.** $\begin{bmatrix} 90 & 101 \\ 252 & 289 \end{bmatrix}$ **29.** $\begin{bmatrix} 27 & 30 \\ 75 & 66 \end{bmatrix}$ **33.** $\begin{bmatrix} 0 & 0 \\ 0 & 0 \end{bmatrix}$; no

35. (a) $\begin{bmatrix} 5 & 2 & 1 & 2 & 2 & 0 \\ 7 & 4 & 2 & 1 & 1 & 1 \end{bmatrix}$ (b) $\begin{bmatrix} \$0.31 \\ 0.79 \\ 0.95 \\ 0.73 \\ 1.89 \\ 1.69 \end{bmatrix}$ (c) $\$9.32$; $\$11.54$

EXERCISES 3.7

1. $x = 1, y = 2$ **3.** $x = 2, y = 3$ **5.** $x = 2, y = 3$ **7.** $x = 1, y = 2, z = -1$
9. $x = 2, y = -3, z = 3$ **11.** $x = 3, y = -1, z = 4$ **13.** $x = 1, y = 2, z = 1$
15. $x = 1, y = -1, z = 2$ **17.** $x = 1, y = 1, z = 2$ **19.** (a) $x = 1, y = 2, z = -1$
(b) $x = 1, y = 2, z = 3$

EXERCISES 3.8

1. No solution   **3.** $x = 3c + 6, y = c$ **5.** $x = 2, y = 3, z = 1$ **7.** $x = 2 + c,$
$y = -3 + 2c, z = c$ **9.** $x = \frac{24}{7} - \frac{3}{7}c, y = \frac{9}{7} + \frac{5}{7}c, z = c$ **11.** $x = \frac{4}{5} - \frac{2}{5}c,$
$y = \frac{3}{5}c - \frac{11}{5}, z = c$ **13.** $x = 3, y = 1, z = 2$

EXERCISES 3.10

1. $\frac{1}{2}\begin{bmatrix} 4 & -2 \\ -5 & 3 \end{bmatrix}$ **3.** $\frac{1}{4}\begin{bmatrix} 7 & 10 \\ 6 & 8 \end{bmatrix}$ **5.** $\frac{-1}{30}\begin{bmatrix} 3 & -1 & -4 \\ 12 & -14 & 4 \\ -12 & -6 & 6 \end{bmatrix}$ **7.** $\frac{1}{26}\begin{bmatrix} -6 & -2 & 12 \\ -7 & 2 & 1 \\ 11 & 8 & -9 \end{bmatrix}$

9. $\frac{1}{10}\begin{bmatrix} -5 & 5 & 5 \\ 5 & -3 & -1 \\ 5 & 1 & -3 \end{bmatrix}$ **11.** $\frac{1}{27}\begin{bmatrix} 8 & 5 & -7 \\ -1 & -4 & 11 \\ -7 & -1 & 23 \end{bmatrix}$ **13.** $A^{-1} = \frac{1}{6}\begin{bmatrix} 1 & 1 \\ -2 & 4 \end{bmatrix}$;

$x = 4, y = -1$ **15.** $A^{-1} = \frac{1}{5}\begin{bmatrix} -1 & 2 \\ 4 & -3 \end{bmatrix}$ (a) $x = 2, y = 3$ (b) $x = 4, y = -3$

(c) $x = -3, y = 10$ (d) $x = 1, y = -1$ **17.** $A^{-1} = \begin{bmatrix} -7 & -5 & 6 \\ 13 & 10 & -11 \\ -1 & -1 & 1 \end{bmatrix}$

(a) $x = 2, y = 3, z = -1$ (b) $x = 3, y = -2, z = 5$ (c) $x = 2, y = 4, z = 7$ (d) $x = 0,$

$y = 2, z = 1$ **19.** $A^{-1} = \left(\dfrac{-1}{55}\right)\begin{bmatrix} 5 & -10 & -5 \\ -15 & -3 & 4 \\ -15 & 19 & -7 \end{bmatrix}$; $x = -2, y = 5, z = 3$

21. $A^{-1} = \left(\dfrac{-1}{5}\right)\begin{bmatrix} -3 & 1 & 4 \\ 1 & -2 & -3 \\ 2 & -4 & -1 \end{bmatrix}$; $x = 1, y = 1, z = 1$

23. $A^{-1} = \left(\dfrac{1}{40}\right)\begin{bmatrix} -16 & 6 & 9 \\ 48 & -38 & 3 \\ -24 & 24 & -4 \end{bmatrix}$; station wagons, 2; full-size cars, 3; intermediate

cars, 4.

CHAPTER 4

EXERCISES 4.2

3. (a) Not reflexive: 3 $\cancel{\mathscr{R}}$ 3, because 3 $\neq$ 3 + 1. Not symmetric: 3 $\mathscr{R}$ 2 (3 = 2 + 1), but
2 $\cancel{\mathscr{R}}$ 3 (2 $\neq$ 3 + 1). Not transitive: 4 $\mathscr{R}$ 3 (4 = 3 + 1) and 3 $\mathscr{R}$ 2 (3 = 2 + 1), but
4 $\cancel{\mathscr{R}}$ 2 (4 $\neq$ 2 + 1).   (b) Reflexive, symmetric, and transitive.   (c) Not reflexive: 7 $\cancel{\mathscr{R}}$ 7
(same sign). Symmetric. Not transitive: 3 $\mathscr{R}$ −5 and −5 $\mathscr{R}$ 8, but 3 $\cancel{\mathscr{R}}$ 8.   (d) Not reflexive:
1 $\cancel{\mathscr{R}}$ 1 (3 is not a factor of 1). Symmetric and transitive.   (e) Not reflexive: 2 $\cancel{\mathscr{R}}$ 2 (because
2 = 2). Symmetric. Not transitive: 2 $\mathscr{R}$ 3 (2 $\neq$ 3) and 3 $\mathscr{R}$ 2 (3 $\neq$ 2), but 2 $\cancel{\mathscr{R}}$ 2.   (f) Not
reflexive: 5 $\cancel{\mathscr{R}}$ 5 (neither 5 is 7). Symmetric. Not transitive: 2 $\mathscr{R}$ 7 and 7 $\mathscr{R}$ 5, but 2 $\cancel{\mathscr{R}}$ 5.
(g) Not reflexive: 5 $\not>$ 5. Not symmetric: 8 > 3, but 3 $\not>$ 8. Transitive.   (h) Reflexive and
transitive. Not symmetric: 4 $\mathscr{R}$ 3 (4 $\not<$ 3) but 3 $\cancel{\mathscr{R}}$ 4 (3 < 4).   (i) Not reflexive: 1 $\cancel{\mathscr{R}}$ 1 (5 is not
a factor of 1 + 1). Symmetric. Not transitive: 3 $\mathscr{R}$ 7 (5 is a factor of 10) and 7 $\mathscr{R}$ 8 (5 is a
factor of 15), but 3 $\cancel{\mathscr{R}}$ 8 (5 is not a factor of 11).   (j) Not reflexive: 1 $\cancel{\mathscr{R}}$ 1 (1 + 1 $\neq$ 0).
Symmetric. Not transitive: 3 $\mathscr{R}$ −3 (3 + (−3) = 0) and −3 $\mathscr{R}$ 3, but 3 $\cancel{\mathscr{R}}$ 3 (3 + 3 $\neq$ 0).
5. 4.2.1 is not; neither reflexive nor symmetric.   4.2.2 is; the classes are the sets of all bricks
of each color—all red bricks, all yellow bricks, and so on.   4.2.3 is not; neither reflexive nor
symmetric.   4.2.4 is; classes are {1, 3} and {2, 4}.   4.2.5 is not; not reflexive, symmetric,
or transitive.

EXERCISES 4.3

1. (a) No. {a, b} and {x, y} are equivalent but not equal.   (b) Yes. If two sets contain
precisely the same elements, then matching each element of one set with itself in the other
produces a one-to-one correspondence, and hence they are equivalent.

3. There are 24 such correspondences; for example,

$$\{a, b, c, d\} \quad \{a, b, c, d\} \quad \{a, b, c, d\} \quad \{a, b, c, d\} \quad \{a, b, c, d\}$$
$$\{*, o, \$, \#\} \quad \{*, o, \$, \#\} \quad \{*, o, \$, \#\} \quad \{*, o, \$, \#\} \quad \{*, o, \$, \#\}$$

5. (a) $\{a, b\}, \{c, d\}$ (b) $\{a, b, c, d, e, f, g, h, i, j\}, \{x, 3, y, 5, A, *, 8, !, q, R\}$
(c) $\{a, b, c, \ldots, y\}, \{0, 1, 2, \ldots, 24\}$ (d) $\{0, 1, 2, \ldots, 99\}, \{x_0, x_1, x_2, \ldots, x_{99}\}$

7. (a) Consider $\{v, w, x, y, z\} \in 5$ and $\{a, b, c, d, e, f, g, h\} \in 8$.
$$\{v, w, x, y, z\}$$
$$\{a, b, c, d, e\}$$
is a one-to-one

correspondence and $\{a, b, c, d, e\}$ is a proper subset of $\{a, b, c, d, e, f, g, h\}$, so $5 < 8$.

(b) Similarly, for $\{a, b, c\} \in 3$ and $\{a, b, c, d, e, f, g\} \in 7$,
$$\{a, b, c\}$$
$$\{c, d, e\}$$
is a one-to-one correspondence

and $\{c, d, e\}$ is a proper subset of $\{a, b, c, d, e, f, g\}$. Thus, $3 < 7$ and hence $3 \leq 7$.

9. Associativity: Let a, b, and c be whole numbers, with reference sets A, B, and C, respectively, such that each of these sets is disjoint from the other two. By definition of sum, $(a + b) + c$ is represented by $(A \cup B) \cup C$, and $a + (b + c)$ is represented by $A \cup (B \cup C)$. But union of sets is associative (if you are not familiar with this fact, it is easy to check from the definition of $\cup$), so $(A \cup B) \cup C = A \cup (B \cup C)$, implying that $(a + b) + c = a + (b + c)$. Commutativity: Let a and b be whole numbers with disjoint reference sets A and B, respectively. By definition of sum, $a + b$ is represented by $A \cup B$, and $b + a$ is represented by $B \cup A$. It is easy to see that $A \cup B = B \cup A$, so $a + b = b + a$, as required.

11. Let A, B, and C be reference sets for a, b, and c, respectively, such that C is disjoint from both A and B. Then $a + c$ and $b + c$ are represented by $A \cup C$ and $B \cup C$, respectively. $a < b$ means that A is equivalent to a proper subset of B. Extend this correspondence to an equivalence between $A \cup C$ and a proper subset of $B \cup C$ by matching each element of C with itself. Thus, $a + c < b + c$, by definition of $<$.

13. Let S be a reference set for n, and recall that $\varnothing$ is the reference set for 0.   (a) By definition of sum, $n + 0$ is represented by $S \cup \varnothing = S$, our reference set for n, so $n + 0 = n$. (b) By definition of product, $n \cdot 0$ is represented by $S \times \varnothing$, which is empty (since there are no second elements for the ordered pairs); that is, $S \times \varnothing = \varnothing$, so $n \cdot 0 = 0$.

EXERCISES 4.4

1. Let A and B be reference sets for 5 and x, respectively. Then $A \cup B$ represents $5 + x$. Since A contains five elements, $A \cup B$ must contain at least five elements and hence cannot represent 3, regardless of what B (or x) is.

3. (a) $(3, 0), (4, 1), (5, 2)$ (b) $(17, 0), (18, 1), (27, 10)$ (c) $(0, 2), (1, 3), (5, 7)$ (d) $(0, 25),$
$(10, 35), (100, 125)$ (e) $(0, 0), (1, 1), (57, 57)$

5. (a) $(-5) \oplus 0 = [0, 5] \oplus [0, 0] = [0 + 0, 5 + 0] = [0, 5] = -5$. $0 \odot (+7) = [0, 0] \odot$

$[7, 0] = [0 \cdot 7 + 0 \cdot 0, 0 \cdot 0 + 0 \cdot 7] = [0, 0] = 0$ (b) Let $x = [a, b]$. $x \oplus 0 = [a, b] \oplus$
$[0, 0] = [a + 0, b + 0] = [a, b] = x$ (c) Let $x = [a, b]$. $0 \odot x = [0, 0] \odot [a, b] =$
$[0 \cdot a + 0 \cdot b, 0 \cdot b + 0 \cdot a] = [0, 0] = 0$

7. $[a, 0] \leqslant [b, 0]$ iff $a + 0 < 0 + b$ iff $a < b$, by definition of $\leqslant$.

9. Let $[a, b], [c, d]$, and $[e, f]$ be arbitrary integers.   (a) $([a, b] \oplus [c, d]) \oplus [e, f] =$
$[a + c, b + d] \oplus [e, f] = [(a + c) + e, (b + d) + f] = [a + (c + e), b + (d + f)]$
(why?) $= [a, b] \oplus [c + e, d + f] = [a, b] \oplus ([c, d] \oplus [e, f])$ (b) $[a, b] \odot [c, d] =$
$[ac + bd, ad + bc]$ and $[c, d] \odot [a, b] = [ca + db, cb + da]$. But $ac + bd = ca + db$ and
$ad + bc = cb + da$ by commutativity of whole number multiplication and addition.
(c) $([a, b] \odot [c, d]) \odot [e, f] = [ac + bd, ad + bc] \odot [e, f]$
$$= [(ac + bd)e + (ad + bc)f, (ac + bd)f + (ad + bc)e]$$
$$= [ace + bde + adf + bcf, acf + bdf + ade + bce]$$
$[a, b] \odot ([c, d] \odot [e, f]) = [a, b] \odot [ce + df, cf + de]$
$$= [a(ce + df) + b(cf + de), a(cf + de) + b(ce + df)]$$
$$= [ace + adf + bcf + bde, acf + ade + bce + bdf]$$
(d) $[a, b] \odot ([c, d] \oplus [e, f]) = [a, b] \odot [c + e, d + f]$
$$= [a(c + e) + b(d + f), a(d + f) + b(c + e)]$$
$$= [ac + ae + bd + bf, ad + af + bc + be]$$
$([a, b] \odot [c, d]) \oplus ([a, b] \odot [e, f]) = [ac + bd, ad + bc] \oplus [ae + bf, af + be]$
$$= [ac + bd + ae + bf, ad + bc + af + be]$$

11. Let $x = [a, b]$ and $y = [c, d]$.
(a) $(-x) \oplus (-y) = [b, a] \oplus [d, c] = [b + d, a + c]$
    $-(x \oplus y) = -[a + c, b + d] = [b + d, a + c]$
(b) $x \odot (-y) = [a, b] \odot [d, c] = [ad + bc, ac + bd]$
    $(-x) \odot y = [b, a] \odot [c, d] = [bc + ad, bd + ac]$
    $-(x \odot y) = -([a, b] \odot [c, d]) = -[ac + bd, ad + bc] = [ad + bc, ac + bd]$
(c) $-(-x) = -[b, a] = [a, b] = x$

13. $(-4) \ominus (-7) = [0, 4] \ominus [0, 7]$ definition of negative integer
$= [0, 4] \oplus (-[0, 7])$ definition of $\ominus$ (4.4.9)
$= [0, 4] \oplus [7, 0]$ definition of additive inverse
$= [7, 4]$ definition of $\oplus$
$= [3, 0]$ Theorem 4.4.2
$= +3$ definition of positive integer

EXERCISES 4.5

1. Definition 4.5.8: $a/b \oslash c/d = a/b \odot (c/d)^{-1}$ 4.5.17: $\frac{2}{5} \oslash \frac{3}{7} = \frac{2}{5} \odot (\frac{3}{7})^{-1} = \frac{2}{5} \odot \frac{7}{3} = \frac{14}{15}$
4.5.18: $\frac{7}{8} \oslash \frac{2}{3} = \frac{7}{8} \odot (\frac{2}{3})^{-1} = \frac{7}{8} \odot \frac{3}{2} = \frac{21}{16}$ 4.5.19: $\frac{10}{9} \oslash \frac{5}{3} = \frac{10}{9} \odot (\frac{5}{3})^{-1} = \frac{10}{9} \odot \frac{3}{5} = \frac{30}{45} = \frac{2}{3}$
4.5.20: $\frac{12}{13} \oslash \frac{4}{1} = \frac{12}{13} \odot (\frac{4}{1})^{-1} = \frac{12}{13} \odot \frac{1}{4} = \frac{12}{52} = \frac{3}{13}$ Theorem 4.5.4: For any two rational
numbers a/b and c/d, $c/d \odot (a/b \oslash c/d) = a/b$.
3. (a) $(5, 9), (10, 18), (15, 27)$ (b) $(15, 6), (5, 2), (-10, -4)$ (c) $(-27, 95), (-270, 950)$,
$(-54, 190)$ (d) $(0, 1), (0, 2), (0, 47)$ (e) $(1, 1), (9, 9), (-13, -13)$
5. (a) $[3, 7] \oplus [0, 1] = [3 \cdot 1 + 7 \cdot 0, 7 \cdot 1] = [3, 7]$
      $[3, 7] \odot [1, 1] = [3 \cdot 1, 7 \cdot 1] = [3, 7]$

$$[0, 1] \oplus [1, 1] = [0 \cdot 1 + 1 \cdot 1, 1 \cdot 1] = [1, 1]$$
$$[1, 1] \odot [0, 1] = [1 \cdot 0, 1 \cdot 1] = [0, 1]$$

Let $r = [a, b]$, where a and b are integers, $b \neq 0$:   (b) $[a, b] \oplus [0, 1] = [a \cdot 1 + b \cdot 0, b \cdot 1] = [a, b]$ (c) $[a, b] \odot [1, 1] = [a \cdot 1, b \cdot 1] = [a, b]$ (d) $[a, b] \odot [0, 1] = [a \cdot 0, b \cdot 1] = [0, b]$

7. By definition of $\lessgtr$, $[x, 1] \lessgtr [y, 1]$ iff $x \cdot 1 < y \cdot 1$; that is, iff $x < y$.

9. Let $[a, b]$, $[c, d]$, and $[e, f]$ be arbitrary rational numbers, and recall that commutativity, associativity, and distributivity hold for $+$ and $\cdot$ in I.

(a) $[a, b] \oplus [c, d] = [ad + bc, bd] = [cb + da, db] = [c, d] \oplus [a, b]$

(b) $([a, b] \oplus [c, d]) \oplus [e, f] = [ad + bc, bd] \oplus [e, f] = [(ad + bc)f + (bd)e, (bd)f]$
$$= [adf + bcf + bde, bdf]$$
$$[a, b] \oplus ([c, d] \oplus [e, f]) = [a, b] \oplus [cf + de, df] = [a(df) + b(cf + de), b(df)]$$
$$= [adf + bcf + bde, bdf]$$

(c) $([a, b] \odot [c, d]) \odot [e, f] = [ac, bd] \odot [e, f] = [(ac)e, (bd)f]$
$$[a, b] \odot ([c, d] \odot [e, f]) = [a, b] \odot [ce, df] = [a(ce), b(df)]$$

(d) $\quad [a, b] \odot ([c, d] \oplus [e, f]) = [a, b] \odot [cf + de, df]$
$$= [a(cf + de), b(df)] = [acf + ade, bdf]$$

$([a, b] \odot [c, d]) \oplus ([a, b] \odot [e, f]) = [ac, bd] \oplus [ae, bf]$
$$= [(ac)(bf) + (bd)(ae), (bd)(bf)] = [b(acf + ade), b(bdf)]$$
$$= [acf + ade, bdf], \quad \text{by Theorem 4.5.2}$$

11. Let $r = [a, b]$ and $s = [c, d]$, where $a, b, c, d \in I$.   (a) $r^{-1} \odot s^{-1} = [b, a] \odot [d, c] = [bd, ac] = [ac, bd]^{-1} = (r \odot s)^{-1}$ (b) $r \odot s^{-1} = [a, b] \odot [d, c] = [ad, bc] = [bc, ad]^{-1} = ([b, a] \odot [c, d])^{-1} = (r^{-1} \odot s)^{-1}$ (c) $(r^{-1})^{-1} = ([a, b]^{-1})^{-1} = [b, a]^{-1} = [a, b] = r$

13. $\frac{5}{2} \div \frac{10}{8} = [5, 2] \div [10, 8]$ rational numbers are quotients of integers
$$= [5, 2] \odot [10, 8]^{-1} \quad \text{definition of } \div \text{ (4.5.8)}$$
$$= [5, 2] \odot [8, 10] \quad \text{definition of reciprocal}$$
$$= [40, 20] \quad \text{definition of } \odot$$
$$= [2, 1] \quad \text{Theorem 4.5.2}$$
$$= 2 \quad \text{description of integers within } Q$$

EXERCISES 4.6

1. (a) $3, 6, 9, 12$ (b) $\frac{1}{3}, \frac{1}{4}, \frac{1}{5}, \frac{1}{6}$ (c) $\frac{1}{2}, \frac{2}{3}, \frac{3}{4}, \frac{4}{5}$ (d) $4, 4, 4, 4$ (e) $0, 2, 0, 2$ (f) $\frac{1}{2}, 1, \frac{9}{8}, 1$

3. Since P may not be rational, the sequence terms $P + \frac{1}{2}$, $P + \frac{1}{3}$, $P + \frac{1}{4}$, and so on, may not be rational numbers. (In fact, if P is not rational, $P + \frac{1}{2}$, and so on, makes no sense at all.)

7. (a) $\{2, 1, \frac{2}{3}, \ldots, 2/n, \ldots\}$, $\{1, \frac{1}{2}, \frac{1}{3}, \ldots, 1/n, \ldots\}$, $\{0, 0, 0, \ldots, 0, \ldots\}$ (b) $\{0, -\frac{1}{2}, -\frac{2}{3}, \ldots, -1 + 1/n, \ldots\}$, $\{-1, -1, -1, \ldots, -1, \ldots\}$, $\{-0.9, -0.99, -0.999, \ldots, -1 + 1/10^n, \ldots\}$ (c) $\{\frac{1}{2}, \frac{2}{3}, \frac{3}{4}, \ldots, n/(n + 1), \ldots\}$, $\{\frac{3}{2}, \frac{5}{4}, \frac{9}{8}, \ldots, 1 + 1/2^n, \ldots\}$, $\{\frac{3}{2}, \frac{4}{3}, \frac{5}{4}, \ldots, 1 + 1/n, \ldots\}$

9. (a) 4th term is $\frac{11}{8}$; 5th term is $\frac{45}{32}$.   (b) Since $1^2 = 1$ and $2^2 = 4$, the number whose square is 3 must lie in $\langle 1, 2 \rangle$. The first four terms of the sequence are $\frac{3}{2}, \frac{3}{2}, \frac{13}{8}$, and $\frac{27}{16}$.

EXERCISES 4.7

1. (a) $(3, 8)$ (b) $(-3, -2)$ (c) $(\frac{19}{3}, 2)$ (d) $(-34, 31)$ (e) $(7, 4)$ (f) $(0, 0)$

3. (a) $(a, b) \oplus (c, d) = (a + c, b + d) = (c + a, d + b) = (c, d) \oplus (a, b)$

(b) $((a, b) \oplus (c, d)) \oplus (e, f) = (a + c, b + d) \oplus (e, f) = ((a + c) + e, (b + d) + f) = (a + (c + e), b + (d + f))$ (why?) $= (a, b) \oplus (c + e, d + f) = (a, b) \oplus ((c, d) \oplus (e, f))$

(c) $(a, b) \odot (c, d) = (ac - bd, ad + bc) = (ca - db, cb + da) = (c, d) \odot (a, b)$

(d) $(a, b) \odot ((c, d) \oplus (e, f)) = (a, b) \odot (c + e, d + f)$
$$= (a(c + e) - b(d + f), a(d + f) + b(c + e))$$
$$= (ac + ae - bd - bf, ad + af + bc + be)$$
$((a, b) \odot (c, d)) \oplus ((a, b) \odot (e, f)) = (ac - bd, ad + bc) \oplus (ae - bf, af + be)$
$$= (ac - bd + ae - bf, ad + bc + af + be)$$

5. $(a + bi) + (c + di) = (a + c) + (bi + di) = (a + c) + (b + d)i$, which corresponds to $(a + c, b + d)$. $(a + bi) \cdot (c + di) = a(c + di) + bi(c + di) = ac + adi + bci + bdi^2 = ac + (ad + bc)i - bd = (ac - bd) + (ad + bc)i$, which corresponds to $(ac - bd, ad + bc)$.

CHAPTER 5

EXERCISES 5.2

1. There are actually 11 more, such as

3. (a) No; for example, $\{a, b\}$ and $\{x, y\}$ can be put in one-to-one correspondence, but do not contain the same elements. (b) Yes; if two sets contain exactly the same elements, then by matching each element of one set with itself in the other, we get a one-to-one correspondence.

EXERCISES 5.3

1. $\{4, 8, 12, \ldots, 4n, \ldots\}$ is a proper subset of N^{even}, and the matching of $2n$ with $4n$ for each n

is a one-to-one correspondence between the two sets. 3.

$$\{\ldots, -3, -2, -1, 0, 1, 2, 3, \ldots\}$$
$$\{\ldots, -6, -4, -2, 0, 2, 4, 6, \ldots\}$$

describes a one-to-one correspondence between I and one of its proper subsets.

EXERCISES 5.4

1.
$$N = \{1, 2, 3, \ldots, n, \ldots\}$$
$$N^{\text{odd}} = \{1, 3, 5, \ldots, 2n - 1, \ldots\}$$

3. (a) 17 (b) 61 (c) -37 (d) 200 (e) $2p$

5.
$$\{1, 3, 5, 7, \ldots, 2n - 1, \ldots\}$$
$$\{30, 60, 90, 120, \ldots, 30n, \ldots\}$$

7. $\frac{4}{5}$ 9. $-\frac{4}{3}$ 11. 27 13. -19

EXERCISES 5.5

1. (a) $0.\overline{1}$ (b) 0.4 (or $0.4\overline{0}$) (c) $2.1\overline{6}$ (d) $0.6\overline{81}$ (e) $1.88\overline{63}$ (f) $6.6\overline{428571}$

3. $0.2020020002\ldots$, $0.5353353335\ldots$, $0.4747747774\ldots$, $0.09199099919999\ldots$, $0.09188099918888\ldots$, and so on.

EXERCISES 5.6

1. (a) Using the scheme of Cantor's diagonalization process, we get $0.121112\ldots$ (b) Using the Cantor diagonalization process with different pairs of digits (instead of 1 and 2) produces different numbers: Using 1 and 3, we get $0.131113\ldots$; using 7 and 8, we get $0.778777\ldots$; using 4 and 9, we get $0.444444\ldots$ (c) As many numbers as there are ordered pairs of distinct digits—that is, $10 \cdot 9$, or 90 such numbers.

EXERCISES 5.7

1. The answers to most of these depend on your choice of point P. Independent of this choice are: (h) 2 inches, and (i) 0 inches.   **3.** Yes. $[0, 1] \leftrightarrow R$ and $R \leftrightarrow R \times R$ so $[0, 1] \leftrightarrow R \times R$.
5. Yes. $[0, 1] \leftrightarrow R$ and (by Exercise 4) $R \leftrightarrow R \times R \times R$, so $[0, 1] \leftrightarrow R \times R \times R$.

EXERCISES 5.8

1. 5 is the collection of all sets that are equivalent to $\{a, b, c, d, e\}$. Two other reference sets are $\{*, o, \#, \$, \&\}$, $\{x, y, z, A, +\}$.   **3.** (a) $\aleph_0$, because Q is equivalent to N. (b) $\mathbf{c}$, because $[0, 1]$ is equivalent to R

5. Let N and $\{a, b, c\}$ be reference sets for $\aleph_0$ and 3, respectively.
$$\begin{array}{c} \{a, b, c\} \\ |\ \ |\ \ | \\ \{1, 2, 3\} \end{array}$$ describes a

one-to-one correspondence between $\{a, b, c\}$ and a proper subset of N, so condition (1) of the definition of $>$ is satisfied. Clearly, $\{a, b, c\}$ and N are not equivalent, so condition (2) is also satisfied.
7. Let R and $\{a, b, c, d, e\}$ be reference sets for $\mathbf{c}$ and 5, respectively. $\{a, b, c, d, e\}$ can be put in one-to-one correspondence with $\{1, 2, 3, 4, 5\}$, a proper subset of R, and cannot be put in one-to-one correspondence with all of R.
9. Let A, B, and C be reference sets for the cardinal numbers $\mathcal{A}$, $\mathcal{B}$, and $\mathcal{C}$, respectively. $\mathcal{A} < \mathcal{B}$ implies that A is equivalent to some proper subset S of B. $\mathcal{B} < \mathcal{C}$ implies that B is equivalent to some proper subset T of C, and hence, by carrying through the two correspondences, A is equivalent to a proper subset T_1 of T, and of course T_1 is also a proper subset of C. Thus, condition (1) of the definition of $<$ is satisfied. We verify condition (2) indirectly: Suppose it is false; that is, suppose A and C are equivalent. Since B is equivalent to a proper subset of C, the combination of these two correspondences makes B equivalent

to a proper subset of A. The Cantor-Schröder-Bernstein Theorem then implies that A and B are equivalent, contradicting the hypothesis $\mathscr{A} < \mathscr{B}$. Hence, condition (2) must be true.

EXERCISES 5.9

1. $\varnothing, \{a\}, \{b\}, \{c\}, \{a, b\}, \{a, c\}, \{b, c\}, \{a, b, c\}$ **3.** (There are many possible correct answers to this question; only one is given here.)   (a)

$$\{\quad u, \quad\quad v, \quad\quad w, \quad\quad x, \quad\quad y, \quad z\,\}$$
$$|\quad\quad\quad|\quad\quad\quad|\quad\quad\quad|\quad\quad\quad|\quad\quad\quad|$$
$$\{\{u, v\}, \{w, x, y\}, \{z\}, \{v, x, z\}, \{w, y\}, \varnothing\}$$

(b) $W = \{v, w, z\}$ **5.** $S = \{\varnothing, \{a\}, \{b\}, \{a, b\}\}$; its size is 4. The set of all subsets of S is $\{\varnothing, \{\varnothing\}, \{\{a\}\}, \{\{b\}\}, \{\{a, b\}\}, \{\varnothing, \{a\}\}, \{\varnothing \{b\}\}, \{\varnothing, \{a, b\}\}, \{\{a\}, \{b\}\}, \{\{a\}, \{a, b\}\}, \{\{b\}, \{a, b\}\}, \{\varnothing, \{a\}, \{b\}\}, \{\varnothing, \{a\}, \{a, b\}\}, \{\varnothing, \{b\}, \{a, b\}\}, \{\{a\}, \{b\}, \{a, b\}\}, S\}$; its size is 16.

EXERCISES 5.10

(There are many different correct answers to these exercises.)
1. (a) By the Continuum Hypothesis, $\aleph_1 = c$, so we can let $A = R$ (the set of all real numbers).
(b) $5, \frac{1}{2}, \sqrt{2}$ (c) $N = \{1, 2, 3, \ldots, n, \ldots\}$ **3.** (a) Let C be the set of all subsets of B (from Exercise 2).   (b) $\{\{1\}, \{\frac{1}{2}\}, \{\frac{1}{3}\}\}, \{\{\sqrt{2}\}\}, \varnothing$

EXERCISES 5.11

1. Since any common elements would only appear once in the union, there might not be enough elements to represent the true sum of the two numbers. For example, if we represent 3 and 4 by $\{a, b, c\}$ and $\{b, c, d, e\}$, respectively, then $\{a, b, c\} \cup \{b, c, d, e\} = \{a, b, c, d, e\}$ would yield the "sum" $3 + 4 = 5$.
3. Let N and $\{a, b, c, d, e\}$ represent $\aleph_0$ and 5, respectively.

$$N \cup \{a, b, c, d, e\} = \{a, b, c, d, e, 1, 2, 3, \ldots, \quad n, \quad \ldots\}$$
$$|\;\;|\;\;|\;\;|\;\;|\;\;|\;\;|\;\;|\quad\quad\quad |$$
$$N = \{1, 2, 3, 4, 5, 6, 7, 8, \ldots, n + 5, \ldots\}$$

is a one-to-one correspondence, so $N \cup \{a, b, c, d, e\}$ is in $\aleph_0$.
5. Let $\{a_1, \ldots, a_{10}\}$ represent 10 and let R represent c. Write R as $N \cup S$, where S is the set of all real numbers not in N. By Exercise 4 we know $\{a_1, \ldots, a_{10}\} \cup N$ is equivalent to N, so extending this correspondence to all of R by matching each element of S with itself we get a one-to-one correspondence between $\{a_1, \ldots, a_{10}\} \cup N \cup S$ and $N \cup S$; that is, between $\{a_1, \ldots, a_{10}\} \cup R$ and R, as required.
7. From previous work we know that c is the size of any interval of real numbers, with or without endpoints. Representing the two c's of the sum by $[0, 1)$ and $[1, 2]$, respectively, we get $[0, 1) \cup [1, 2] = [0, 2]$, which also represents c, as required.
9. Let x be the cardinal number of the set of all subsets of R. Then, by Cantor's Theorem, $x > c$, so $c + x > c$.

EXERCISES 5.12

1. No. The definition of product depends on the Cartesian product of the reference sets, which distinguishes the elements of one set from those of the other. Hence, common elements cannot "get lost" in the process.

3. Mimicking the example $2 \cdot \aleph_0$ given in this section, we can start by using $\{0, 1, 2\}$ as a reference set for 3 and $T = \{0, 3, 6, \ldots, 3(n - 1), \ldots\}$ as a reference set for $\aleph_0$. (Is it clear that T has size $\aleph_0$?) Then

$$\{0, 1, 2\} \times T = \{(0, 0), (1, 0), (2, 0), (0, 3), (1, 3), (2, 3), \ldots\}$$
$$W = \{ \quad 0, \quad 1, \quad 2, \quad 3, \quad 4, \quad 5, \ldots\}$$

where this correspondence results from adding the terms in each pair. But it is easy to see that the set W of whole numbers has size $\aleph_0$ (why?), so $3 \cdot \aleph_0 = \aleph_0$.

5. Consider $\{\frac{1}{2}, \frac{1}{3}, \frac{1}{4}, \ldots, 1/(n + 1), \ldots\}$, where n runs through all the natural numbers, and let S be everything else in $[0, 1)$. Then

$$[0, 1) = \{\tfrac{1}{2}, \tfrac{1}{3}, \tfrac{1}{4}, \ldots, 1/(n + 1), \ldots\} \cup S$$
$$[0, 1] = \{1, \tfrac{1}{2}, \tfrac{1}{3}, \ldots, \quad 1/n, \quad \ldots\} \cup S$$

describes an obvious one-to-one correspondence between $[0, 1)$ and $[0, 1]$, where it is understood that each element of S is matched with itself.

7. No. By Exercise 4, for example, $3 \cdot \aleph_0 = 17 \cdot \aleph_0$ (since they both equal $\aleph_0$), but $3 \neq 17$. Similarly, $5 \cdot \mathbf{c} = 249 \cdot \mathbf{c}$ [by Exercise 6(c)], but $5 \neq 249$.

CHAPTER 6

EXERCISES 6.3

3. Let the letters H and T denote, respectively, a "head" and a "tail." Then the sample space is $\{HH, HT, TH, TT\}$.   **5.** $S = \{(H, 1), (H, 2), (H, 3), (H, 4), (H, 5), (H, 6), (T, 1), (T, 2), (T, 3), (T, 4), (T, 5), (T, 6)\}$

7. Let the six students be labeled $A, B, C, D, E,$ and F and let the ordered pair (x, y) represent the event that x is the president and y is the secretary. The sample space consisting of 30 possible elections is $S = \{(A, B), (A, C), (A, D), (A, E), (A, F), (B, A), (B, C), (B, D), (B, E), (B, F), (C, A), (C, B), (C, D), (C, E), (C, F), (D, A), (D, B), (D, C), (D, E), (D, F), (E, A), (E, B), (E, C), (E, D), (E, F), (F, A), (F, B), (F, C), (F, D), (F, E)\}$.

9. Let the letters B and G denote, respectively, a "boy" and a "girl." The sample space of six possible arrangements is $S = \{(B, B, G, G), (B, G, B, G), (B, G, G, B), (G, G, B, B), (G, B, B, G), (G, B, G, B)\}$.

11. Let G_1, G_2, G_3 and $L_1, L_2,$ and L_3 denote, respectively, the three gentlemen and the three ladies. The sample space of six dancing possibilities is

G_1	G_2	G_3
L_1	L_2	L_3
L_1	L_3	L_2
L_2	L_1	L_3
L_2	L_3	L_1
L_3	L_1	L_2
L_3	L_2	L_1

13. $S = \{$HHH, HHT, HTH, HTT, THH, THT, TTH, TTT$\}$ (a) $\{$THH, THT, TTH, TTT$\}$
(b) $\{$HHT, HTH, THH$\}$ (c) $\{$HHT, HTH, THH, HHH$\}$ (d) $\{$TTT$\}$
15. (a) $\{(1, 5), (2, 5), (3, 5), (4, 5), (5, 5), (6, 5), (5, 1), (5, 2), (5, 3), (5, 4), (5, 6)\}$ (b) $\{(1, 1), (2, 1),$
$(3, 1), (4, 1), (5, 1), (6, 1), (1, 3), (2, 3), (3, 3), (4, 3), (5, 3), (6, 3), (1, 5), (2, 5), (3, 5), (4, 5), (5, 5), (6, 5)\}$
(c) $\{(1, 5), (2, 4), (3, 3), (4, 2), (5, 1)\}$ (d) $\{(1, 1), (2, 2), (3, 3), (4, 4), (5, 5), (6, 6)\}$

EXERCISES 6.4

1. (a) $\frac{3}{6}$ (b) $\frac{3}{6}$ (c) $\frac{2}{6}$ (d) $\frac{4}{6}$ (e) $\frac{4}{6}$ (f) 0 **3.** (a) $\frac{5}{10}$ (b) $\frac{3}{10}$ (c) $\frac{4}{10}$ (d) $\frac{4}{10}$
5. (a) $\frac{4}{15}$ (b) $\frac{10}{15}$ (c) $\frac{9}{15}$ (d) $\frac{11}{15}$ **7.** (a) $\frac{6}{25}$ (b) $\frac{11}{25}$ (c) $\frac{15}{25}$ (d) $\frac{18}{25}$ **9.** (a) $\frac{1}{8}$ (b) $\frac{3}{8}$
(c) $\frac{3}{8}$ (d) $\frac{7}{8}$ (e) $\frac{1}{2}$ **11.** (a) $\frac{5}{36}$ (b) $\frac{8}{36}$ (c) $\frac{6}{36}$ (d) $\frac{6}{36}$ **13.** (a) $\frac{5}{15}$ (b) $\frac{7}{15}$ (c) $\frac{5}{15}$
15. $\frac{2}{6}$ **17.** $\frac{2}{6}$

EXERCISES 6.5

1. Yes   **3.** A and B are mutually exclusive; B and C are not mutually exclusive; A and C
are mutually exclusive.   **5.** (a) Yes (b) Yes (c) Yes (d) $P(A \cup B) = \frac{8}{36}$; $P(A \cup C) = \frac{24}{36}$;
$P(B \cup C) = \frac{20}{36}$ **7.** (a) The sum of probabilities is less than 1. (b) The sum of
probabilities is more than 1. (c) The probability that Mr. Carlson should register for at
least two courses should be at least 0.67.   **9.** (a) 0.23 (b) 0.94 (c) 0.51 (d) 0.77
11. (a) 0.84 (b) 0.37 (c) 0.48 (d) 0.35   **13.** (a) 0.45 (b) 0.63 (c) 0.36 (d) 0.18
(e) 0.73 (f) 0.64 (g) 0.27 (h) 0.81   **15.** (a) 0.28 (b) 0.72 (c) 0.64   **17.** (a) $\frac{262}{500}$
(b) $\frac{238}{500}$ (c) $\frac{206}{500}$ **19.** (a) 0.28 (b) 0.49 (c) 0.72   **21.** (a) 0.40 (b) 0.60 (c) 0.43
23. 0.55

EXERCISES 6.6

1. (a) $\frac{13}{52}$ (b) $\frac{13}{26}$ (c) $\frac{13}{39}$ **3.** (a) $\frac{1}{7}$ (b) $\frac{1}{4}$ **5.** (a) 0.29 (b) 0.52 (c) $\frac{23}{40}$ (d) $\frac{23}{35}$
(e) $\frac{23}{52}$ **7.** (a) $\frac{26}{35}$ (b) $\frac{26}{43}$ (c) $\frac{35}{52}$ **9.** (a) $\frac{11}{24}$ (b) $\frac{11}{35}$ **11.** (a) $\frac{15}{30}$ (b) $\frac{15}{60}$ (c) $\frac{30}{75}$
13. (a) $\frac{225}{500}$ (b) $\frac{145}{500}$ (c) $\frac{100}{500}$ (d) $\frac{100}{225}$ (e) $\frac{100}{145}$ **15.** (a) $\frac{7}{20}$ (b) $\frac{3}{20}$ **17.** (a) $\frac{1}{30}$
(b) $\frac{1}{6}$ **19.** (a) $\frac{14}{55}$ (b) $\frac{1}{55}$

EXERCISES 6.7

1. $S = \{HH, HT, TH, TT\}$; $A = \{HH, HT\}$, $B = \{HT, TT\}$, $A \cap B = \{HT\}$, $P(A) = \frac{2}{4}$, $P(B) = \frac{2}{4}$, $P(A \cap B) = \frac{1}{4}$. Since $P(A \cap B) = P(A) \cdot P(B)$, it follows that A and B are independent. **3.** $P(A) = \frac{13}{52}$, $P(B) = \frac{4}{52}$, $P(A \cap B) = \frac{1}{52}$. Since $P(A \cap B) = P(A) \cdot P(B)$, we conclude that A and B are independent. Also, $P(C) = \frac{8}{52}$ and $P(A \cap C) = \frac{2}{52}$. The events A and C are independent, since $P(A \cap C) = P(A) \cdot P(C)$. **5.** (a) 0.56 (b) 0.06 (c) 0.94 (d) 0.38 **7.** (a) 0.6375 (b) 0.0375 (c) 0.9625 (d) 0.3250 **9.** (a) 0.8550 (b) 0.0050 (c) 0.9950

EXERCISES 6.8

1. $\{0, 1, 2\}$ **3.** $\{-4, -2, 0, 2, 4\}$ **5.** $\{3, 4, 5, \ldots, 16, 17, 18\}$

EXERCISES 6.9

1. Yes **3.** Yes **5.** The proposition is unfavorable to the newspaper boy since he receives 22 cents for the paper on the average. **7.** One employee **9.** $8550 **11.** Expected gain, $500 **13.** $35 **15.** 0.0657

CHAPTER 7

EXERCISES 7.2

1.

Class boundaries	Frequency
15.5–18.5	8
18.5–21.5	7
21.5–24.5	9
24.5–27.5	16
27.5–30.5	8
30.5–33.5	1
33.5–36.5	1

3.

Class boundaries	Frequency
1.5–7.5	1
7.5–13.5	9
13.5–19.5	12
19.5–25.5	12
25.5–31.5	11
31.5–37.5	7
37.5–43.5	5
43.5–49.5	3

5.

Class boundaries	Frequency
21.45–24.45	10
24.45–27.45	21
27.45–30.45	20
30.45–33.45	16
33.45–36.45	6
36.45–39.45	3
39.45–42.45	3
42.45–45.45	1

EXERCISES 7.3

1. (a) $x_1 + x_2 + x_3$ (b) $x_1y_1 + x_2y_2 + x_3y_3 + x_4y_4$ (c) $(x_1 + 4) + (x_2 + 4) + (x_3 + 4) + (x_4 + 4) + (x_5 + 4)$ (d) $(x_1 + y_1 + 4) + (x_2 + y_2 + 4) + (x_3 + y_3 + 4) + (x_4 + y_4 + 4)$
3. (a) 12   (b) -5 (c) 50   (d) 110   (e) 42   (f) 19

EXERCISES 7.4

1. Mean = 10.6, median = 11, mode = 11   **3.** Mean = 32, median = 31.5, mode does not exist   **5.** Mean = 70.7, median = 75, mode = 76   **7.** Mean = 20.333, median = 20, mode = 20   **9.** Mean = 78.857, median = 79.5, mode = 78   **11.** Mean = 16,000, median = 15,000   **13.** Mean = 76.36, median = 75   **15.** Mean = 57.862, median = 55   **17.** (a) Arithmetic mean   (b) 346

EXERCISES 7.5

1. $x = 9, s = 1.732$ **3.** $x = 12, s = 2.449$ **5.** $x = 21, s = 3.651$ **7.** $x = 2.7, s = 0.510$ **9.** $\mu = 12,000, \sigma = 2828.427$ **11.** $s = 20.499$

EXERCISES 7.7

1. (a) 0.9207   (b) 0.3907   (c) 0.1635   (d) 0.1456   (e) 0.95   (f) 0.8731   **3.** (a) 0.6826   (b) 0.8354   (c) 0.9544   (d) 0.9826   (e) 0.9974   (f) 0.9980   **5.** (a) 1.81   (b) 1.76   (c) -1.65 (d) -1.67 **7.** (a) 0.0668   (b) 0.0228   (c) 0.6915   (d) 0.6826   (e) 0.3830   (f) 0.9544   **9.** (a) 0.3830   (b) 0.1587   (c) 0.1587   **11.** $\mu = 69$

EXERCISES 7.8

1. 0.1359   **3.** 0.3085   **5.** (a) 0.1587   (b) 0.1359   (c) 0.0082   **7.** (a) 0.9066   (b) 0.1359   (c) 0.6826   **9.** 0.0668   **11.** After 85.36 minutes   **13.** 7.301 ounces   **15.** 72   **17.** 0.0062   **19.** 0.0668

EXERCISES 7.9

1. $\hat{y} = -0.3 + 2.3x$; 13.5; 22.7   **3.** (a) $\hat{y} = 310.7575 + 4.7105x$ (b) 367.2835; 381.415   **5.** (a) $\hat{y} = 1.494 + 0.1624x$ (b) 7.99   **7.** (b) $\hat{y} = 31.4286 + 5.2381x$ (c) 70.7143   **9.** $\hat{y} = 13.8827 + 1.1647x$; 24.5979

CHAPTER 8

EXERCISES 8.1

1. 199   **3.** 10,000   **5.** 19,929   **7.** $n(n + 1)/2$. Because they record the number of points in an equilateral triangle array.   **9.** The sum is always a perfect square. Specifically, the sum of the nth and the $(n + 1)$st triangle number is $(n + 1)^2$.

EXERCISES 8.2

1. (a) True (b) False (c) True (d) True (e) False (f) False (g) False (h) True
3. $K + 5$ **5.** If $b|c$, then $bx = c$ and $c/b = x$. But $xb = c$, so $x|c$ **7.** If $a|b$ and $a|c$, then $b = am$ and $c = an$, so $b - c = am - an = a(m - n) = ax$, where $x = m - n$. Therefore, $a|(b - c)$.   **9.** Addition

EXERCISES 8.3

1. $6 = 2 \cdot 3, 8 = 2^3, 10 = 2 \cdot 5, 14 = 2 \cdot 7, 15 = 3 \cdot 5, 21 = 3 \cdot 7, 22 = 2 \cdot 11, 26 = 2 \cdot 13$, $27 = 3^3$. Must be the cube of a prime or the product of two distinct primes.
3. (a) $7 \cdot 13$ (b) $2^3 \cdot 3 \cdot 5$ (c) $13 \cdot 23$ (d) 313 (e) 347 (f) $2^4 \cdot 31$ (g) $2^2 \cdot 11 \cdot 19$
(h) $2 \cdot 3 \cdot 313$ (i) $3 \cdot 13 \cdot 107$ (j) $2^2 \cdot 5^2 \cdot 7 \cdot 11 \cdot 13$ (k) $7 \cdot 11 \cdot 13^2 \cdot 23$ (l) $2^6 \cdot 5^6$
5. If n is composite, then n has a prime divisor p, and $n = p \cdot x$. p and x cannot both be greater than $\sqrt{n}$, or their product would be greater than n. Then either $p \leq \sqrt{n}$ or $x \leq \sqrt{n}$, and n must have prime divisors less than or equal to $\sqrt{n}$.
7. $11^2 - 2 = 119 = 7 \cdot 17$ **9.** 60; 180

EXERCISES 8.4

1. $2^4 \cdot 3^2 \cdot 5, 2^4 \cdot 3^2 \cdot 7, 2^4 \cdot 3 \cdot 5 \cdot 7, 2^4 \cdot 3 \cdot 7^2, 2^4 \cdot 5 \cdot 7^2, 2^4 \cdot 7^3, 2^3 \cdot 3^2 \cdot 5 \cdot 7, 2^3 \cdot 3^2 \cdot 7^2,$
$2^3 \cdot 3 \cdot 5 \cdot 7^2, 2^3 \cdot 3 \cdot 7^3, 2^3 \cdot 5 \cdot 7^3, 2^2 \cdot 3^2 \cdot 5 \cdot 7^2, 2^2 \cdot 3^2 \cdot 7^3, 2^2 \cdot 3 \cdot 5 \cdot 7^3, 2 \cdot 3^2 \cdot 5 \cdot 7^3$
3. (a) 1, 2, 3, 4, 5, 6, 10, 12, 15, 20, 30, 60 (b) 1, 3, 5, 9, 15, 27, 45, 81, 135, 405 (c) 1, 2, 3, 4, 6, 7, 9, 12, 14, 18, 21, 27, 28, 36, 42, 54, 63, 84, 108, 126, 189, 252, 378, 756
5. $n \cdot x = n + n^2 + n^3 + \cdots + n^k + n^{k+1}$, so $n \cdot x - x = n^{k+1} - 1$. Thus, $(n - 1) \cdot x = n^{k+1} - 1$ and $x = \dfrac{n^{k+1} - 1}{n - 1}$.

EXERCISES 8.5

1. 0, 1, 1, 3, 1; 6, 1, 7, 4, 8; 1, 16, 1, 10, 9; 15, 1, 21, 1, 22; 11, 14, 1, 36, 6; 16, 13, 28, 1, 42
3. (a) 21 (b) 240 (c) 37 (d) 1 (e) 1 (f) 496 (g) 844 (h) 1890 (i) 1875 (j) 191,548

(k) 122,333   (l) 1,480,437   **5.** If p is prime, $P(p) = 1$, which must be less than p, so p is deficient. The set of prime numbers is infinite, so the set of deficient numbers is infinite.
7. $P(2^k) = 1 + 2 + \cdots + 2^{k-1} = 2^k - 1$ by Theorem 8.4.1.
9. Let $D(n) = k$. Then $P(n)$ is the sum of $k - 1$ divisors, and is at least as large as $1 + 2 + 3 + \cdots + (k - 1) = (k^2 - k)/2$ (see Exercise 8.1.7). Since n is composite, $k \geq 3$, so $k^2 \geq 3k$, $k^2 - k \geq 2k$, and $(k^2 - k)/2 \geq k$; that is, $P(n) \geq (k^2 - k)/2 \geq k = D(n)$.

EXERCISES 8.6

1. $2^5 \cdot 63 = 2016$; $P(2016) = 4536$ **3.** False; $110 = 2 \cdot 5 \cdot 11$, and 110 is deficient.
5. $(2^k - 1)(2^{k-1}) = [(2^k - 1)(2^k)]/2$, a triangle number.

EXERCISES 8.7

1. $M_{11} = 2047 = 23 \cdot 89$ **3.** 31; 23

EXERCISES 8.8

1. $672 = 2^5 \cdot 3 \cdot 7$; $S(672) = 2016 = 3 \cdot 672$ **3.** This follows easily from Definition 8.8.2 and the fact that $S(k) = P(k) + k$.   **5.** $P(24) = 36$; $P(36) = 55$; $P(55) = 17$; $P(17) = 1$; $P(1) = 0$ **7.** If n is semiperfect, then some set of divisors, $d_1, d_2, \ldots, d_k$, adds up to n. Then the multiple $x \cdot n$ has divisors $x \cdot d_1, x \cdot d_2, \ldots, x \cdot d_k$, which sum to $x \cdot n$.
9. (a), (b), (c), (e), (g) are semiperfect; (d) and (h) are weird.

CHAPTER 9

EXERCISES 9.2

1. $a * b = b, b * c = d, d * d = a, c * a = c.$

3.

*	1	2
1	1	1
2	1	1

*	1	2
1	2	1
2	1	1

*	1	2
1	1	2
2	1	1

*	1	2
1	1	1
2	2	1

*	1	2
1	1	1
2	1	2

*	1	2
1	2	2
2	2	2

*	1	2
1	1	2
2	2	2

*	1	2
1	2	1
2	2	2

	1	2
1	2	2
2	1	2

	1	2
1	2	2
2	2	1

	1	2
1	1	1
2	2	2

	1	2
1	2	2
2	1	1

	1	2
1	1	2
2	1	2

	1	2
1	2	1
2	2	1

	1	2
1	1	2
2	2	1

	1	2
1	2	1
2	1	2

EXERCISES 9.3

1. (a) $(60 \div 6) \div 2 = 10 \div 2 = 5$; $60 \div (6 \div 2) = 60 \div 3 = 20$ (b) $(1 * 2) * 3 = (1 + 2 \cdot 2)$
$* 3 = 5 * 3 = 5 + 2 \cdot 3 = 11$; $1 * (2 * 3) = 1 * (2 + 2 \cdot 3) = 1 * 8 = 1 + 2 \cdot 8 = 17$

3.

	p	q	r
p	q	q	q
q	q	q	q
r	q	q	q,

	p	q	r
p	p	q	r
q	q	r	p
r	r	p	q,

others

EXERCISES 9.4

1 (a) q (b) 2 (c) s (d) 1 (e) s (f) 0 (g) q (h) 4 (i) $x = q$ (j) $x = 1$ (k) $p \circ q = r$, but $q \circ p = t$ (l) $(q \circ r) \circ t = p \circ t = q$, but $q \circ (r \circ t) = q \circ p = t$ (m) Table I, because $\circ$ is not associative   **3.** (a) 3 (b) $7' = -1, 2' = 4$ (c) $a' = 6 - a$

EXERCISES 9.5

1. (a)

	1	2	3	4
1	1	2	3	4
2	3	4	1	2
3	2	1	4	3
4	4	3	2	1

(b)

	1	2	3	4
1	1	2	3	4
2	1	2	3	4
3	1	2	3	4
4	1	2	3	4

(c)

	1	2	3	4
1	1	1	1	1
2	2	2	2	2
3	3	3	3	3
4	4	4	4	4

(d)

	1	2	3	4
1	1	2	3	4
2	2	2	4	3
3	3	4	1	2
4	4	3	2	1

(There are other possible examples of each.)

3.

	1	2	3	4		1	2	3	4
1	1	2	3	4	1	1	2	3	4
2	2	3	4	1	2	2	1	4	3
3	3	4	1	2	3	3	4	1	2
4	4	1	2	3	4	4	3	2	1

Both are groups, but associativity is not obvious; a case-by-case check is necessary to prove it.
5. At most one: The element a must have an inverse a' in G, so the given equation implies that $a' * (a * x) = a' * b$. Thus,

$$(a' * a) * x = a' * b \qquad \text{by associativity}$$
$$z * x = a' * b \qquad \text{by definition of inverse}$$
$$x = a' * b \qquad \text{by definition of identity}$$

Thus, if a solution exists, it must be $a' * b$. At least one: $a' * b$ is a solution, since

$$a * (a' * b) = (a * a') * b = z * b = b$$

7. Suppose not. Then every element other than z has a different inverse. By Exercise 6, elements that are inverses of each other can be paired up without "overlap" among pairs. Since every element besides z must have an inverse, all those elements of the group can be paired, implying that there is an even number of them. Counting z, then, we have an odd number of elements in the group; contradiction.
9. Right cancellation does not: $1 * 3 = 498 * 3$, because both equal $2 \cdot 3\,(=6)$, but clearly $1 \neq 498$. Left cancellation does: If $a * b = a * c$, then $2b = 2c$, so $b = c$.

EXERCISES 9.6:

1.

	b	c			b	c
b	b	c	and	b	c	b
c	c	b		c	b	c

are the only two group operations.

3. (a) $\{j\}, \{j, f\}, \{j, g\}, \{j, k\}, \{j, h, i\}, \{j, f, g, h, i, k\}$ (b) They are all divisors of 6, the total number of elements in B (c) $(B, \circ)$ is not commutative.
5. Let S be a subset of a finite group $(G, *)$ which satisfies properties (1) and (2), and let a be an arbitrary element of S. For any finite whole number n, let na denote $a * a * \cdots * a$, where there are n a's in the string. Since S is closed under $*$, na is an element of S for any whole number n. Since G is finite, there must be two different "multiples" of a which are equal; say $na = ka$, for $k < n$. Then (by cancellation) $(n - k)a = z$, so $(n - k - 1)a = a'$ and must be in S (by closure), as required.
7. (a) By definition of inverse, it suffices to show that $(a * b) * (a' * b') = z$.

$$(a * b) * (a' * b') = (a * (b * a')) * b' \qquad \text{by associativity}$$
$$= (a * (a' * b)) * b' \qquad \text{by commutativity}$$
$$= (a * a') * (b * b') \qquad \text{by associativity}$$
$$= z * z \qquad \text{by definition of inverse}$$
$$= z \qquad \text{by definition of identity}$$

(b) $(f * k)' = h' = i$, but $f' * k' = f * k = h$; $(a * b)' = b' * a'$

EXERCISES 9.7

1. $\{0\}$, $\{0, 3, 6\}$, and S itself   **3.** (a) 90 elements because that is the smallest common multiple of 2, 5, and 9   (b) 30 elements   **5.** (a) There are many ways to do this; one is

$$\{b, d, g, j\}, \quad \{a, c, e, f\}, \quad \{h, i, k, l\}, \quad \{m, n, o, p\}, \quad \{q, r, s, t\}$$

(b) Yes   (c) $\{a, b, c\}$, or any subset whose number of elements is not a divisor of 20.   (d) Yes. Any subset containing 1, 2, 4, 5, 10, or 20 elements will work.

EXERCISES 9.8

1. (a) $\{5, 8, 11, 2\}$ (b) $\{5, 9, 1\}$ (c) $\{2, 5, 8, 11\}$ (d) $\{2, 6, 10\}$ (e) $\{0, 3, 6, 9\}$
(f) $\{9, 0, 3, 6\}$ (g) $\{4, 8, 0\}$ (h) $\{4, 7, 10, 1\}$ (i) $\{3, 7, 11\}$.
3. (a) $p \circ L = \{p, q, r\}$ $q \circ L = \{q, p, s\}$ $r \circ L = \{r, s, p\}$
$s \circ L = \{s\}$ $t \circ L = \{t, u\}$ $u \circ L = \{u, t, q\}$
(b) Six. No.   (c) No. If it were, all distinct cosets would have to be disjoint and contain the same number of elements.
5. Suppose that $x * H$ and $y * H$ both contain some element g. Then $g = x + h_1 = y + h_2$ for some h_1 and h_2 in H. Since H is a group, h_2' is in H and we may write $(x * h_1) * h_2' = (y * h_2) * h_2'$. By associativity, $x * (h_1 * h_2') = y * (h_2 * h_2') = y$. Now let $y * h$ be an arbitrary element of $y * H$. Then $y * h = (x * (h_1 * h_2')) * h = x * ((h_1 * h_2') * h)$. But h_1, h_2', and h are all in H, which is closed under $*$, so $((h_1 * h_2') * h)$ is in H. Thus, $y * h = x * $ (some element of H) and hence is in $x * H$, implying that $y * H$ is a subset of $x * H$.

Review Exercise.   (a) No. No identity.   (b) Yes. Identity is 1; $2' = 3$.   (c) No. Not closed: 0 is not in the set.   (d) No. Cancellation does not hold: There are two 1's in the first row, and so on.   (e) No. If it were, $\{1, 3\}$ would be a two-element subgroup of a five-element group, contradicting Lagrange's Theorem.   (f) No. 2 has no inverse. (3 works as an inverse for 2 from the right but not from the left; 4 works from the left but not from the right.)   (g) Yes. Identity is 6; $2' = 4$.   (h) No. If it were, $\{1, 2, 3\}$ would form a three-element subgroup of an eight-element group, contradicting Lagrange's Theorem.

EXERCISES 9.9

3. Denote the eight rigid motions as follows:

r_0—"rotation" of $0°$ v—reflection about vertical median
r_1—rotation of $90°$ h—reflection about horizontal median
r_2—rotation of $180°$ d_1—reflection about upper left/lower right diagonal
r_3—rotation of $270°$ d_2—reflection about lower left/upper right diagonal

$\circ$	r_0	r_1	r_2	r_3	v	h	d_1	d_2
r_0	r_0	r_1	r_2	r_3	v	h	d_1	d_2
r_1	r_1	r_2	r_3	r_0	d_1	d_2	h	v
r_2	r_2	r_3	r_0	r_1	h	v	d_2	d_1
r_3	r_3	r_0	r_1	r_2	d_2	d_1	v	h
v	v	d_2	h	d_1	r_0	r_2	r_3	r_1
h	h	d_1	v	d_2	r_2	r_0	r_1	r_3
d_1	d_1	v	d_2	h	r_1	r_3	r_0	r_2
d_2	d_2	h	d_1	v	r_3	r_1	r_2	r_0

No, this group is not commutative; for example, $v \circ r_1 \neq r_1 \circ v$.

5. (a) There are only two possible rigid motions—the $0°$ "rotation" and a flip about the bisector of the right angle; call them r and f, respectively.

	r	f
r	r	f
f	f	r

(b) Yes   (c) If we let $r = 0$ and $f = 1$, this table matches the first table of Example 9.2.5.
(d) Yes, any of the subgroups $\{j, f\}, \{j, k\}, \{j, g\}$.

CHAPTER 10

EXERCISES 10.2

1. Undefined terms, definitions, axioms, and theorems   **3.** No. There are none.   **5.** Yes. Some words can be used to facilitate the definition of others. "Parallelogram" before "square"; "trapezoid" could fit anywhere.

EXERCISES 10.3

1. Given line L, it contains points p and q, by A3. By A4, there is a line $M \neq L$ containing p. M contains another point, r, by A3; and r cannot be in L, by A2.
3. Given parallel lines L and M, they contain points p and q, r and s, respectively, by A3. Since L and M have no points in common, these points are distinct.
5. The proof is the same as that in the text up through the parenthetical comment. Now if $q = r$, we have $p, q \in L$ and $p, q \in M$. Since the lines are distinct, one of them must contain a third point.

EXERCISES 10.4

1. Axioms 1, 3, and 5 are obviously satisfied. There are exactly six possible pairs of points and each is in one, and only one, line, so A2 is satisfied. And A4 is satisfied, since each point is in three different lines.
3. Axioms 1, 3, and 5 are obviously satisfied. Each pair of points is in exactly one line; the fact that $\{1, 2\}$, $\{1, 3\}$, and $\{2, 3\}$ determine the same line is no problem. A4 is easily seen to be satisfied.
5. (a) Points $= \{1, 2, 3, 4, 5, 6, 7, 8\}$. Lines are the 28 doubleton subsets.
(b) 1 1 2 2 3 3   (c) 1 1 2 3
    2 4 4 5 4 5   2 4 4 4
    3 5   3

EXERCISES 10.5

1. Axioms 3, 4, and 5 are easily seen to be satisfied.   **3.** Yes, it is consistent; no, it is not complete. See Examples 10.4.1 and 10.4.2.   **5.** Yes. A model of four points and six two-point lines shows this.   **7.** Independent: B1, B2, B5, B6, B7, B8, B9. Dependent: B3, B4.

APPENDIX A

1. Negations: (a) The flower is not purple.   (b) Not all freshmen are flowers.   (c) No birds are fire engines.   (d) The development of the number systems is not motivated exclusively by a consideration of algebraic equations.   (e) No more than one meeting has been held to discuss the renovation of the schedule.   (f) Not all dogs have fleas *or* Some dogs do not have fleas.
(g) There are not three trees in the front yard.   (h) $8 + 3 \neq 11$.   (i) Some textbooks are useful.   (j) For some x, there does not exist a y such that y is less than x.
Contraries: (Many other possible answers are correct.)   (a) The flower is white.   (b) All freshmen are students *or* No freshmen are flowers.   (c) All birds are sailboats.   (d) The development of the number systems is motivated exclusively by a consideration of inequalities.   (e) No meetings have been held to discuss the renovation of the schedule.
(f) No dogs have fleas.   (g) There is only one tree in the front yard.   (h) $8 + 3 = 15$.   (i) All textbooks are useful.   (j) For each x, there does not exist a y such that y is less than x.
3. (a) If roses are red, then snowmen like carrots.   (b) If snowmen like carrots, then roses are red.   (c) Roses are red if and only if snowmen like carrots.   (d) If roses are not red, then snowmen like carrots.   (e) If roses are red, then snowmen do not like carrots.   (f) If snowmen do not like carrots, then roses are not red.

APPENDIX B

EXERCISES B.1

1. (d), (e), (f), (g), (h), (i), and (j) are well-defined sets.   **3.** (d) {Alaska, Maine, Michigan, Montana, New York, North Dakota, Washington, Wisconsin, Vermont}   (e) {15, 16, 17, 18, 19, 20}   **5.** (a), (b), (d), (f), and (g) are true statements.   **7.** (b) and (f) are equal to A. (c) and (d) are equal to B.   **9.** (a) {Bob, Charles}, {Bob, Dick}, {Bob, Edward}, {Charles, Dick}, {Charles, Edward}, {Dick, Edward}   (b) {Bob, Charles, Dick}, {Bob, Charles, Edward}, {Bob, Dick, Edward}, {Charles, Dick, Edward}

EXERCISES B.2

1. {1, 7, 8, 9, 10}   **3.** {2, 3, 4, 5, 6, 7, 8, 10}   **5.** {1, 2, 3, 5, 7, 8, 9, 10}   **7.** {1, 9}
9. {2, 4, 6, 8}   **11.** {1, 2, 4, 5, 8}   **13.** {2, 3, 4, 5, 6, 7, 8, 9}   **15.** Universal set
17. {1, 3, 5, 6, 7, 9}   **19.** {3, 5, 7, 9}   **25.** The set of secretaries or the stockholders of an insurance company   **27.** The set of stockholders and keypunch operators in an insurance company   **29.** The set of secretaries or keypunch operators in an insurance company
31. The set of either the secretaries and the stockholders *or* the secretaries and keypunch operators in an insurance company   **33.** The set of students not registered in chemistry
35. The set of students in botany or chemistry or possibly both   **37.** The set of students in botany, chemistry, and physics   **39.** The set of students in botany or chemistry or physics.

INDEX